RESONANCE

Edited by **Jan Awrejcewicz**

Resonance
http://dx.doi.org/10.5772/intechopen.68248
Edited by Jan Awrejcewicz

Contributors

Betül Çalişkan, Ali Cengiz Çalişkan, Karlo Queiroz Costa, Nadson Souza, Jefferson Costa, Rafael Santos, Andre Cruz, Koichi Narahara, Klara Kalinova, Jiri Naprstek, Cyril Fischer, Vladimir L. Kalashnikov, Yannick Le Bras, Jean-Marc Greneche, Xianghong Kong, Lina Qiu, Gaobiao Xiao, Di He, Ken-Ichi Hino, Juan Manuel Velazquez Arcos, Jaime Granados-Samaniego, Alejandro Pérez-Ricárdez, Ricardo Teodoro Páez Hernández, Lukas Kurmann

Published by InTech
Janeza Trdine 9, 51000 Rijeka, Croatia

Publishing Process Manager Ana Vais
Technical Editor SPi Global
Cover InTech Design team

First published November, 2017
Printed in Croatia
Legal deposit, Croatia: National and University Library in Zagreb

Additional hard copies can be obtained from orders@intechopen.com

Resonance, Edited by Jan Awrejcewicz
p. cm.
Print ISBN 978-953-51-3633-0
Online ISBN 978-953-51-3634-7

For EU product safety concerns:
IN TECH d.o.o., Prolaz Marije Krucifikse Kozulić 3, 51000 Rijeka, Croatia,
info@intechopen.com or visit our website at intechopen.com.

Meet the editor

Jan Awrejcewicz is a professor and the head of the Department of Automation, Biomechanics and Mechatronics of Lodz University of Technology. He is a recipient of multiple honorary degrees and the Humboldt Award. He is author/coauthor of over 730 scientific papers and 48 monographs. He has edited 22 books and 26 journal special issues. He is a principal investigator in 38 research grants by the Polish Ministry of Education and Science (21), Committee for Scientific Research (15), and National Science Centre (2). He is a member of the editorial boards of 74 journals (15 with IF). He is an organizer and the head of a series of 14 international conferences Dynamical Systems: Theory and Application (Lodz, 1992–2017), and others. His papers and research cover various disciplines of mathematics, mechanics, biomechanics, automatics, physics, and computer-oriented sciences, with the main focus on nonlinear processes.

Contents

Preface

Resonance is a common phenomenon, which is observed both in nature and in numerous devices and structures. It occurs in literally all types of vibrations. To mention just a few examples, acoustic, mechanical, or electromagnetic resonance can be distinguished. In the present book, several chapters dealing with different aspects of resonance phenomena have been collected.

Out of numerous submitted chapters, the following 12 have been eventually selected for publication in this book. A brief overview of them is as follows.

In the chapter by *Çalişkan and Çalişkan,* the fundamentals of magnetic resonance are described. The authors deal with both electron spin (paramagnetic) resonance and nuclear magnetic resonance.

Le Bras and Greneche give an overview of magnetostrictive resonators, principles related to such resonators, and their respective performances including sensing application domains and limitations. They review different magnetostrictive materials to design resonators. Moreover, they propose an analytical model, which can be used to determine the magnetoelastic coupling factor.

Náprstek and Fischer focus on the stochastic resonance (SR) and related topics. Aside from classical definition and basic features of SR, they discuss the most important methods of investigation of SR as well as recommend an experimental procedure to verify the results of stochastic simulation, among others.

Kalashnikov deals with chaotic and stochastic resonance as well as stochastic antiresonance (SAR). The chapter presents SAR in a Raman fiber amplifier, chaotic resonance between a dissipative soliton and linear waves, and stochastic resonance and antiresonance in mode-locked lasers.

The chapter authored by *He* considers the optimization method of double-well bistable stochastic resonance system. Besides the optimization method, the corresponding analysis results are given especially under low SNR circumstances. Also, an example of application of the proposed method in cognitive radio networks is given.

Narahara studies resonances in left-handed (LH) waves developed in nonlinear electrical lattices. To investigate resonances involving LH waves, nonlinearity is introduced to composite right-handed and left-handed (CRLH) transmission lines. Head-on collision, three-wave mixing, harmonic resonance, and soliton decay are considered.

In the chapter authored by *Kurmann*, an introduction to parametric and autoparametric resonance is given. The chapter is supplemented with numerous examples of literature to the topic. Also, examples with numerical simulations and analytical methods are presented.

In the chapter by *Kalinova*, results of investigations of the resonance effect of nanofibrous membranes for sound absorption applications are presented. The chapter comprises the theoretical basics of membrane and Helmholtz's resonators, followed by the description of the design of an acoustic element used for the study.

Souza et al. present the results of the investigation of the influence of a dielectric shell on metallic spherical nanoparticles in the resonant modal response of an SPR-type sensor. In their research, they compare analytical solutions with those obtained with the use of the finite element method and experimental data.

Kong et al. use Mie theory to analyze Fano resonances in simple high-permittivity structures such as spheres or core-shell particles. The chapter includes also the investigations for arbitrary-shaped objects as well as for periodic structures. For each structure, different theoretical methods are presented together with numerical analysis.

Hino et al. review the laser-induced Fano resonance in condensed matter systems. They study two physical processes, i.e., a Floquet excitor in semiconductor super lattices driven by a strong continuous-wave laser and the coherent phonon induced by an ultrashort pulse laser in bulk crystals.

Last but not least, *Velázquez-Arcos* et al. present their approach to the reasons of a sudden loss of signal propagation due to significant changes in the broadcasting regime. Furthermore, they propose a method to avoid such loss of signal and thus to enhance the broadcasting process.

I sincerely hope that any reader of this book will find at least some of the investigated topics interesting and inspiring for his or her research.

Jan Awrejcewicz
Department of Automation, Biomechanics and Mechatronics
Lodz University of Technology
Lodz, Poland

Chapter 1

Magnetic Resonance

Betul Çalişkan and Ali Cengiz Çalişkan

Additional information is available at the end of the chapter

http://dx.doi.org/10.5772/intechopen.70516

Abstract

Magnetic resonance is divided into electron spin resonance (ESR) [electron paramagnetic resonance (EPR)] and nuclear magnetic resonance (NMR) according to the working region in the electromagnetic spectrum. If the studied region is in the microwave region, this resonance type is electron spin resonance. If the region studied is the radio frequency region, then nuclear magnetic resonance is mentioned. ESR and NMR are similar in terms of their basic theorem.

Keywords: electron spin resonance (ESR), electron paramagnetic resonance (EPR), nuclear magnetic resonance (NMR), microwave frequency, radio frequency

1. Introduction

Nuclear magnetic resonance (NMR) spectroscopy examines the interaction of nuclear spins forming an atom with the magnetic field applied to them. Electron spin (paramagnetic) resonance (ESR, EPR) spectroscopy studies the interaction of the electron spins with the applied magnetic field.

The resonance term is used to determine that an external factor is in harmony with the natural frequency of the magnetic system. The natural frequency is the radio frequency (RF) or microwave (MD) frequency, which is in agreement with the Larmor rotation frequency of the magnetic moments in the magnetic field.

The magnetic moment referred to NMR is a nonzero nuclear moment. In other words, NMR deals with nuclei whose spin value is nonzero. The magnetic moment referred to EPR is the magnetic moment of the electron. EPR studies magnetic systems with unpaired electrons.

2. Magnetic resonance spectroscopy (NMR and EPR spectroscopy)

Nuclear magnetic resonance (NMR) was first observed by F. Bloch in 1946. In the same period, the electron spin resonance (ESR) experiment was first performed by YK Zavoysky in 1944.

Magnetic resonance spectroscopy is similar to other types of absorption spectroscopy. Magnetic resonance is based on the interaction of matter with electromagnetic radiation. Electromagnetic radiation for NMR is in the radio frequency domain. For the EPR, it is in the microwave area. As a result of this interaction, the transition from the high energy state to the low energy state leads to an energy release in the amount of (ΔE) [1].

$$\Delta E = hv \tag{1}$$

Within the external magnetic field, the magnetic moment of the nucleus or electron makes a precession movement with the Larmor frequency (ω) around the magnetic field. The Larmor precession movement tries to orient the spins or magnetic dipole moments in the direction of the magnetic field. This process is called **the relaxation process** [2].

The state of the system reaching the thermal equilibrium is called relaxation time. Relaxation times are divided into two. The first is T_1 spin-lattice (longitudinal) relaxation time. It affects the z-component of the magnetization vector. T_1 determines the energy flow rate of neighboring molecules (lattice) from the nuclear spin system. It is the time to reach the thermal equilibrium of the neighboring molecules with the nuclear spin system. The second is T_2 spin-spin (transverse) relaxation time. T_2 affects the x and y components of the magnetization vector. The process of reaching thermal equilibrium as the result of the interaction between the spins without transferring energy to the neighboring molecules is called the spin-spin relaxation process. T_2 is related to **the full width at half maximum (fwhm)** ($\Delta v_{1/2}$) of the NMR signal.

The orientations of the magnetic moments are in the form of different spin populations at different energy levels. Boltzmann expression is used for low energy state (N_{-}) and high energy state (N_{+}) spin populations with temperature effect.

$$\frac{N_-}{N_+} = e^{hv/kT} \tag{2}$$

where k is Boltzmann's constant and T is the temperature in K.

The ratio of the magnetic moment to the spin angular momentum is **called the gyromagnetic ratio** (γ). This expression is also equal to the ratio of the Larmor precession frequency to the magnetic field.

$$\vec{\omega} = \gamma \vec{H} \tag{3}$$

where ω is 2π times the precession frequency ($2\pi v$) and H is the applied magnetic field.

This expression contains both the resonance condition and the magneton concept. For EPR studies, Bohr magneton is valid whereas for NMR studies nuclear magneton is applied.

http://dx.doi.org/10.5772/intechopen.70516

Magneton is related to the concept of spin magnetic moment (μ_I: the nuclear spin magnetic moment, μ_S: the electron spin magnetic moment). It is a concept related to the ratio between the mass and the charge of a particle having a different spin from zero.

$$\beta = \frac{e\hbar}{2m_e} \tag{4}$$

where β is the Bohr magneton, β=9.2741 × 10^{-24} J/T, e is the charge of the spinning particle, and m_e is the mass of the electron. For the nuclear magneton β_n, this would be:

$$\beta_n = \frac{e\hbar}{2m_p} \tag{5}$$

β_n (nuclear magneton) is much smaller than β (Bohr magneton) since the proton mass (m_p) is 1836 times as great as that of the electron, β_n=5.05 × 10^{-27} J/T. The magneton relates to the basic equation above because:

$$\gamma = \frac{g\beta}{\hbar} \tag{6}$$

where g is a proportionality constant usually referred to as the g-value or g is the spectroscopic splitting constant, and is Planck's constant divided by 2π. Hence:

$$\frac{\omega}{H} = \frac{g\beta}{\hbar} \tag{7}$$

or

$$\omega\hbar = g\beta H \tag{8}$$

Since ω is $2\pi v$, then:

$$hv = g\beta H \tag{9}$$

The above expression is called resonance condition in both NMR and ESR [1].

Although many processes are similar in the EPR and NMR experiments, the tools used in the experiments differ. In EPR, it is used in microwave components, such as wave-guide, cavities, and klystron tubes. In NMR, inductances, capacitors, conductors that transmit radio frequency energy, and vacuum tubes are used [2].

2.1. NMR spectroscopy

Magnetic dipole moment of the nucleus:

$$\vec{\mu}_I = \gamma_I \vec{I} \tag{10}$$

$$\vec{\mu}_I = g_n \frac{\beta_n}{\hbar} \vec{I} \tag{11}$$

$$\mu_I = g_n \left(\frac{e\hbar}{2m_p}\right) \frac{1}{\hbar} \left(\sqrt{I(I+1)}\hbar\right) \tag{12}$$

$$\mu_I = g_n \beta_n \sqrt{I(I+1)} \tag{13}$$

Here, I is nuclear spin and g_n is the nuclear g-factor (g_n=5.5855).

The interaction between the external magnetic field and the nuclear magnetic moment is given as follows:

$$E = -\vec{\mu}_I \cdot \vec{H_0} \tag{14}$$

$$E = -\mu_I H_0 cos\theta \tag{15}$$

where θ is the angle between the dipole and the magnetic field. There are two orientations for a proton with a nuclear spin 1/2. This indicates the quantum number of magnetic spin, m_I. For $m_I = \pm\frac{1}{2}$, energy takes values $-\frac{1}{2}g_n\beta_n H_0$ and $+\frac{1}{2}g_n\beta_n H_0$ as shown in **Figure 1**.

The nuclear magnetic resonance transition occurs between two energy levels. The transition between the two energy levels constitutes the resonance condition.

$$\Delta E = hv = g_n \beta_n H_0 \tag{16}$$

is called **the resonance condition for NMR**.

Nuclear magnetic resonance stays on two important interactions. The first one is **the chemical shift** and the other is **the spin-spin coupling**. A third interaction can also be mentioned. This is **the exchange interaction**. Thus, we can list three important interactions in NMR as follows:

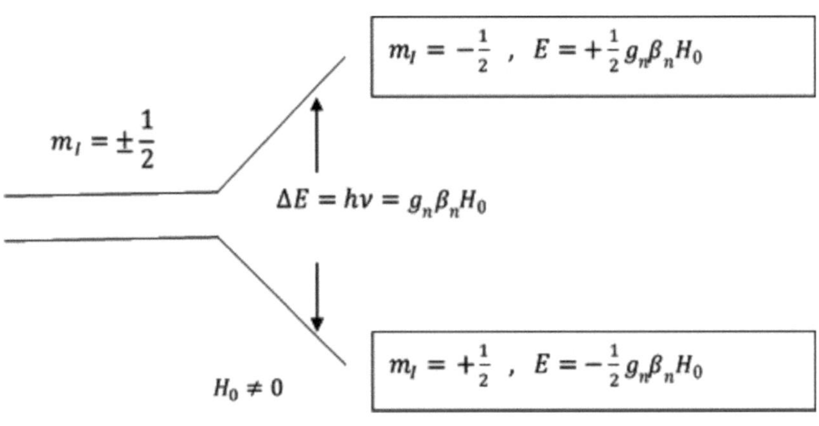

Figure 1 . Nuclear magnetic resonance transition.

http://dx.doi.org/10.5772/intechopen.70516

1. The chemical shift
2. The spin-spin coupling
3. The exchange interaction
   a. Slow exchange interaction
   b. Fast exchange interaction

However, the third influence is not taken into consideration. So, we will focus on two interactions.

The effective Hamiltonian expression for NMR consists of the sum of **the nuclear Zeeman Hamiltonian** and **the nuclear spin-spin interaction Hamiltonian** terms:

$$\mathcal{H} = -g_n\beta_n \vec{H} \cdot \vec{I} \cdot \vec{S} + \vec{I} \cdot \vec{\vec{Q}} \cdot \vec{I} \tag{17}$$

Here, $\vec{\vec{Q}}$ is **the quadrupole interaction tensor** (interaction between two nuclear spins) interaction tensor [3].

2.1.1. Chemical shift

The electrons surrounding the nucleus of a molecular system show a spherical distribution. The external magnetic field applied on the system creates polarity in the electron distribution in the spherical structure. That is, a current flows through the molecule. This current induces a magnetic field by induction where the core is located. This field is called **the internal magnetic field** (**Figure 2**). The internal magnetic field is opposite to the external magnetic field. The total

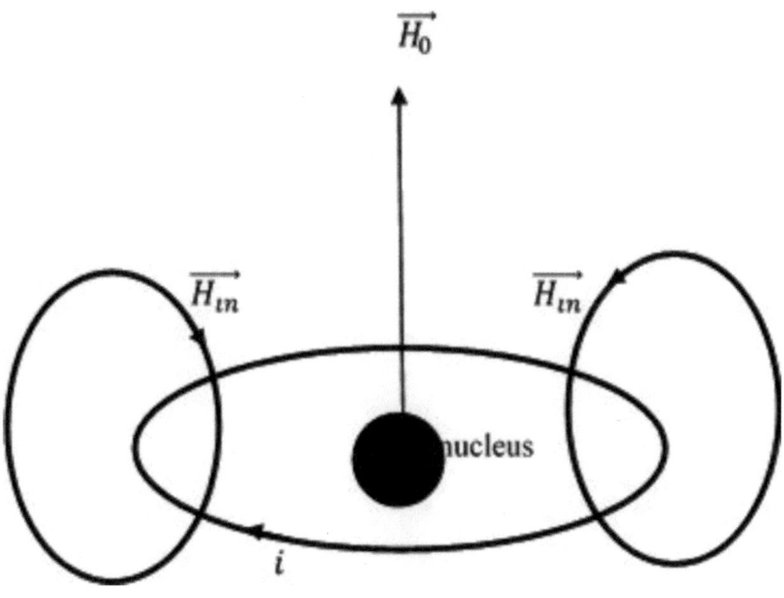

Figure 2. Internal magnetic field and external magnetic field orientation.

magnetic field seen by the nucleus is different from the external magnetic field. This brings about a shift in the resonance frequency of the nucleus. This is called the chemical shift. That is, the electron-nucleus interaction originating from the magnetic field created by moving charges is the chemical shift (**Figure 3**).

Accordingly, the nucleus sees the effective magnetic field given by the formula:

$$\overrightarrow{H_{eff}} = \vec{H}_0 - \vec{H}_{in} \tag{18}$$

The internal magnetic field is connected to the external magnetic field (Eq. (18))

$$\vec{H}_{in} = \sigma \vec{H}_0 \tag{19}$$

The internal magnetic field is connected to the external magnetic field by **the diamagnetic shielding coefficient** (σ). In NMR, tetramethylsilane, $Si(CH_3)_4$, is generally used as a standard sample for comparison. The chemical shift is shown as δ. Its scale is parts per million (ppm).

$$\delta = (\sigma_T - \sigma_X).10^6 \text{ ppm} \tag{20}$$

$$\delta = \left(\frac{\omega_X - \omega_T}{\omega_0}\right).10^6 \text{ ppm} \tag{21}$$

$$\delta = \left(\frac{H_X - H_T}{H_0}\right).10^6 \text{ ppm} \tag{22}$$

2.1.2. Spin-spin coupling

Contrary to the dipole-dipole interaction, it is a new type of interaction that is not dependent on the orientation of the molecule. It is the indirect spin-spin interaction period that occurs through the electrons that form chemical bonds in the molecule. In other words, the interaction of a nucleus with another nucleus through an electron cloud is a spin-spin coupling. The spin of an electron near the A nucleus is S_A and the spin of an electron near the B nucleus is S_B. In

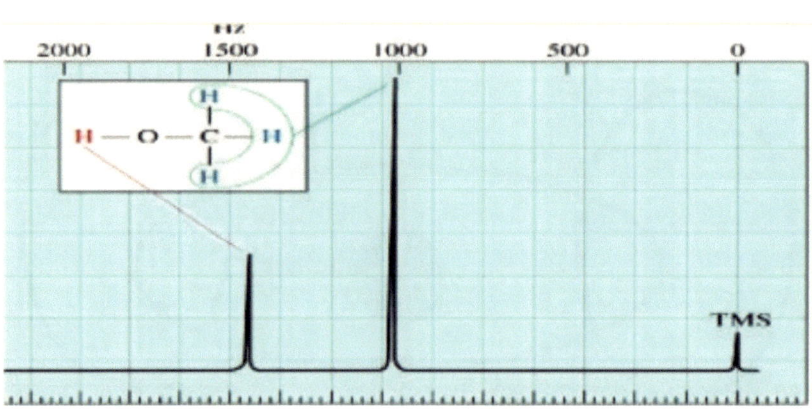

Figure 3. The chemical shift.

http://dx.doi.org/10.5772/intechopen.70516

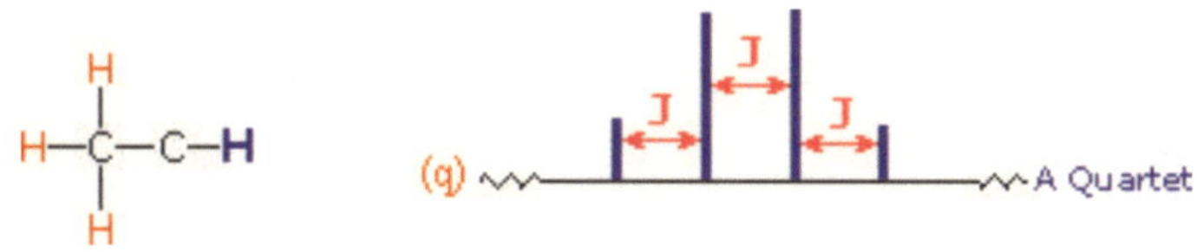

Figure 4. A spin-spin coupling example.

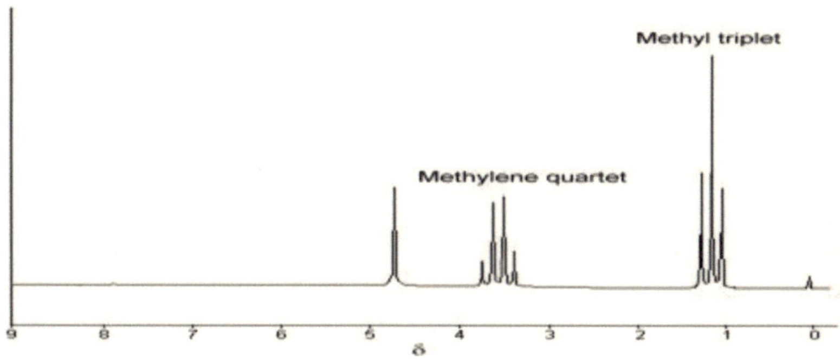

Figure 5. An example of an NMR spectrum.

external magnetic field, the opposite or the same direction of orientation of A and B nuclei with $I = 1/2$ is the spin-spin coupling state of A and B nuclei. The energy of the spin-spin coupling is given as

$$E = \hbar J_{AB} \vec{I}_A \cdot \vec{I}_B \tag{23}$$

J is the spin-spin coupling coefficient.

When a nucleus or nucleus group interacts with n magnetically equivalent nuclei with spin quantum number I, the observed number of splits is $(2nI + 1)$. **Figure 4** shows a spin-spin coupling example and **Figure 5** shows an NMR spectrum example.

2.2. EPR (ESR) spectroscopy

EPR is a magnetic resonance method such as NMR. EPR deals with substance that contains unpaired electrons. These substances are free radicals, triplet excited states, and most transition metal and rare earth species. Among the parameters found in the EPR experiments are the g-factor, the hyperfine structure constant (*hf*), the nuclear quadrupole coupling constant, and the zero-field splitting constant. However, mostly **the *g*-factor** and **the hyperfine structure constant** are among the more studied parameters.

For EPR analysis, the sample is placed in a strong magnetic field. The applied electromagnetic radiation is in the microwave area. Due to the interaction between the magnetic moment of the free electron and the external magnetic field, the spin of the electron is directed parallel or antiparallel to the magnetic field. The energy difference between the two orientations gives the resonance condition for EPR.

$$\Delta E = hv = g_e \beta H_0 \tag{24}$$

Here, g_e is the free electron g-factor (g_e=g_S=2.0023) and β is Bohr magneton, β=9.2741 $\times 10^{-24}$ J/T.

Magnetic dipole moment of the free electron:

$$\vec{\mu}_S = \gamma_S \vec{S} \tag{25}$$

$$\vec{\mu}_S = -g_S \frac{\beta}{\hbar} \vec{S} \tag{26}$$

$$\mu_S = -g_S \left(\frac{e\hbar}{2m_e}\right) \frac{1}{\hbar} \left(\sqrt{S(S+1)}\hbar\right) \tag{27}$$

$$\mu_S = -g_S \beta \sqrt{S(S+1)} \tag{28}$$

where S is the electron spin.

The interaction between the external magnetic field and the magnetic moment of the free electron is given as follows:

$$E = -\vec{\mu}_S \cdot \vec{H_0} \tag{29}$$

$$E = -\mu_S H_0 cos\theta \tag{30}$$

where θ is the angle between the dipole and the magnetic field. There are two orientations for a electron spin 1/2. This indicates the quantum number of magnetic spin, m_S. For $m_S = \pm\frac{1}{2}$, energy takes values $+\frac{1}{2} g_e \beta H_0$ and $-\frac{1}{2} g_e \beta H_0$ as shown in **Figure 6**.

The electron spin resonance transition occurs between two energy levels. The transition between the two energy levels constitutes the resonance condition.

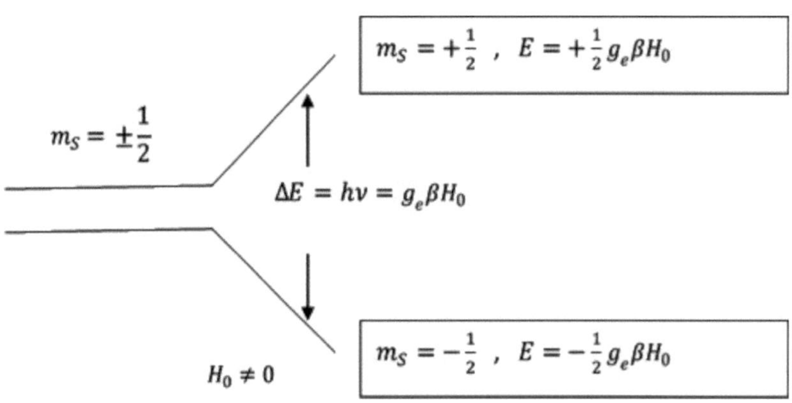

Figure 6. Electron spin resonance transition.

$$\Delta E = h\nu = g_e\beta H_0 \tag{31}$$

is called **the resonance condition for EPR**.

2.2.1. *g-factor*

The unpaired electrons can cause a slight shift in the resonance line due to the internal magnetic field effect. This effect is expressed as a g-value shift in the EPR. The largest shifts occur in paramagnetic transition metal ions. The g-value parameter in the EPR is identical to the chemical shift parameter in NMR.

To calculate the g-value, the values of the microwave frequency and the magnetic field must be known. g-value is obtained from the resonance condition.

$$g_e = \frac{h\nu}{\beta H_0} \tag{32}$$

The g-value calculation can also be performed using a sample with the known g-value as a reference. The reference material for EPR is **diphenylpicrylhydrazyl**. The g-value of the standard sample is 2.0036.

$$g = g_{ref}\frac{H_{ref}}{H} \tag{33}$$

2.2.2. *Hyperfine coupling*

The interaction between the unpaired electron and the nucleus is called the hyperfine structure interaction. For hyperfine structure interaction, the nuclear spin value must be different from zero ($I \neq 0$). The hyperfine structure interaction is divided into **the isotropic hyperfine structure interaction (Fermi contact)** and **the anisotropic hyperfine structure interaction (dipole-dipole interaction)**. While the anisotropic interaction is dependent on the orientation of the molecule, the isotropic effect is not dependent on the orientation of the molecule. The symbol of the isotropic hyperfine structure interaction constant is "*a*," whereas the symbol of the anisotropic hyperfine structure interaction constant is "*A*." Usually the Gauss unit is used for hyperfine structure constant. In addition to the Gauss unit, the unit of MHz is also used (1 G ≈ 2.8 MHz). The value of a is expressed as:

$$a = \frac{8\pi}{3} g_e\beta g_n\beta_n|\psi(0)|^2 \tag{34}$$

where $|\psi(0)|^2$ is the probability of finding the electron in the s-sphere for the hydrogen atom. For isotropic hyperfine structure interaction, the Hamiltonian is expressed as follows:

$$\mathcal{H} = a\,\vec{S}\cdot\vec{I} \tag{35}$$

For the anisotropic hyperfine structure interaction, it is expressed as:

$$\mathcal{H} = \vec{S} \cdot \vec{\vec{A}} \cdot \vec{I} \tag{36}$$

The effective Hamiltonian expression for EPR consists of the sum of **the electron Zeeman Hamiltonian** and **the hyperfine structure interaction Hamiltonian** terms:

$$\mathcal{H} = \beta \vec{H} \cdot \vec{\vec{g}} \cdot \vec{S} + \vec{S} \cdot \vec{\vec{A}} \cdot \vec{I} \tag{37}$$

where $\vec{\vec{g}}$ and $\vec{\vec{A}}$ are in the tensor form. In the resonance case, the average of the diagonal elements of the g-tensor gives the isotropic value of the g-factor.

$$g = \frac{1}{3} Trace\left(\vec{\vec{g}}\right) \tag{38}$$

In the same way, the average of the diagonal elements of the A-tensor gives isotropic value "a."

$$a = \frac{1}{3} Trace\left(\vec{\vec{A}}\right) \tag{39}$$

An example of an EPR spectrum is shown in **Figure 7**.

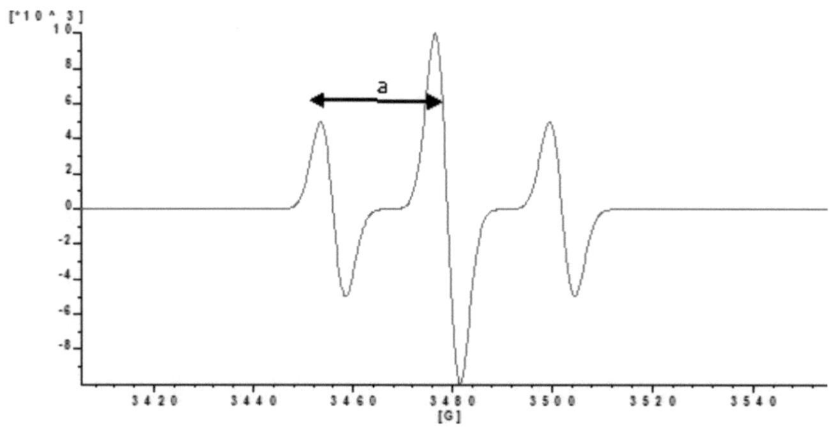

Figure 7. An example of an EPR spectrum and the measurement of the "a" value. If there are n equivalent nuclei, the spectrum shows $2nI + 1$ splits. Here, the nuclei with the same hyperfine structure constant as the equivalent nuclei expression are meant. If the nuclear spin is 1/2, the number of lines and the relative intensity are given by the binomial theorem (**Figure 8**). The line-to-line spacing gives a hyperfine structure constant [4–6].

n° of H											
1					1		1				
2				1		2		1			
3			1		3		3		1		
4		1		4		6		4		1	
5	1		5		10		10		5		1
.....	.....	.....	.....	.....	.....	.....	.....	.....	.....	.....	.....

Figure 8. The Pascal triangle.

The difference between the hyperfine structure splitting of two inequivalent protons and the hyperfine structure splitting of two equivalent protons is shown in **Figures 9** and **10**, respectively.

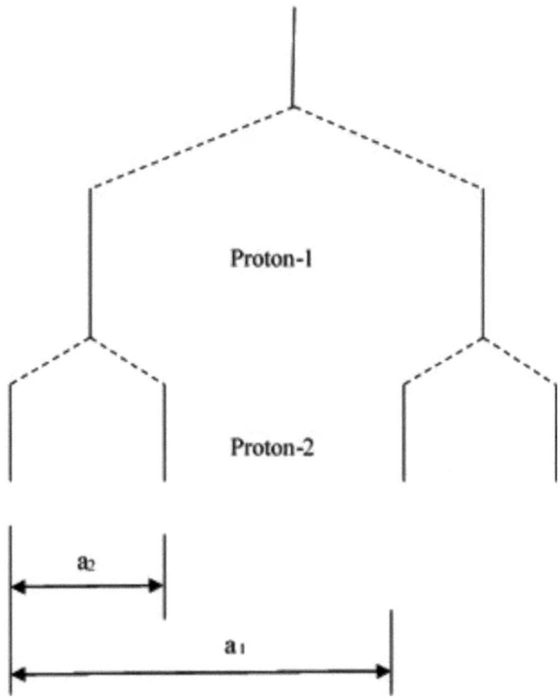

Figure 9. The hyperfine structure splitting of two inequivalent protons ($a_1 \neq a_2$).

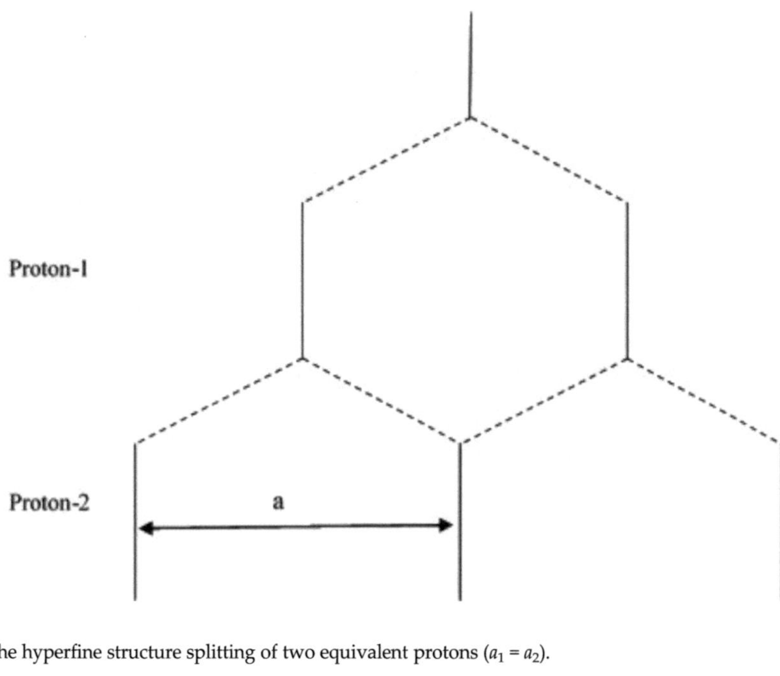

Figure 10. The hyperfine structure splitting of two equivalent protons ($a_1 = a_2$).

The hyperfine structure interaction in the EPR is identical to the spin-spin coupling interaction in NMR.

3. Conclusion

EPR and NMR form the magnetic resonance spectroscopy. EPR and NMR depend on the same basic principles. However, these two experimental methods differ because of the differences in the physical quantities between the electron and the nucleus. These differences stand out in terms of charge, mass, and magnetons (Bohr magneton or nuclear magneton).

Author details

Betul Çalişkan[1]* and Ali Cengiz Çalişkan[2]

*Address all correspondence to: bcaliska@gmail.com

1 Department of Physics, Faculty of Arts and Science, Pamukkale University, Kinikli, Denizli, Turkey

2 Department of Chemistry, Faculty of Science, Gazi University, Ankara, Turkey

References

[1] Foster MA. Magnetic Resonance in Medicine and Biology. 1st ed. Oxford, OX, England; New York: Pergamon Press; 1984. p. 1-3

[2] Varian Associates. Instrument Division. NMR and EPR Spectroscopy. Papers presented at Varian's Third Annual Workshop on Nuclear Magnetic Resonance and Electron Paramagnetic Resonance, held at Palo Alto, California. Oxford: Pergamon Press; 1960. pp. 1-3

[3] Poole CP, Jr., Farach HA. The Theory of Magnetic Resonance. New York: Wiley-Interscience; 1972. pp. 1-270

[4] Weil JA, Bolton JR, Wertz JE. Electron Paramagnetic Resonance: Elementary Theory and Practical Applications. New York: Wiley Interscience; 1994. p. 1-392

[5] Atherton NM. Electron Spin Resonance: Theory and Applications. New York: Wiley Interscience; 1993. p. 1-297

[6] Gordy W. Theory and Applications of Electron Spin Resonance. New York; Chichester [etc.]: Wiley; 1980. p. 1-197

Chapter 2

Magneto-Elastic Resonance: Principles, Modeling and Applications

Yannick Le Bras and Jean-Marc Greneche

Additional information is available at the end of the chapter

http://dx.doi.org/10.5772/intechopen.70523

Abstract

The magnetostriction effects are first discussed in the frame of the magneto-elastic resonance to define important values mainly the magneto-elastic coupling factor, k_{33}. We review the different magnetostrictive materials according to their developments, with a special attention to amorphous ribbons to design magnetostrictive resonators. Furthermore, we focus on the current instrumental setups including their limitations, and then on the usual measurement procedures of the resonators, particularly the frequency domain measurement. In addition, an innovative approach based on the magneto-elastic impedance is reported, together with an analytical model which establishes the complete transfer function between the input and output voltages. This model is applied to ribbon-shaped materials, particularly to determine the magneto-elastic coupling factor. These resonators are suitable to sensing applications, i.e., to estimate the influential quantities such as the temperature, magnetic fields and mass stuck on the resonating surface.

Keywords: resonant frequency, magnetostrictive resonators, magneto-mechanical coefficient, analytical model, sensor

1. Introduction

This chapter deals with the magneto-elastic resonance: this form of mechanical resonance involves magneto-mechanical properties of some ferromagnetic materials. Consequently, it presents some similarities to other types of resonance, such as the existence of resonant and anti-resonant frequencies. The behaviors of magnetostrictive resonators, which also result from magneto-mechanical properties, give rise to some specific particularities. After introducing the main features on magneto-elastic resonance, we first report on the magnetostriction effects and the relevant characteristics of subsequent materials in order to design magnetostrictive resonators. Then, we detail an analytical model in the case of a ribbon-shaped resonator allowing the magneto-elastic coupling factor to be estimated. Finally, we propose some resonator based on a magnetostrictive amorphous ribbon, which behaves as a good platform for sensing applications while we report on the emblematic example of freezing-rain sensor.

2. Magnetostriction effects

Resonance could be defined as a phenomenon that occurs when a vibrating system has greater amplitude at some specific resonant frequencies. In case of magneto-elastic resonance, the vibrating system can be established from a magnetostrictive material. One can consider, in a first approach, this mechanical resonance occurs with a magnetic cause. To obtain the excitation, i.e., a mechanical vibrating strain, one applies a vibrating magnetic field and the magnetostriction converts magnetic variation into strain variation. In return, when resonance occurs, strain is at a maximum. As a consequence of magnetostrictive effects, magnetic values are also at maxima resulting from maximum of mechanical values. Thus, magneto-elastic resonances result from mechanical and magnetic resonances. Consequently, studying magneto-elastic resonance requires knowing of magnetostriction effect, as presented in the next section.

2.1. Magnetostriction

2.1.1. Definition

Magnetostriction can be defined, in a first approach, as the property of some ferromagnetic materials to modify their shape due to change in magnetization [1, 2]. In practice, only some ferromagnetic materials have significant shape and magnetization correlated changes. This phenomenon was discovered by James Prescott Joule in 1842 studying a sample of iron. Since Joule's discovery, many magnetostriction effects have been highlighted, such as bending, torsion, density changes, or Young's modulus variations. We still use the term of magnetostriction for all magneto-elastic properties. This chapter is only concerned by changes in shape. More precisely, two effects are involved in common magnetostrictive resonators, the Joule and Villari effects; the last one corresponds to the inverse magneto-elastic effect.

2.1.2. Joule and Villari effect

A ferromagnetic material with parallelepiped shape elongates or shrinks under a magnetic field according to the longitudinal Joule effect. The reversal effect, change of magnetization while submitted to a mechanical stress is known as Villari effect. These effects are depicted by the material magnetostriction curves which describe the variation of the relative deformation $\lambda = \frac{dL}{L}$, where L is the length of the sample, versus H the magnetic field applied to the material.

A typical curve is characterized by a maximum, as illustrated in **Figure 1**; in addition, one clearly observes some hysteresis also commonly called "butterfly loop," because of its symmetrical shape. In the next part, curves are restricted to the positive fields. The asymptotic

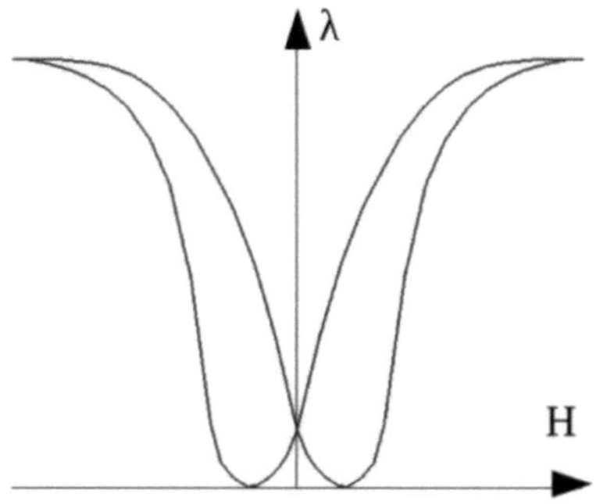

Figure 1. Typical magnetostriction curves in form of butterfly loop.

elongation or shrinkage gives rise to magnetostriction at saturation: the maximum value λ_s corresponds to the saturation magnetostrictive coefficient, i.e., the strength of the magneto-elastic coupling, which can be thus either positive or negative, respectively. It is usually found in the order of 10^{-6} but could rise up to 10^{-3} in the case of Terfenol-D ($Tb_xDy_{1-x}Fe_2$, x ~ 0.3), which behaves as the best magnetostrictive material and is commonly applied as engineering magnetostrictive material. The magnetostriction can be described by a quadratic function and the sign is strictly dependent on the material, not on the direction of the applied magnetic field. Note that it differs from piezomagnetism (analogously piezoelectric effect), which is characterized by a linear coupling between the mechanical strain and the magnetic polarization.

It is important to emphasize that function λ(H) is not linear and could be more rugged than the typical one illustrated in **Figure 1**. Indeed, λ(H) is not strictly monotonous (as for Fe), for which one can distinguish two regimes with positive and negative values of λ, corresponding to an elongation and shrinkage of material for small and larger fields, respectively.

In addition, the deformation is not exclusively dependent on the magnetic field. Indeed, among the other parameters, the temperature plays an important role: when the temperature increases, the elongation decreases as the magnetization is reduced. The magnetostriction curves also depend on the direction of the applied field respect to the easy magnetization axes, i.e., the shape and the chemical purity of the sample and also on its thermomagnetic history.

2.1.3. Causes

The physical mechanisms of the magnetostrictive effects have not been yet described successfully at the atomic scale, to the best of our knowledge. But they are not necessary for our current topic. On the opposite, we would only keep in mind a simple picture, as schematized in **Figure 2**: the main idea is based on the rotation of magnetization in presence of an external magnetic field, which may originate some new arrangements of magnetic domains causing either elongation or shrinkage of the magnetostrictive material [3].

One can distinguish different contributions to the magnetic energy from nano to microscale: exchange interactions, dipolar interactions, magneto-crystalline anisotropy, shape, interface,

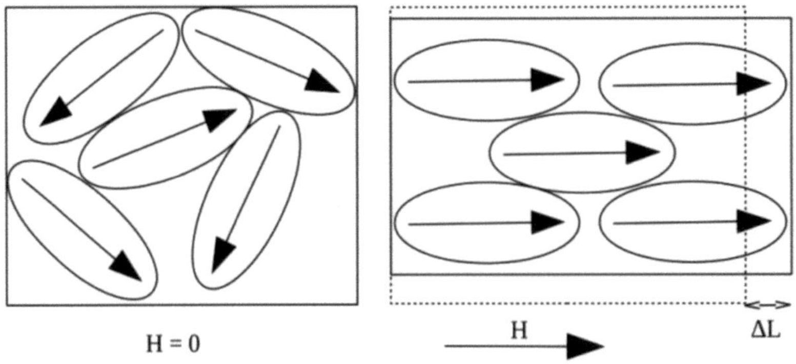

Figure 2. Schematic picture depicting the magnetostriction caused by rotation of magnetization.

and magneto-elastic anisotropy energies. When the material is submitted to mechanical stresses and/or an external magnetic field, the equilibrium of the deformations corresponds to the minimum of the total energy. In the case of a crystalline magnet, the application of uniaxial mechanical stress originates a magnetostrictive contribution to the magnetic anisotropy. It is clear that this magneto-elastic contribution results from the magneto-crystalline term: indeed, under stress crystals can be considered as stressless crystals with a slightly different crystalline structure. In the case of polycrystalline materials, the saturation magnetostriction is an average over different crystal orientations.

Magnetostrictive materials are suitable to convert magnetic into kinetic energy and vice-versa: they can be thus applied to design actuators and sensors. They are implemented in sonars, generation of ultrasound for medical, industrial uses, or for active control of noise and vibration, using simultaneously the opposite effect for vibration measurement and the direct one to carry out the corrective action.

2.2. Characteristic quantities, magneto-elastic coupling factor

2.2.1. Curves, magnetostriction at saturation and slope

The useful information expected by engineers and technicians is the magnetostriction curve, as plotted in **Figure 1**, which characterizes the magnetostrictive material [4]. Indeed, one could easily define the maximum of deformation and estimate λ_s which is usually reported by the manufacturer in the literature. This value presents the advantage to be unequivocal and weakly dependent on further physical parameters except temperature.

This value is able to predict the maximum change in length as $\Delta L_{max} = \lambda_s \cdot L$ but does not describe the sensitivity of the magneto-mechanical conversion. But the slope $d = \left(\frac{\partial\lambda}{\partial H}\right)_\sigma$ is a useful representation of materials properties, since it indicates how rapidly the strain changes with the relevant applied field, according to Jiles [5]. The largest slope, d_{max}, corresponds to the best operating point. But, it is important to emphasize that literature does not report on d_{max}

but on λ_s, because of some dependencies on influence quantities (especially d_{max}, sometimes noted d_{33}, depends on the direction of the magnetic field).

2.2.2. *Magneto-mechanical coupling coefficient k_{33}*

The more relevant characteristic of a resonator is obviously the magneto-mechanical coupling coefficient k_{33}, a dimensionless parameter, which describes the energy conversion as k_{33}^2 is the energy conversion ability from magnetic into elastic energy and inversely. The values of k_{33} which can be estimated from the slope of the curve are expected to be theoretically ranged from 0 up to 1. The larger value which is 0.97 has been observed for an amorphous metallic ribbon.

Du Trémolet de Lacheisserie has proposed an equation to estimate the effective magneto-mechanical coupling k_{33} coefficient from the calculation of Gibbs free energy, as

$$k_{33} = d_{33}\sqrt{\frac{Y^H}{\mu_{33}{}^{\sigma}}} \qquad (1)$$

where, $\mu_{33}{}^{\sigma}$ is the permeability at constant stress and Y^H the Young's modulus at constant field (certain conditions are reported in a next section). Indeed, the effective value of k_{33} depends on the boundary conditions (geometry and fixation of the magnetostrictive material acting as resonator) and the mode of induction of the magnetic field.

2.3. Materials

Since Joule and his discovery of magnetostriction on an iron sample, many new magnetostrictive materials have been identified [3, 5, 6]. Hartemann proposed [1] to classify them into four main categories: nickel and metallic crystalline alloys, the first materials to be used, ferrites, iron-rare-earth alloys, and amorphous alloys. But, this classification has to be updated with nanocrystalline alloys as obtained from subsequent annealing on as-quenched alloys on one hand, and the newer Fe-Ga based alloy (Galfenol) on the other hand.

2.3.1. *Nickel, metallic alloys, and magnetostrictive ferrites*

Polycrystalline nickel was the first magnetostrictive material to be used as a transductor. **Figure 3** compares the magnetostriction curves characteristics of Ni (thick) and Fe (thin), revealing negative and positive magnetostriction coefficient, respectively.

Nickel which is semi-soft (or semi-hard) magnetic material, gives clear evidence for a quite large linearity range with a magnetostriction at saturation λ_s of −35 ppm and a magneto-mechanical coefficient k_{33} of 0.3. In addition to a significant hysteresis, Ni characteristics are strongly dependent on its chemical purity and the annealing conditions to get polycrystalline structure: nevertheless Ni remains an excellent standard. As abovementioned, λ_s (Fe) depends on the external field, giving rise to positive and then negative magnetostrictive behavior.

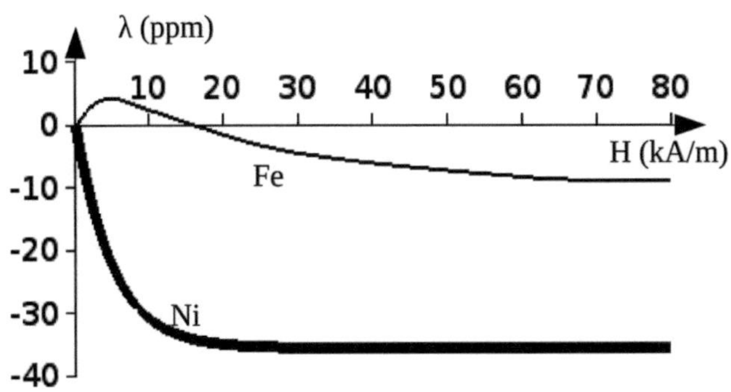

Figure 3. Magnetostriction curves characteristic of Ni (thick) and Fe (thin).

The magnetostrictive properties of Fe-Ni alloys result from a combination of their respective positive and negative magnetostrictive and magneto-crystalline anisotropies: it allows different magnetostrictive characteristics to be tuned as a function of the chemical content. Thus, Permalloy has high permeability and magnetostriction near zero for Permalloy 78 (78% Ni) but Permalloy 45 is greatly magnetostrictive ($\lambda_s = 27 \times 10^{-6}$, $k_{33} = 0.3$). **Table 1** lists some physical characteristics of iron-aluminum (Alfenol), nickel-cobalt, and iron-cobalt alloys.

In the case of ferrites with spinel structure, the magnetic properties are not only dependent on the nature of cations, but also on their distributions into the tetrahedral and octahedral sites giving rise to either direct, inverse, or mixed structures. Consequently, the conditions of elaboration using the ceramic route, the chemical nature, and content of their atomic elements provide large varieties of materials. Co-based ferrites are excellent candidates as magnetostrictive materials (see characteristics listed in **Table 1**, in addition to their high resistivity compared to those of

Material	λ_s **(ppm)**	k_{33} **max ()**	d_{33} **max [6]** $(10^{-9}$ m/A)
Fe	−9		0.3
Ni	−35	0.3	−3
Co	−62		−0.2
Permalloy 45 (Ni45-Fe55)	27	0.3	
Permalloy 80 (Ni80-Fe15-Mo5)	<1.2		
Alfer 13 (Al 13-Fe 87)	40	0.3	
Co 4.5-Ni 95.5	−36	0.5	
Fe 30-Co 70 laminated	130		
Ferrites Fe_3O_4	40	0.36	
Ferrites $CoFe_2O_4$	−110		
Terfenol ($TbFe_2$)	1753	0.35	
Terfenol-D ($Tb_{0.3}Dy_{0.7}Fe_2$)	1100	0.75	57
Galfenol	250	0.7	

Table 1. Specific characteristics (λ_s, k_{33max}, d_{33max}) of some selected magnetostrictive materials.

metal alloys). Microferrites can then be used at higher frequencies, but their mechanical fragility remains a serious weakness.

Nickel and metal alloys are used mainly as actuators for applications requiring small displacements with a large force like for ultrasound emission.

2.3.2. Iron-rare-earth compounds

As noted by du Trémolet de Lacheisserie, studies developed on iron-rare-earth compounds are exemplary in the field of magnetic materials. The aim of the researchers was to combine advantages of 3d metals and/or alloys able of operating at room temperature and under relatively small magnetic fields, but with poor magnetostrictive effects and 4f metals which exhibit high values of magnetostriction coefficient but very low Curie temperatures. Those studies developed in the 70's led to significantly improved magnetostrictive materials with deformations 50–100 times larger. Thus, Clark first developed the $TbFe_2$ alloy named Terfenol (TERbium, FEr, Naval Ordnance Laboratory), which exhibits a relatively high Curie temperature with giant magnetostriction but with a great magneto-crystalline anisotropy. Then, he elaborated a mixed alloy combining two different rare-earth species, giving rise to $Tb_{0.3}Dy_{0.7}Fe_2$ which exhibits rather similar advantages than Terfenol but is easier to be magnetically saturated ($\lambda_s = 1100 \times 10^{-6}$, $k_{33} = 0.75$). Terfenol compounds which behave as hard magnets are brittle and expensive. According to its characteristics, Terfenol-D remains currently an excellent magnetostrictive material. Indeed, it is suitable to be applied as magnetostrictive actuator at room temperature, but with restriction in use as resonator. We report magnetostriction curve under preload (**Figure 4**): it appears different curves and in particular, the maximum slope of the curves reported are 15, 80, and 40 10^{-9} A/m for pressures of 0, 20, and 40 MPa, respectively. Such values are greater than those predicted by du Trémolet de Lacheisserie [4].

It is important to emphasize that, as observed in its website [7], Etrema™ reports the curve established without load which does not allow correct values of k_{33} to be extrapolated; indeed, as illustrated in **Figure 4**, the values of the magneto-mechanical coefficient can be well estimated providing that the material (Terfenol-D) is submitted to important loads. Thus, one has

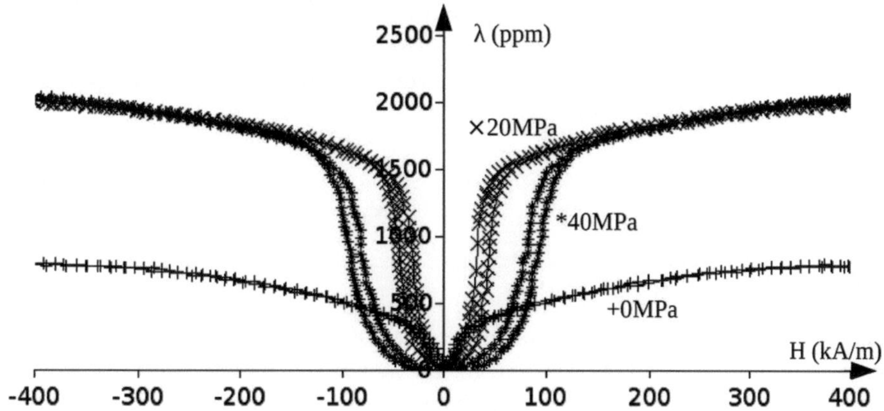

Figure 4. Magnetostriction curve characteristic of Terfenol-D with and without preloading.

to be very careful to estimate the value of k_{33}. In addition, most of resonators work without load, making that Terfenol cannot act as an excellent magnetostrictive material, contrarily to Ni which possesses a large k_{33} value with a small polarization.

2.3.3. Amorphous and nanocrystalline material

In the case of usual resonators, the highest values of magnetostriction coefficient are not strictly necessary but the key parameter does result from the largest possible magnetostrictive effect obtained in presence of a magnetic field as small as possible, i.e., large d and k_{33} values [6, 8]. Some ribbon-shaped amorphous glasses possess very good magneto-elastic properties (great k_{33} for small field) associated to excellent mechanical properties. Let us remember that the metallic amorphous ribbons, also called metallic glasses, are obtained by rapid quenching from the induction melt (10^6 K/s) using the roller technique: a molten alloy is ejected by a flume onto a cooled rotating wheel. The experimental conditions (temperature of the melt, size of capillary, distance capillary-wheel, nature and surface state of the wheel, protective gas, etc.) have to be optimized to get regular ribbons over large lengths (up to several km). Their thicknesses—typically ranged from 20 to 40 μm—favor some mechanical brittleness which depends on quenching conditions. The amorphous ribbons are usually soft magnets with relative permeability more than 10^5 and coercive field near 1 A/m, very low magneto-crystalline anisotropy while their magnetostrictive properties are strongly correlated to their chemical composition (particularly that of Fe, Ni, and Co). The magnetic properties can be improved by annealed under a magnetic field, transverse to increase magnetostriction (longitudinal to annihilate). The largest magneto-mechanical coupling coefficient k_{33} was measured on Metglas 2605SC ribbon annealed at 390°C under a transverse in-plane magnetic field of 400 kA/m for approximately 10 min. The magneto-mechanical coupling coefficient k_{33} is close to 1. We report technical properties of two ribbons of metallic glasses, the best 2605SC and the most used 2826 MB.

As listed in **Table 2**, the main characteristics of metallic amorphous ribbons make them good candidates as magnetostrictive resonators (soft magnet, mechanically soft, large k_{33}), despite their weak thicknesses. Nanocrystalline alloys (such as FINEMET, NANOPERM, and HITP ERM) which result from a subsequent annealing of the amorphous precursor do not exhibit better magnetostrictive characteristics. An alternative is related to bulk amorphous glasses (BMG) which could be obtained as cylindrical rods by mold casting and suction casting techniques: some of them possess excellent soft or hard magnetic properties with saturation magnetostriction values ranged up to 40×10^{-6}.

3. Magnetostrictive resonator

3.1. Structure

Resonators consist of a magnetostrictive material, one or two exciting coils, one or two pick-up coils and eventually a support (a schematic view is given in **Figure 5**) [9, 10]. Exciting coil, either Helmholtz type or a rather long cylindrical coil, aims to produce a homogeneous magnetic field with an alternating component and a DC component. Exciting coil converts vibrating current into vibrating field. This field induces vibrations in the ribbon-shaped material with a resonant frequency which is dependent on its size, usually length L.

Properties	Units	Metglas 2605SC	Metglas 2826 MB
Composition		$Fe_{81}B_{13.5}Si_{3.5}C_2$	$Fe_{40}Ni_{38}B_4Mo_{18}$
Thickness	(μm)	17	29.2
Density	$kg.m^{-3}$	7320	7900
Magnetostriction at saturation	(ppm)	30	12
Magneto-mechanical coupling coefficient	—	0.97 (H = 50 A/m)	0.3
Crystallization temperature	K	480	410
Young's modulus	GPa	25	100-110
Résistivité électrique	μ.Ω.m	1.35	1.38
Perméabilité relative maximum	—	300,000	800,000

Table 2. Specific physical parameters characteristic of two ribbon-shaped metallic glasses.

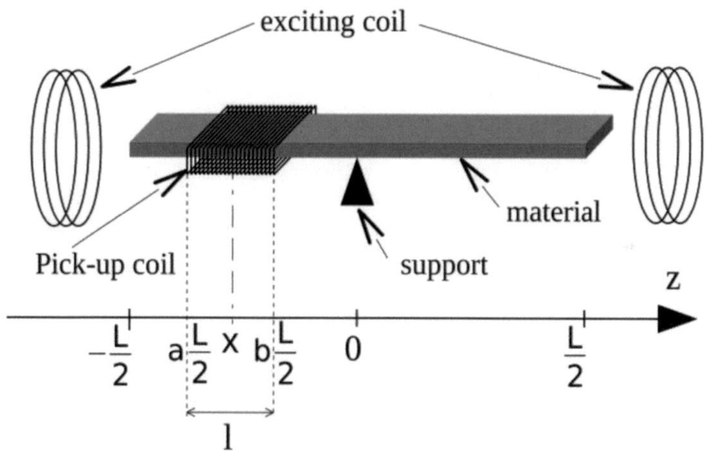

Figure 5. Principle of a ribbon-shaped resonator.

As described by Grimes [10], vibrations "can be detected magnetically with a pick-up coil, acoustically with a microphone, or optically with a laser emitter and a photo-transistor." Next, we focus on resonators with vibration detection based on the Villari effect, resulting from the pick-up coil as previously mentioned.

Ideally, the best configuration would be to study a free ribbon, thus in practice, the ribbon does simply lye on a flat and smooth surface or be centered in its middle on a support.

3.2. Operating principle, elementary model

A generator delivers a current with an alternating component to the exciting coil which in turn generates a magnetic field proportional to the current. The magnetostrictive material subjected to the field, is thus deformed. By applying a sinusoidal current component, the material vibrates according to a sinusoidal mode. The excitation is the strains originated from field's variation, and the resonator acts thus as a mechanical resonator.

The resonator with parallelepiped shape could be modeled as a plate, where ultrasound waves propagate without losses. Assuming a one-dimensional problem and choosing as system a slice of thickness dx, its mass is

$$dm = \rho \cdot h \cdot edx \tag{2}$$

where h, e, and ρ are the thickness, depth, and the density of the plate, respectively.

The balance of forces represented in **Figure 6** provides $\overrightarrow{f(x)} = \sigma(x)\cdot S\cdot \vec{x}$ and $\overrightarrow{f(x+dx)} = -\sigma(x+dx)\cdot S\cdot \vec{x}$, where S is the cross-section of the resonator.

When applying fundamental principle of dynamics, it comes out:

$-\sigma(x+dx)S + \sigma(x)S = \rho S dx \frac{\partial^2 u}{\partial t^2} \Rightarrow -\frac{\partial \sigma}{\partial t} = \rho \frac{\partial^2 u}{\partial t^2}$, where u is the displacement.

From both the Hooke's law establishing the proportionality between relative elongation and constraint, $\lambda = \frac{-1}{Y}\cdot \sigma$, and the expression of the elongation is given as

$$\lambda = \frac{dL}{L} = \frac{u(x+dx) - u(x)}{dx} = \frac{\partial u}{\partial x} \tag{3}$$

one gets $\frac{\partial^2 u}{\partial t^2} - \frac{1}{c^2}\frac{\partial^2 u}{\partial x^2} = 0$ where $c = \sqrt{\frac{Y}{\rho}}$.

This equation can be solved using harmonic solutions $u(t,x) = e^{j2\pi ft}\cdot u(x)$ with

$u(t,x) = U_1 e^{j\frac{2\pi f}{c}x} + U_2 e^{-j\frac{2\pi f}{c}x}$ and assuming constraint at x = 0 with u(0)=0 and boundary conditions $\sigma\left(\frac{L}{2}\right) = \sigma\left(-\frac{L}{2}\right) = 0$.

Consequently, $\sigma(x) = -j2Y\frac{U\cdot 2\pi f}{c}\cos\left(\frac{2\pi f}{c}x\right)$, to satisfy $\sigma\left(x = \frac{L}{2}\right) = 0$ involving $\frac{2\pi f}{c}L = \pi(2p+1)$ where p is a positive integer.

Resonances are established at frequencies $f_p = (2p+1)\frac{c}{2L} = \frac{(2p+1)}{2L}\cdot\sqrt{\frac{Y}{\rho}}$, and particularly the fundamental frequency corresponding to that characteristic of the material:

$$f_0 = \frac{1}{2L}\cdot\sqrt{\frac{Y}{\rho}}. \tag{4}$$

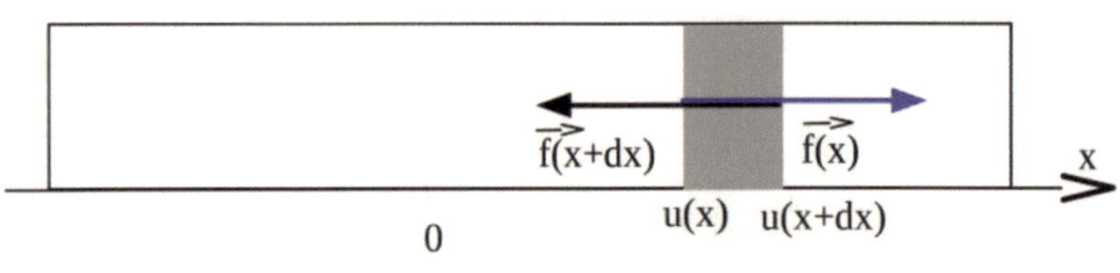

Figure 6. Resonator modeling, with the stress forces.

3.3. Measurement

By Villari effect, resonator's vibration generates a time-varying magnetic flux, measured by the pick-up coils. Three different routes as time domain measurement, frequency domain measurement, and magneto-elastic impedance can convert this variation to estimate f_0.

3.3.1. *Time domain measurement*

The strategy consists in exciting the resonator to its natural frequency. Then, one could apply to the exciting coil a rectangular wave-train pulse, or even better, a sinusoidal wave-train, as the current variation is limited by the inductance. Then the response, the pick-up coil's voltage, is a damped sine wave-train (**Figure 7**). The natural frequency can be thus determined by fast Fourier transform (FFT), frequency counting or demodulation. FFT gives the spectrum of the voltage which maximum corresponds to the natural frequency. Furthermore, frequency counting consists in the determination of a number of oscillations. Thus, a comparator converts the voltage into a rectangular shape voltage whose frequency can be easily determined by a counter, according to the definition of a frequency. The last technique consists in demodulating the pick-up's signal: a phase-locked loop replaces the counter and gives voltage corresponding linearly to the frequency. The frequency counting and demodulation techniques require less high-performance instrumentation but remain more difficult to be well achieved. On the contrary, FFT gives *a priori* better results and particularly the quality factor characteristics of the resonance.

3.3.2. *Frequency domain measurement*

The resonant frequency results from the transfer function. The excitation coil is connected to a function generator and the pick-up coil to a voltage measurement system. The generator delivers a sinusoidal voltage as a function of frequency (sweeping mode) giving rise to V(f) corresponding to the amplitude of the pick-up coil (as illustrated in **Figure 9b**). The resonant frequency corresponds to the maximum. This measurement can be carried out using only a spectrum analyzer that delivers the magnitude of the input signal versus frequency: such an approach allows the resonant frequency, the anti-resonant frequency, and the resonance quality to be obtained.

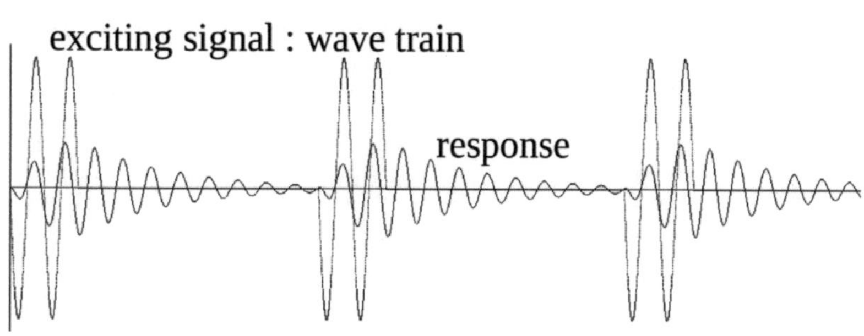

Figure 7. Time domain measurement signals: wave train and response as continous and dashed line, respectively.

3.3.3. Magneto-elastic impedance

The experimental setup for measuring the evolution of the impedance is close to that of frequency domain measurement. An analyzer measures real and imaginary parts of voltage as a function of the frequency, allowing thus the variation of impedance which is experimentally similar to the transfer function versus frequency. Let us note that the instrumentation is similar for the two techniques, but physicists prefer the later one which also gives the evolution of the permeability.

3.4. Development

The development of a resonator requires a setup comprising polarization and excitation coils (possibly one for both), two pick-up coils, a continuous power supply, and an analyzer, in addition to the magnetostrictive material. Consequently, its achievement is not a difficult task providing some rules to be satisfied.

3.4.1. The resonator

As concluded in Section 2.2, the main criterion of choice is the magneto-mechanical coefficient or the slope of curves $\lambda(H)$, i.e., a material with a k_{33} at least more than 10%, for a DC field easy to obtained in the lab. For preliminary tests, Ni foil or amorphous 2826MB would be a good choice according to Section 2.3, but the optimal choice depends on the application.

The output is often the resonant frequency that is inversely proportional to the length of the resonator with parallelepiped shape. To get an acute resonance, the resonator requires a strictly constant length L: consequently, the cutting has to be done with extreme caution. Different techniques such as paper guillotine, laser beam, diamond wire saw, or electrical discharge machining have to be optimized according to the brittleness of the material and preventing from contamination and from crystallization in the case of amorphous ribbons. After cutting, the material may undergo subsequent treatment under field in neutral atmosphere to improve magnetostriction.

3.4.2. Coils and electrical setup

Polarization and excitations coils such as Helmholtz or long cylindrical solenoid, do create a uniform field. In addition, as the field produced by a coil is proportional to the current, it is easier to get only one coil, using links capacitor and inductor to discriminate the DC and AC component voltage (see **Figure 8**). The advantage of Helmholtz coils is that the resonator is placed outdoor, but as the field decreases with the square of the coil diameter, a large coil creates a greater field than Helmholtz type with the same current.

The coil picks up the time derivative of flux $\Phi_n = \mu_0 H \cdot S_n + \mu_0 M \cdot S_{rib}$ resulting from n loops of surface S mounted around the material. Then the flux is image of the magnetization M, but also of the magnetic field H. The field component is removed using a differential measurement from two pick-up coils. Indeed, the second coil is identical to the first one and placed, out of the material, symmetrically centered in the excitation field, what measured is the voltage of the

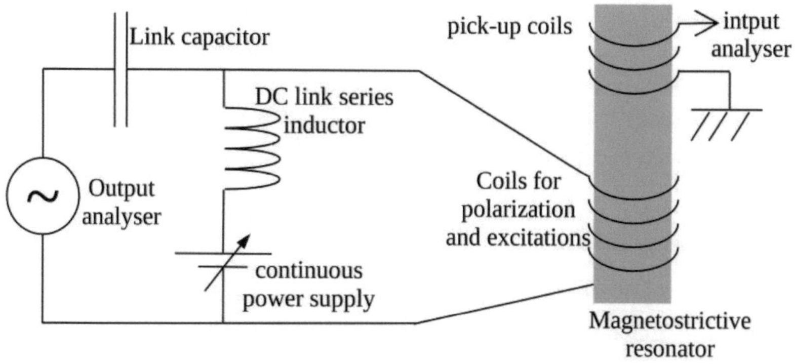

Figure 8. Coils and electrical setup.

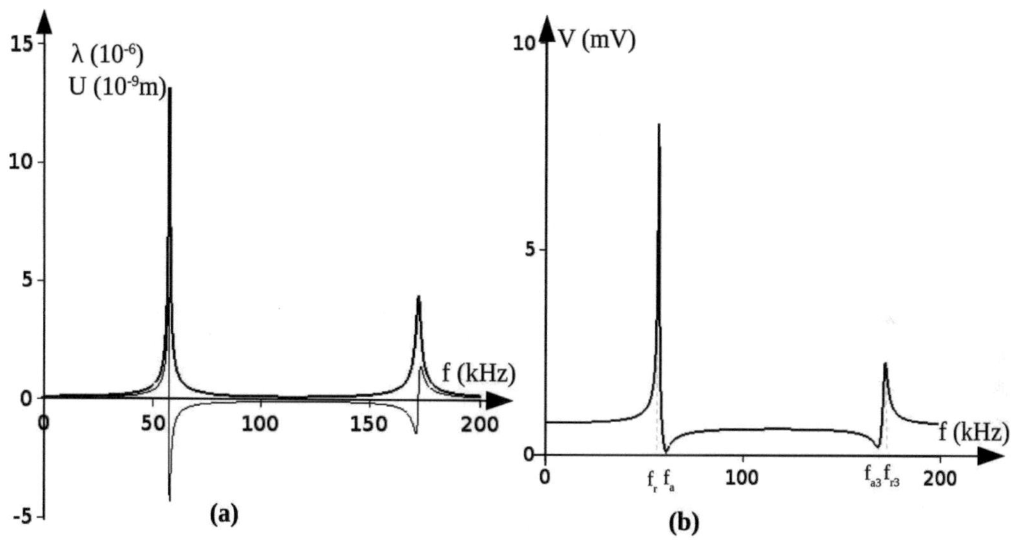

Figure 9. a: Examples of frequencies responses of the maximum strains in the middle (thick,) and displacement at the ends (thin). b: Examples of frequencies responses of the pick-up coil's voltage.

serial coils: $u = n\frac{d(\mu_0 M \cdot S_{rib})}{dt}$. The voltage u is thus proportional to the derivative of the magnetization change with the sinusoidal strain.

3.4.3. Setting

The first step consists in determining the place of the pick-up coil corresponding to a minimum of voltage without resonator in order to optimize the compensation. Then, the measuring range has to be refined from an approximate value of the resonant frequency. The resonance is obtained by adjusting first the amplitude of the excitation at around 1 V and scanning the DC voltage. The final refinement of the position of the pick-up coil and the amplitude of the AC voltage gives rise to a curve similar to that displayed in **Figure 9b**.

4. Analytical model

An analytical model is thus necessary to estimate the frequency dependence of the output/input coils voltage ratio when submitted to an electrical excitation [11]. It consists in establishing the equations coupling mechanical and magnetic quantities.

4.1. Modeling and assumptions

3D general equations can be derived into the two following 1D Eqs. (5) and (6), assuming a low AC component of the magnetic field

$$\widetilde{B} = d\cdot\widetilde{\sigma} + \mu^{\sigma}\cdot\widetilde{H} \tag{5}$$

$$\widetilde{\lambda} = \frac{1}{Y}\cdot\widetilde{\sigma} + d\cdot\widetilde{H} \tag{6}$$

where $\widetilde{X}$ refers to a low level AC quantity, B, H, ε, σ, Y, d, and μ^{σ} are the magnetizing flux density, the magnetic field, the strain, the stress, the Young modulus, the slope of the magnetostriction curve, and the magnetic permeability at constant stress, respectively.

The ribbon is assumed to be set in the middle. The magnetic field is uniform and has two components: H_{DC} and alternating $\widetilde{H}$ with $H_{DC} \gg \widetilde{H}$. In addition, $\widetilde{H}$ is low enough to neglect the effects of hysteresis. The complex Young modulus is expressed as $\overline{Y} = Y(1 + j\eta)$, where the imaginary part takes into account mechanical and magnetic losses with η, the damping factor characteristic of the resonator.

The boundaries conditions are related to the two ends of the ribbon mechanically free, except strains due to magnetostriction:

$$\lambda\left(z = -\frac{L}{2}\right) = \widetilde{\varepsilon_0} = d\cdot\widetilde{H} \tag{7}$$

$$\lambda\left(z = \frac{L}{2}\right) = \widetilde{\varepsilon_0} = d\cdot\widetilde{H} \tag{8}$$

The vibrations resulting from the excitation field are described by the wave propagation equation derived from Newton's second law:

$$\frac{\partial\sigma}{\partial t} = \rho\,\frac{\partial^2 u}{\partial t^2} \tag{9}$$

where u is the longitudinal displacement and ρ the density.

From Eqs. (3), (9), and (6), assuming that $\frac{\partial\widetilde{u}}{\partial z} = 0$ (uniform field), one gets

$$\overline{Y}\frac{\partial^2\tilde{\lambda}}{\partial z^2} = \rho\,\frac{\partial^2\tilde{\lambda}}{\partial t^2} \tag{10}$$

4.2. The strain expression

Eq. (10) can be solved by using harmonic oscillations expressed as

$$\widetilde{\lambda}(z,t) = e^{j\omega t} \cdot E \cdot e^{jKz} \tag{11}$$

From Eqs. (10) and (11) $K = \pm(k_r + jk_i)$ with $k_r = \frac{\sqrt{\frac{\rho}{Y}}\omega}{\sqrt[4]{1+\eta^2}} cos\left(\frac{tg^{-1}(\eta)}{2}\right)$ with $k_i = \frac{-\sqrt{\frac{\rho}{Y}}\omega}{\sqrt[4]{1+\eta^2}} sin\left(\frac{tg^{-1}(\eta)}{2}\right)$.

Considering above boundary conditions, one finally obtains from Eq. (12):

$$\lambda(z) = d{\cdot}\widetilde{H}\left(E_{r0} + jE_{i0}\right)\left(e^{(-k_i+jk_r)z} + e^{(k_i-jk_r)z}\right) \tag{12}$$

With

$$E_{r0} = \frac{e^{\frac{3k_iL}{2}} + 3e^{\frac{k_iL}{2}} + 3e^{-\frac{k_iL}{2}} + e^{-\frac{3k_iL}{2}} - 4e^{\frac{k_iL}{2}}cos^2\left(\frac{k_rL}{2}\right) - 4e^{-\frac{k_iL}{2}}cos^2\left(\frac{k_rL}{2}\right)}{e^{2k_iL} - 2 + e^{-2k_iL} + 16cos^2\left(\frac{k_rL}{2}\right) - 16cos^4\left(\frac{k_rL}{2}\right)} cos\left(\frac{k_rL}{2}\right)$$

$$E_{i0} = \frac{e^{\frac{3k_iL}{2}} + e^{\frac{k_iL}{2}} - e^{-\frac{k_iL}{2}} - e^{-\frac{3k_iL}{2}} - 4e^{\frac{k_iL}{2}}cos^2\left(\frac{k_rL}{2}\right) + 4e^{-\frac{k_iL}{2}}cos^2\left(\frac{k_rL}{2}\right)}{e^{2k_iL} - 2 + e^{-2k_iL} + 16sin^2\left(\frac{k_rL}{2}\right)cos^2\left(\frac{k_rL}{2}\right)} sin\left(\frac{k_rL}{2}\right).$$

At this stage, the frequency variation of the strain can be plotted but the displacement at ends of the resonating ribbon is preferred. This second curve can be obtained by means of a contactless measurement as laser vibrometer or microphone.

The expression of the motion is determined from Eqs. (3) and (12):

$$u(z) = d{\cdot}\widetilde{H}\left(E_{r0} + jE_{i0}\right)\left(\frac{e^{(-k_i+jk_r)z}}{(-k_i + jk_r)} + \frac{e^{(k_i-jk_r)z}}{(k_i - jk_r)}\right)$$

Figure 10a reports the frequencies' responses of the maximum strain (in the middle) and displacement (at the ends) for a ribbon taken from an anti-theft which is Vitrovac 4040 ($Fe_{39}Ni_{39}Mo_4Si_6B_{12}$; $\rho = 7400$ kg.m^{-3} and L = 37 mm. The refined characteristics are $k_{33} = 0.312$, $d = 20 \times 10^{-9}$ m/A, $H_{ACmax} = 4$ A/m and $\eta = 0.012$.

The mechanical responses are quite different from the measurement based on the inverse magnetostrictive (Villari) effect (**Figure 9b**).

4.3. Expression of the frequency response

As coils associated with the magnetostriction convert mechanical quantities into electrical quantities, the next step consists thus in substituting mechanical by magnetic quantities.

From Eqs. (12), (5), and (1):

$$\mu = \frac{\widetilde{B}}{\widetilde{H}} = \mu^{\sigma}\left(1 - {k_{33}}^2 + {k_{33}}^2\left(E_{r0} + jE_{i0}\right)\left(e^{(-k_i+jk_r)z} + e^{(k_i-jk_r)z}\right)\right) \tag{13}$$

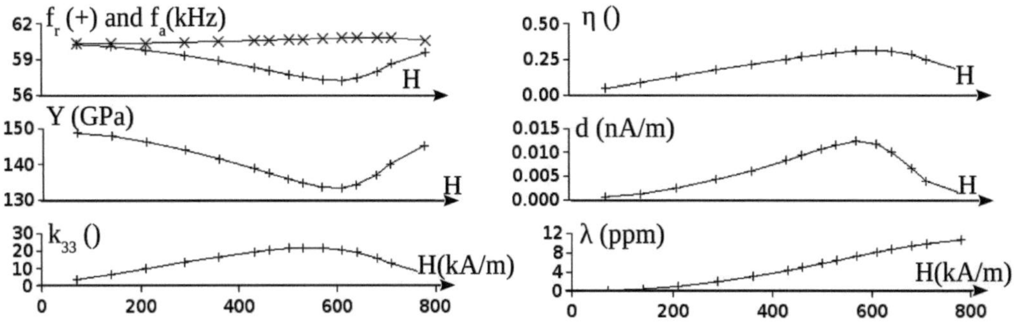

Figure 10. Evolution of resonant and anti-resonant frequency f_r (+) and f_a (x), Young's modulus Y, magneto-mechanical coefficient k_{33}, damping η, slope of the magnetostriction curves d, and strain λ as function of DC field for Vitrovac ribbon.

The current applied to the exciting coil assumed to be only inductive is

$$i(t) = \frac{V_{effexc}}{L_{exc}\omega}\sqrt{2}sin(\omega t), \tag{14}$$

where V_{effexc} and L_{exc} correspond to the rms voltage and the inductance of the excitation coil.

As the B magnetic field can be neglected out of the resonating ribbon, it is expressed as

$$B = \mu \cdot n_l \cdot i \tag{15}$$

assuming an infinitely long solenoid coil where n_l: number of loops per unit length

From Eqs. (13)–(15), one gets than $B(z) = 2\sqrt{2}\cdot\frac{n}{l(b-a)}\frac{V_{effexc}}{L_{exc}\omega}\sqrt{2}sin(\omega t)\cdot\mu(z)$

The output voltage is calculated as:$v(t) = \frac{d\Phi}{dt} = \frac{d\left(\int_{a\frac{l}{2}}^{b\frac{l}{2}} n_l\cdot dz\cdot B(z)S\right)}{dt}$

The frequency response defined as $T = \frac{V_M}{V_{effexc}}$ where V_M corresponds to rms value of V(t) is then:

$$T = T_0\left(1 + 2\frac{{k_{33}}^2}{1-{k_{33}}^2}\cdot\frac{E_{r0}+jE_{i0}}{l(b-a)}\cdot\frac{\left(e^{(-k_i+jk_r)a\frac{l}{2}} + e^{(k_i-jk_r)b\frac{l}{2}} - e^{(-k_i+jk_r)b\frac{l}{2}} - e^{(k_i-jk_r)a\frac{l}{2}}\right)}{k_i - jk_r}\right)$$

$$(\text{With})T_0 = \frac{S}{L_{exc}}\cdot\frac{2{n_b}^2}{l(b-a)}\cdot(1-{k_{33}}^2)\mu^{\sigma}$$

The frequency response T is function of a, b, L, n_b, L_{exc}, S, μ^{σ}, k_{33}, η, l, Y, ρ, and f. While considering T function of T_0, which then becomes a parameter, T is function of T_0, a, b, L, k_{33}, ρ, l, Y, η, and f.

It is important to emphasize that the gain depends on not only the material parameters (L, k_{33}, η, Y), but also on the size and position of the pick-up coil (a, b, and l).

A typical response is plotted in **Figure 9b**. The resonant frequencies, noted f_{rk} and f_0 or f_r for the fundamental one can be estimated from Eq. (4). The anti-resonant frequencies, f_{ak}, are not observable when studying strain response (**Figure 9a**). One observes in **Figure 9b** reversal of some anti-resonant harmonic frequencies when they are smaller than the resonant frequency: details are reported in [12].

5. Applications of magnetostrictive resonator's characterization

This model allows interestingly to estimate the values of k_{33}, η, Y, T_0, from a frequency response, providing that a, b, L, l, and ρ are known [12]. The strategy consists first in saving couples of data (f, T) from a classic analyzer and then to fit them using a least squares method to determine the set (k_{33}, η, Y, T_0). In addition, from Eq. (1), $k_{33} = d_{33}\sqrt{\frac{Y^H}{\mu_{33}{}^{\sigma}}}$, it becomes possible to estimate the value of d and then by integration that of strain λ. **Figure 10** displays the different data characteristic of Vitrovac sample. It is important to emphasize that the present contactless and cheap method is well suitable to characterize soft magnetic resonators.

6. Influence quantities

The frequency response of a resonator is strongly dependent on its geometry such length L, its physical properties (μ^{σ},Y, ρ), and the operating conditions. But, some particular quantities may influence the sensing response of the magnetostrictive resonators.

6.1. Effects of field and temperature

6.1.1. Effect of field

The magnetostrictive effects depend obviously on magnetic field which is an unavoidable influence quantity, but easily quantified. The thicker line in **Figure 11** describes the variation of resonant frequency, at 20°C, versus the applied field. From Eq. (4), resonant frequency

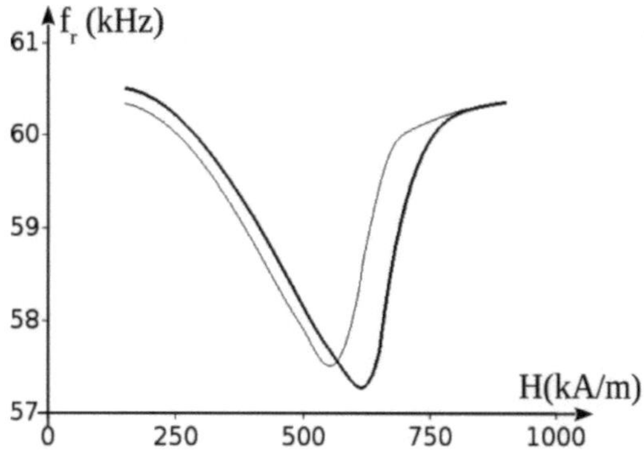

Figure 11. Evolution of magnetostriction curves for 20°C and 100°C (thick and thin line, respectively).

appears as a function of $Y^{1/2}$ assuming the density constant, also labeled as ΔY effect [13]. The variations of d or k_{33} versus the field are sources of perturbation and can be considered as influence quantities which are intrinsic to the material: consequently, one does control that d remains constant in a wide range, i.e., Ni is a better choice than amorphous ribbon.

6.1.2. *Effect of temperature*

As magnetization decreases with temperature, magnetostriction does the same. **Figure 11** compares frequency response versus magnetic field for two temperatures for our ribbon (thicker line corresponds to 20°C and the thinner one to 100°C) [14]. One concludes that increasing the temperature increases the minimum of resonant frequency, but decreases k_{33}. The temperature proves as an important influence quantity with change of resonant frequency up to 15%. This effect can be reduced by tuning the DC field lower than the anisotropy field while that of the thermal expansion can be neglected.

6.2. Effect of a mass stuck on the surfaces

Any mass coated on the resonator tends to absorb vibrations: the effect of inertial mass Δm coating the resonator has to be studied, assumed to be uniformly applied. In Eq. (2), the mass of the system, a slice of thickness dx, changes from $dm = \rho \cdot e \cdot d \cdot dx$ to $\rho \cdot e \cdot d \cdot dx + \Delta m \frac{dx}{L}$. This change acts equivalent to that of density from ρ to $\rho\left(1 + \frac{\Delta m}{L\rho e h}\right)$. Then the resonant frequency becomes:

$$f_0 = \frac{1}{2L}\sqrt{\frac{Y}{\rho\left(1 + \frac{\Delta m}{L\rho e h}\right)}} \tag{16}$$

It is expected only a decrease of f_0. Indeed, one observes a decrease of the maximum due to the losses generated by the friction between the resonator and the coating mass.

6.3. Effects of operating conditions

Any cause of frictions originates from an influence quantity, among them the viscosity and the density of the fluid wrapping the resonator [10]. An increase of viscosity increases the losses of the resonator and then affects damping ratio η. Such an effect can be quantified by comparing the frequency responses corresponding to two different values of damping: one expects a decreasing of the maximum concomitant to the decrease of the resonant frequency which significantly differs when the damping ratio increases (see Eq. (4)). Consequently, the attachment of the resonator disturbs strongly the frequency response.

7. Magnetostrictive sensors

7.1. Freezing-rain sensor, an emblematic example

Freezing-rain sensors which are the emblematic examples of magnetostrictive resonators, particularly because of the non-contact measurement, are used to detect the icing conditions from the mass deposition of ice layer stuck on their surface as well as its growth [15]. The aerospace manufacturer Goodrich™ proclaims that "its sensor detects the presence of icing conditions so that appropriate actions can be taken to prevent damage to power and communication lines, to warn of road hazards, or to keep ice off wind turbine blades or a plane's wings." It is also announced that "surfaces are automatically defrosts itself when ice accumulation reaches 0.5 mm." Technical available information (including natural resonant frequency: 40 kHz, frequency decreases to 130 Hz, strut height of 2.54 cm, and strut diameter: 3.10 cm, the material is a nickel alloy rod) allows us to make some calculations. But boundaries' conditions differ from our model: free at both ends, here fixed-free than the frequency is $f_0 = \frac{1}{4L} \cdot \sqrt{\frac{Y}{\rho}}$, in our case the displacement in the middle is zero, it appears the ribbon fixed in the middle. Taking Young's modulus and density of nickel, respectively 200 GPa and 8908 kg.m^{-3}, from a length of 2.54 cm gives a frequency of 47 kHz.

The frequency, for a cylindrical resonator, with a thickness of ice e_{ice} is given by

$$f_0 = \frac{1}{4L}\sqrt{\frac{Y}{\rho\left(1+\frac{\rho_{ice}\left((d+e_{ice})^2-d^2\right)}{\rho\, d^2}\right)}} \tag{17}$$

Thus, the values of the resonant frequency and its shift are estimated at 47 kHz and 85 Hz using Ni resonator, which are rather consistent with those given the manufacturer (40 kHz and 130 Hz) obtained with Ni based alloy resonator. Finally, we do emphasize that the sensitivity of the sensor is essentially due to the strut diameter (see Eq. (17)).

7.2. Chemical sensor

Magneto-elastic sensors can be used to detect chemicals such as carbon dioxide [16] or ammonia [17], biological cells [18], or to measure pH [19]. The principle is to detect the mass of chemical or biomass stuck on the surfaces. Using amorphous ribbon, the sensibility is excellent. The difficulty is to functionalize the surface in such a way that the product to be detected sticks the surface and no other contaminating elements. Ruan et al. describe the functionalization process to design a sensor for measuring ricin in solution [19]. The sample is first cut using a computer controlled laser cutter. Then it is coated with a 10 nm Cr layer and a 140 nm protective Au layer, with appropriate annealing treatment before functionalization. Magnetostrictive sensors act as excellent platforms to detect very low mass while the wireless and passive nature of these devices allows remote measurements.

7.3. Electronic article surveillance

Magnetostrictive resonators are involved in anti-theft tags which are fixed to merchandise. Tags consist of two mechanically independent free strips, one of a magnetostrictive amorphous ferromagnetic ribbon, and the second one of a magnetically semi-hard film acting as biasing magnet and switch to activate and deactivate the sensor. The good magneto-elastic coupling of the first strip originates the conversion of magnetic energy into mechanical vibrations. Detection gantries emit bursts at frequency close to that of the resonator (58 kHz) inducing thus longitudinal vibrations, which continue even after the burst is over. It results some change in magnetization of the amorphous strip and induces thus an AC voltage to activate the detection gantry's antenna. These tags which are thicker than electromagnetic ones are cheaper and have better detection rates, but vibration and therefore detection can be deleted, when the sensor is submitted to a mechanical pressure (that of the robber!).

7.4. Magnetic sensor, thermometer, and others

The resonant frequency depends on both magnetic field and temperature which can be also measured by a resonator. Garcia-Ambas et al. [14] have investigated the possibility of temperature measurements from the temperature dependence of the magneto-elastic resonance frequency: it occurs when the magnetic biasing field applied to the resonator is close to its anisotropy field. But the sensitivity of the measurement is dependent on the temperature dependence of the magneto-elastic, which is self-correlated to that of the anisotropy constants; these low prize magnetostrictive sensors have the great advantage to make remote measurements. In addition, they can be also involved for differential and multiple measurements. Literature reports several possible applications such as stress [20] and strain [21] measurements or environmental parameters such as viscosity [22].

8. Conclusion

The aim of this chapter deals with an overview of magnetostrictive resonators, their own principles and their respective performances including sensing application domains and limitations. The development of some analytical model allows the characteristics of magnetostrictive to be estimated and the main influential quantities to be defined: thus, it does facilitate the design of new ribbon-shaped resonator suitable for specific applications.

Author details

Yannick Le Bras* and Jean-Marc Greneche

*Address all correspondence to: yannick.le_bras@univ-lemans.fr

Institut des Molécules et Matériaux du Mans, IMMM - UMR CNRS 6283, Université du Maine, Le Mans, France

References

[1] Hartemann P. Effets et matériaux magnétostrictifs. Techniques de L'ingénieur. 1999;**E1880**: 1-13

[2] du Trémolet de Lacheisserie E. Magnétisme, Tome 1, Fondements. Édition Diffusion Presse Sciences; 1999. 496 p. ISBN: 2706108312

[3] Hristoforou E, Ktena A. Magnetostriction and magnetostrictive materials for sensing applications. Journal of Magnetism and Magnetic Materials. 2007;**316**:372-378. DOI: 1016/j.jmmm.2007.03.025

[4] du Trémolet de Lacheisserie E. Magnetostriction: Theory and Applications of Magnetoelasticity. Amsterdam: CRC Press; 1993. 408 p. ISBN: 0849369347, 9780849369346

[5] Jiles DC. Recent advances and future directions in magnetic materials. Acta Materialia. 2003;**51**:5907-5939. DOI: 10.1016/j.actamat.2003.08.011

[6] du Trémolet de Lacheisserie E.: Magnétisme, Tome 2, Matériaux et Applications. Édition Diffusion Presse Sciences, 1999. 510 p. ISBN: 2706108320

[7] TdVib. Terfenol-D [Internet]. 2016. Available from: http://tdvib.com/terfenol-d/# [Accessed: June 06, 2017]

[8] Lupu N, Chiriac H, Ababei G, Grigoras M, Valeanu M, Buruiana T, Harabagiu V, Mihalca I. New 3-D bulk shaped polycrystalline and nanocomposite magnetostrictive materials. In: European School on Magnetism/Workshop on Advanced Magnetic Materials; 9–18 September 2007; Cluj-Napoca, Romania; 2007

[9] Hathaway KB, Spano ML. Measurement of high magnetomechanical coupling factors by resonance techniques. Journal of Applied Physics. 1984;**84**:1765-1767. DOI: 10.1063/1.333470

[10] Grimes CA, Mungle C, Zeng K, Jain M, Dreschel W, Paulose M, Ong K. Wireless magnetoelastic resonance sensors: A critical review. Sensors. 2002;**2**(7):294-313. DOI: 10.3390/s20700294

[11] Le Bras Y, Mazaleyrat F, Greneche JM. Analytical model of the magnetostrictive ribbon shape resonator frequency response. Sensor Letters. 2011;**9**:1801-1806. DOI: 10.1166/sl.2011.1714

[12] Le Bras Y, Lasheras A, Gutierrez J, Mazaleyrat F, Greneche JM. A new magneto-elastic resonance based technique to determine magneto-mechanical parameters of amorphous ferromagnetic ribbons. The Review of Scientific Instruments. 2013;**84**(4):043904. DOI: 10.1063/1.4799177

[13] Gutiérrez J, García-Arribas A, Garitaonaindia J, Barandiarán JM, Squire PT. ΔE effect and anisotropy distribution in metallic glasses with oblique easy axis induced by field annealing. Journal of Magnetism and Magnetic Materials. 1996;**157–158**:543-544. DOI: 10.1016/0304-8853(95)01119-6

[14] Garcia-Ambas A, De Cos D, Gutierrez J, Barandiaran JM. Selectable temperature sensitivity of the magnetoelastic resonance. Sensors and Actuators A. 2003;**106**:111-116. DOI: 10.1016/S0924-4247(03)00146-8

[15] Campbell Scientific. 0871LH1 Freezing-Rain Sensor [Internet]. 2017. Available from: https://www.campbellsci.com/0871lh1 [Accessed: June 06, 2017]

[16] Cai QY, Cammers-Goodwin A, Grimes CA. A wireless remote query magneto-elastic CO_2 sensor. Journal of Environmental Monitoring. 2000;**2**:556-560. DOI: 10.1039/b004929h

[17] Cai QY, Mahaveer KJ, Grimes CA. A wireless, remote query ammonia sensor. Sensors and Actuators. 2001;**B77**:614-619. DOI: 10.1016/S0925-4005(01)00766-3

[18] Ong K, Wang J, Singh R, Bachas L, Grimes CA. Monitoring of bacteria growth using a wireless, remote query resonant-circuit sensor: Application to environmental sensing. Biosensors & Bioelectronics. 2001;**16**:305-312. DOI: 10.1016/S0956-5663(01)00131-2

[19] Ruan C, Varghese OK, Grimes CA. A magneto-elastic ricin immunosensor. Sensor Letters. 2004;**2**:138-144. DOI: 10.1166/sl.2004.041

[20] Barandiaran JM, Gutierrez J. Magneto-elastic sensors based on soft amorphous magnetic alloys. Sensors and Actuators A. 1997;**59**:38-42. DOI: 10.1016/S0924-4247(97)80145-8

[21] Kouzoudis D, Mouzakis DE. A 2826MB Metglas ribbon as a strain sensor for remote and dynamic mechanical measurements. Sensors and Actuators A. 2006;**127**:355-359. DOI: 10.1016/j.sna.2005.12.014

[22] Grimes CA, Kouzoudis D. Remote query measurement of pressure, fluid-flow velocity, and humidity using magneto-elastic thick-film sensors. Sensors and Actuators. 2000;**84**: 205-212. DOI: 10.1016/S0924-4247(00)00306-X

Chapter 3

Stochastic Resonance and Related Topics

Jiří Náprstek and Cyril Fischer

Additional information is available at the end of the chapter

http://dx.doi.org/10.5772/intechopen.70178

Abstract

The stochastic resonance (SR) is the phenomenon which can emerge in nonlinear dynamic systems. In general, it is related with a bistable nonlinear system of Duffing type under additive excitation combining deterministic periodic force and Gaussian white noise. It manifests as a stable quasiperiodic interwell hopping between both stable states with a small random perturbation. Classical definition and basic features of SR are regarded. The most important methods of investigation outlined are: analytical, semi-analytical, and numerical procedures of governing physical systems or relevant Fokker-Planck equation. Stochastic simulation is mentioned and experimental way of results verification is recommended. Some areas in Engineering Dynamics related with SR are presented together with a particular demonstration observed in the aeroelastic stability. Interaction of stationary and quasiperiodic parts of the response is discussed. Some nonconventional definitions are outlined concerning alternative operators and driving processes are highlighted. The chapter shows a large potential of specific basic, applied and industrial research in SR. This strategy enables to formulate new ideas for both development of nonconventional measures for vibration damping and employment of SR in branches, where it represents an operating mode of the system itself. Weaknesses and empty areas where the research effort of SR should be oriented are indicated.

Keywords: stochastic resonance, post-critical processes, dynamic stability, Fokker-Planck equation, Galerkin approach

1. Introduction

The stochastic resonance (SR) is a phenomenon, which can be observed at certain nonlinear dynamic systems under combined excitation including mostly deterministic periodic force and random noise. The phenomenon of this type has been first observed and reported by Kramers, see [1], investigating the interwell hopping in the Brownian motion. Some allusions can also be found in older resources devoted to stochastic processes and theory of stability (Lyapunov, Kolmogorov, Planck, and others).

The genuine phenomenon of SR has been discovered in early 1980s. The initiation point was probably two papers by Nicolis [2, 3] dealing with problems of climatic evolution. Other scientific and application areas followed that inspiration in due time, since it came to light that SR is a generic phenomenon. The idea of SR initiated remarkable cross-disciplinary interest bringing together nonlinear dynamics, statistical physics, information and communication theories, data analysis, life and medical sciences. Individual areas came to the use of SR phenomenon rather independently, and therefore, they introduced slightly different definitions and particular strategies in the first period. This transition time passed and many cross disciplines overlapping in their activities have been built at the unifying background developed by mathematics and theoretical physics. Despite this evolution, the historical aspects are still visible, due to fact that every branch still focuses on different needs, working in different scale and parameter intervals.

The term stochastic resonance was introduced probably in 1981 in informatics to describe the annoying noise in contemporary communication equipment that prevented to detect the weak useful signal. However, researchers recognized soon that under certain conditions, the noise can be helpful to enhance the device sensitivity.

The opportunity to employ SR in mechanics emerged only recently. SR approved to be promising for modeling of certain post-critical effects in nonlinear dynamics, active vibration damping, feedback systems, biomechanics, etc. Therefore, it is worthy of presentating a certain overview to the community of rational and applied dynamics concerning strengths, weaknesses, and application possibilities of SR occurred in theoretical and applied disciplines.

The phenomenon itself manifests in the simplest case by a stable periodic hopping between two nearly constant limits perturbed by random noises. The occurrence of this phenomenon depends on certain combinations of input parameters, which can be determined theoretically and verified experimentally. The classical mathematical definition of SR follows from properties of the Duffing equation with the negative linear part of the stiffness (bistable system) under excitation by a Gaussian white noise together with a deterministic harmonic force with a fixed frequency. It should be highlighted that also more general definitions of SR exist and will be also briefly reported in this chapter. In particular, it considers various types of the random noise, shapes of the deterministic excitation component, types of oscillator nonlinearity (potential of internal forces), and finally also number of stable positions, which can exceed two or drop to one.

In terms of classical Engineering Dynamics, SR can be assumed as a dangerous effect accompanying a post-critical system response. Therefore, it should be eliminated by appropriate selection of parameters and operating conditions (plasma physics, aeroelasticity, rotating machines, etc.) in order to ensure the reliability of the system. On the other hand, SR can characterize the mode of a natural system we are observing, and therefore, it serves as a tool of its investigation (e.g., Brownian motion mentioned above). It can also represent an intentional operating mode of the artificial system, and therefore, it should be considered as a useful state (special excitation or vibration damping devices, energy harvesting, etc.).

Nevertheless, many disciplines predominantly consider SR as a mechanism by which a system embedded in a noisy environment acquires an enhanced sensitivity toward small external signal, when the noise intensity reaches certain finite level. This phenomenon of boosting undetectable signals by resonating with added noise extends to many other systems, whether electromagnetic, physical, or biological, and is an area of intense research. This interpretation of SR shows that noise can play a positive role in systems either designed artificially or observed as a natural systems. Furthermore, SR and its variants can serve to understand many processes in various scales and temperature domains to understand various effects in solid state physics, biophysics, and electronics with possible application to design SR-inspired devices.

The study tries to mimic some excellent review studies published mainly in the areas of physics, informatics, and physiology with emphasis on Engineering Dynamics. See, for instance, papers [4–10], etc. Although their style is quite different, adequately with the branch they represent, they are full of valuable information and worthy to be studied. For reading are recommended problem-oriented monographs, e.g., [11, 12] or books including SR-devoted chapters, e.g., [13–15]. Additional information can be found also at numerous web sites, like popular Wikipedia, Scholarpedia, American Physical Society Sites, Encyclopedia of Maths, or Mathworks, see [16]. Doubtlessly, the largest source of primary information are leading journals edited by world societies of physics, electronics, informatics, and neurosciences. Moreover, lot of conference proceedings are available as well organized, e.g., by IEEE, APS, AIP, SIAM, or OSA.

Apart from this introductory remarks, the chapter consists of six sections (2–6). They have general or specific character oriented to particular disciplines. Section 2 introduces some overview of classical SR definitions, solution methods, and ways of its quantification. The following Section 3 estimates a possible future SR position in mechanics accompanied by a digest of a particular study performed in area of aeroelastic stability. Section 4 is devoted to SR-assisted energy harvesting as a discipline being very close to mechanics and having many joint features with that. Section 5 is unavoidably included for historical reasons dealing with climatology, where the modern SR appeared in the contemporary meaning of the term in early 1980. It gave an inspiration for all other branches, which are commonly discussed. Section 6 pays attention to nonconventional SR definitions dealing with alternative differential operators providing for instance, a possibility to abandon the bistable interwell hopping and to build SR on a monostable system. The use of nonGaussian driving noise is mentioned as well.

Concluding part No. 7 attempts to evaluate position of SR strategy and its strengths and weaknesses. With respect to the area of potential readers, it concentrates to a possible SR involvement in Engineering Dynamics. It means to eliminate dangerous SR-based phenomena occurring in industrial aerodynamics, dynamics of vehicles, and in whatever system endangered by dynamic stability loss and subsequent post-critical emergency regime. In the same time, SR can become the basis for the development of active equipment for vibration damping, earthquake resistance improvement, vehicle stabilization, etc. Let us take a note that SR phenomenon appears in many additional disciplines of theoretical and applied physics, data mining, chemistry, neurophysiology, pattern recognition, etc., where many inherent extensions beyond the classical definition of SR have been developed and used. For more information, see review papers, e.g., [4, 17], where extensions into quantum stochastic resonance with specific applications are outlined.

2. Classical definition of stochastic resonance

In early 1980s, SR has been discovered as a generic phenomenon and the first classical definition has been introduced. Some modifications appeared in due time, but the basic version is still alive serving as the basis of SR mathematical modeling. There is a lot of resources reporting about SR from the viewpoint of the definition in a rigorous or loose interpretation, see for example well-known overview article [4] by Gammaitoni et al. and also review paper [17] by authors of this chapter. Note that although vast majority of cases use the classical definition, a number of problems need the special definition of the SR phenomenon regarding its basic philosophy or individual components. Such settings extending the classical definition will be briefly outlined in Section 6.

2.1. Phenomenon of stochastic resonance

In classical meaning, SR occurs in bistable systems with single degree of freedom (SDOF), when a small periodic force is applied together with a large broad band random noise, see **Figure 1**. The system response is driven by two excitation components resulting in a "system switch" between two stable states. Their positions are given by two wells of the system potential $V(u)$. Wells are separated by a barrier. Its height decisive for the switching is considered as a difference between maximum and minimum of the potential, see **Figure 1**.

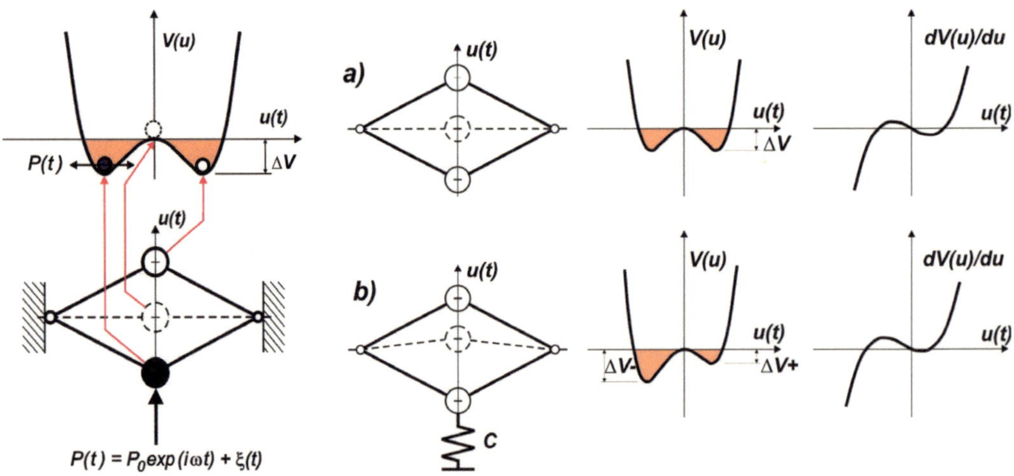

Figure 1. Bistable nonlinear system: (a) Symmetric potential; (b) Nonsymmetric potential.

http://dx.doi.org/10.5772/intechopen.70178

In the absence of periodic forcing, the approximate frequency of escape from one well into the second is given by the following estimate published in the comprehensive study [1]:

$$\omega_e = \sqrt{2} \cdot \exp\left(-\Delta V / \sigma^2\right) \tag{1}$$

where σ^2 is the variance of the noise, and ΔV means the barrier separating potential minima (symmetric potential), see **Figure 1a**. For nonsymmetric potential, the symbols $\Delta V_{-}(u)$,$\Delta V_{+}(u)$ in **Figure 1b** denote the left and right minima, respectively. In classical setting of SR, the Gaussian white noise is taken into account (for a couple of other variants, see Section 6).

If both component are acting, then the degree of switching is related with the noise intensity σ^2, see a sample response in **Figure 2**. When the periodic force is small enough being unable to make the system response switch, the presence of a nonnegligible random component is required for it to happen. When the noise is small (small variance σ^2) very few switches occur, mainly at random with no significant periodicity in the system response—**Figure 2(a)**. When the noise is too strong, a large number of switches occur for each period of the periodic component, and the system response does not show remarkable periodicity—**Figure 2(c)**. Between these two conditions, there exists an optimal value of the noise intensity σ_0^2 that cooperatively concurs with the periodic forcing in order to make almost exactly one switch per period (a maximum in the signal-to-noise ratio)—**Figure 2(b)**. Amplitude of the response alternating component as a function of the noise level is outlined in **Figure 3**. Peakness of the maximum is given by the damping factor. If the damping is too high, the peak can completely disappear and SR vanishes.

The optimum of the noise level σ_0^2 is quantitatively determined by matching of two time scales:

i. the period of the sinusoid (the deterministic time scale); and

ii. the Kramers rate, Eq. (1)—average switch rate induced by the sole noise, which is the inverse of the stochastic time scale. It implicates the denomination "stochastic resonance".

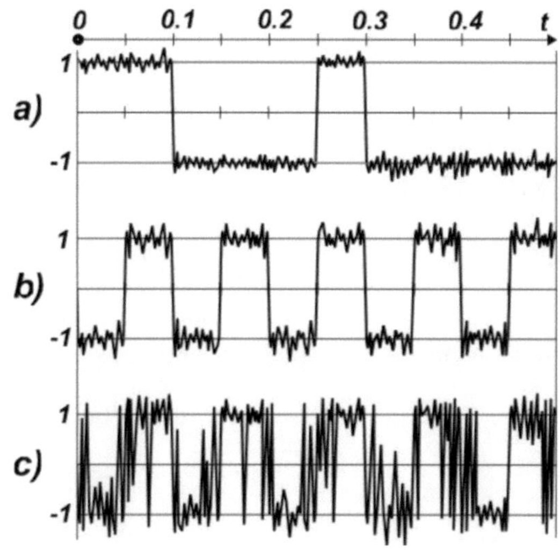

Figure 2. Time history of the system response for various noise variance: (a) low level; (b) optimal level σ_0^2; (c) high level.

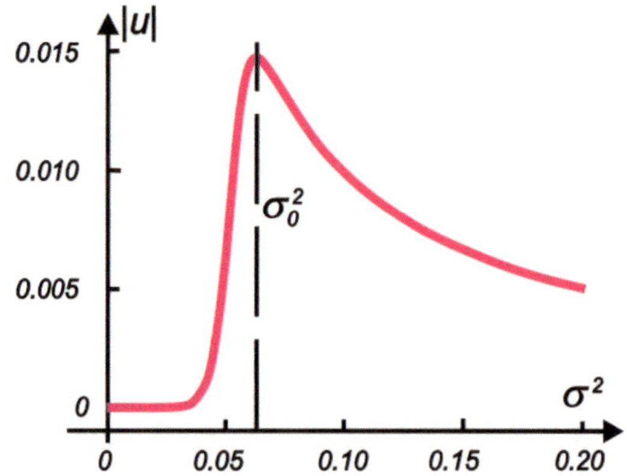

Figure 3. Amplitude of the system response alternating component due to simultaneous excitation by a weak periodic force and a random noise.

The Kramers formula, Eq. (1), is a result of theoretical and empirical investigation motivated by problems of nonlinear optics. Note that, in original resources, the absolute temperature T instead of the variance σ^2 is considered. The formula Eq. (1) is widely used and works very well. During the last decades, a number of areas of optics, quantum mechanics, chemistry, neurophysiology, etc., investigated this formula attempting to use the phenomenon of SR for the description of various effects arising in their branches using both experimental and theoretical ways of investigation, see, e.g., [18, 19].

The mathematical basis of the classical SR definition is related to the Duffing equation with negative linear part of the stiffness (in terms of mechanics). It is the most simple variant and it corresponds together with Gaussian white noise and deterministic harmonic force with a fixed frequency to the classical setting of SR. This configuration will be treated mostly throughout this chapter. Nevertheless, some generalizations and extensions beyond the classical formulation will be introduced in section 6 and furthermore at other remarked places.

Let us assume the nonlinear mass-unity SDOF oscillator written in a normal form:

$$\dot{u} = v; \quad \dot{v} = -2\omega_b \cdot v - V'(u) + P(t) + \xi(t). \tag{2}$$

$V(u)$–potential commonly introduced in a form providing the Duffing equation:

$$V(u) = -\frac{\omega_0^2}{2}u^2 + \frac{\gamma^4}{4}u^4 \quad \Rightarrow \quad V'(u) = dV(u)/du = -\omega_0^2 \cdot u + \gamma^4 \cdot u^3 \tag{3}$$

$\xi(t)$–Gaussian white noise of intensity $2\sigma^2$ respecting conditions:

$$\mathcal{E}\{\xi(t)\} = 0, \quad \mathcal{E}\{\xi(t)\xi(t')\} = 2\sigma^2 \cdot \delta(t - t'), \tag{4}$$

$\mathcal{E}\{\bullet\}, \delta(t)$—operator of the mathematical mean value in Gaussian meaning and Dirac function, respectively,

$P(t) = P_o \exp(i\Omega t)$—external harmonic force with frequency Ω. Amplitude P_o should be understood per unit mass.

Symbols ω_0 and ω_b have a usual meaning of the circular eigenfrequency and circular damping frequency of the associated linear system. The linear part of the $V'(u)$ is negatively making the system metastable in the origin, while the cubic part acts as stabilizing factor beyond a certain interval of displacement u. The system is drafted in **Figure 1** in two versions: (a) system with symmetric potential typical by an equivalent energy needed for hopping from the left into the right potential well and backwards; (b) system with asymmetric potential due to the supplementary linear string, which could be able (when rising its stiffness) to bring the oscillator to monostable state, see Section 6.1, where we will see that also the monostable system under certain circumstances is able to exhibit SR phenomenon.

2.2. Methods of stochastic resonance investigation

Theoretical approaches, either analytical or numerical, are mostly based on an assumption that random processes ruling inside the investigated system are of the Markov type. The primary requirement, namely the dependence of the process on its value only in one previous moment, is usually accomplished. In such a case, a large variety of methods are applicable for the investigation of SR phenomena.

Basically three type of solution procedures can be regarded:

(*i*) *Fokker-Planck* (*FP*) *equation*. It is the equation for cross probability density function (PDF) of the system response. Solution of this equation serves subsequently for the evaluation of various stochastic parameters like mean value, stochastic moments of adequate order, auto and cross correlation functions, probability flow, signal to noise ratio, mutual information etc. Concerning SR itself, the main indicators and parameters of this phenomenon can be evaluated and discussed in relation with the physical character of the problem, see subsection 2.3. So that, PDF is a certain "source function" to obtain all information needed.

Taking into account that random noise in the governing physical differential system, Eq. (2), has an additive character, no Wong-Zakai correction terms emerge, see, e.g., [20–22]. Then, the relevant FP equation, e.g., [23], can be easily written out:

$$\frac{\partial p(u,v,t)}{\partial t} = -\kappa_u \frac{\partial p(u,v,t)}{\partial u} + \frac{\partial}{\partial v}(\kappa_v p(u,v,t)) + \frac{1}{2}\kappa_{vv}\frac{\partial^2 p(u,v,t)}{\partial v^2}, \tag{5}$$

$$\begin{aligned} &\kappa_u,\ \kappa_v \text{ - are drifft coefficients}: \quad \kappa_u = v\ ; \quad \kappa_v = \kappa_v(t) = -2\omega_b \cdot v - V'(u) + P(t)\ , \\ &\kappa_{vv} \text{ - is a diffusion coefficient}: \quad \kappa_{vv} = 2\sigma^2, \end{aligned} \tag{6}$$

together with boundary and initial conditions:

$$\lim_{u,\,v\to\pm\infty} p(u,v,t) = 0(a), \qquad p(u,v,0) = \delta(u,v)(b). \tag{7}$$

Solution of the above FP equation can be conducted using one of the following procedures:

(i-a) Variational solution of Galerkin type. In principle, it is a procedure of decomposition into stochastic moments (or cumulants) with Gaussian closure, e.g., [24]. The demonstration of this procedure is presented in subsection 3.2, where an application to stability analysis of the TDOF aeroelastic system is roughly outlined.

In general, for details of the Galerkin method on the basis of functional analysis rules, see, e.g., [25]. For details of particular solution, see [26–28], and other papers and monographs. The method is suitable namely for stationary solutions, but quasiperiodic solutions can be investigated as well, see, e.g., [29], where detailed procedure outlined above is presented.

(i-b) Generalized Fourier method. Decomposition into a series following eigen functions and values of FP operator.

$$p(u,v,t) = p_o(u,v)\cdot\varphi(t) \quad \Rightarrow \quad p(u,v,t) = \sum_{j=0}^{N} p_j(u,v)\cdot\varphi_j(t) \tag{8}$$

The series Eq. (8) can be substituted into the FPE Eq. (5). Due to the independency of $p_j(u, v)$ or $\varphi_j(t)$ on time or space variables, respectively, the part dependent on time only can be separated on the left side and that dependent on space variables on the right side. They can be equivalent only if both of them equal the same constant λ_j for each part of the series. It can be shown that λ_j are eigen values of the FP operator part, which is on the right side of Eq. (5). Subsequently, $p_j(u, v)$ are relevant eigen functions of this operator and finally $\varphi_j(t)$ are the simple exponential functions with the negative real part. Take a note that the $\lambda_0 = 0$, as the first part of the series Eq. (8) for $j = 0$ represents the stationary part of the FPE solution, provided the stationary part exists. In general, the occurrence of one or more positive real parts of λ_j can reveal positive, which would indicate an instable solution of FPE. However, it is not the case when investigating FPE used for modeling the SR phenomenon.

This approach is applicable rather in special cases with easy searching of eigen functions, when transition process is looked for. For example, see [30]. In general, searching for eigen functions of FP operator is a complex task, and it can prevent application of this method when more than SDOF system is analyzed.

(i-c) Floquet theory. Application of the Floquet theorem:

$$p(u,v,t) = p(u,v,t+T) \tag{9}$$

Suitable for equations with periodically variable coefficients, when transition nonperiodic process is investigated. See [30].

(i-d) Finite element method (FEM) and other numerical procedures. The FEM can be considered as a general numerical solution method of partial differential equation. It can be proved that FEM is well applicable for this purpose under certain circumstance, which are fulfilled regarding FPE. When constructing adequate elements, a care should be taken due to special properties of the FP operator. Significant problem originates from the fact of multi-dimensionality of space we are working with and a delicate character of initial conditions. Moreover, the non-self-adjointness

of the FP operator, special configuration of boundary conditions, etc., should be taken into account. These factors shift application of FEM in this case into a special area where a number of nonconventional problems should be solved.

The FPE is analyzed in an original evolutionary form which enables an analysis of transition effects starting the (nearly) Dirac type initial conditions. The FEM efficiency when solving FPE, which follows from the Duffing stochastic differential equation without external harmonic forces was already studied by the authors in [31]. With the periodic force taken into account, certain difficulties arise due to the time inhomogeneity of the corresponding stochastic process. Many results regarding FEM application on FP equation analysis can be found in [32] or [33]. For the most recent results concerning FEM application to SR problem, see [31], and additional details together with demonstrating examples, see [34].

The method is based on the approximaltion solution of Eq. (5) in the Galerin-Petrov meaning on the piecewise smoothly bounded domain $\Psi \in u \times v$, in $\mathbb{R}^d$, $d = 2$. The initial conditions at $t = 0s$ for PDF are considered in a form of the Gauss distribution function with an initial system position at the point $u_0 = 0$, $v_0 = 0$. For a small value of standard deviation, it approaches the Dirac function as it is primarily requested.

After a spatial discretization of Ψ onto the rectangular finite elements using the bilinear approximation functions and implying boundary condition $p(\partial\Psi, t) = 0$, the system of ordinary differential equations emerges with global matrices $\mathbf{M}$, $\mathbf{S}(t)$ and vector of probability density values $\mathbf{P}(t)$ in nodes of the mesh.

Final differential system has the form as follows:

$$\mathbf{M} \cdot \dot{\mathbf{P}}(t) = \mathbf{S}(t) \cdot \mathbf{P}(t) \tag{10}$$

The matrix $\mathbf{S}(t)$ is time-dependent due to the periodic perturbation entering the drift term of FPE, and in the result, the solution oscillates periodically between the potential wells. In the regime of SR, the switchings are in phase with the external periodic signal $\mathbf{P}(t)$ and the mean residence time is closest to half the signal period $2\pi/\Omega$. Comparing the results obtained by means of FME with those following from the analytical investigation outlined above shows a good compatibility.

The efficiency of FEM is obvious as usual. It enables to investigate details, which are inaccessible using other methods. It applies especially to transition processes starting the excitation and response processes nearby the stability loss, when the Lyapunov exponent is floating around zero and boundary between local and global stability are ambiguous.

(ii) Stochastic simulation–digital and analog. Stochastic simulation is one of the most important methods of SR investigation. The basic idea is straightforward, the governing system Eq. (2) is subdued to numerical integration and subsequently probabilistic parameters including PDF are evaluated. However, extreme caution should be taken, as the differential system is stochastic. Because the system Eq. (2) includes only an additive noise, no Wong-Zakai correction terms are necessary, see [20–22]. However, the strategy of integration should be carefully controlled [35, 36], due to fact that we manipulate with the Ito system. In principle, the time

increment can be neither too long in order to prevent information loss, nor too short to keep the stochastic character of the output. Hence, the care should be taken during manipulations in the corrector phase of one step.

Results obtained in this manner are very important. They serve as a verification of semi-analytical results obtained using one of the procedures mentioned in the previous paragraph (i), and furthermore, the simulation is able to enter into small details, which remain hidden to methods mentioned in (i). It applies particularly to transition process if there is a need of their investigation. On the other hand, like every fully numerical method or simulation, it provides result for one set of parameters only. Like in experiments, it is difficult and laborious to obtain a broader overview.

Analog simulations have been very popular in the past wherever nonlinear differential equations were to be solved. However, they are still very attractive for researchers as they lie at the frontier between digital simulation and experiment. Their advantage is that the parameters can be easily and quickly tuned over a wide range of values and the response can be followed straightforwardly. Many review and technical papers have been published as for instance [37, 38], where the comparison of analog simulation of stochastic resonance with adiabatic theory has been performed. It should be appreciated now that a genuine analog simulation can be effectively emulated at digital computers using commercial software packages, see for instance McSimAPN package, visit <http://www.edn.com>. Moreover, actually whatever hybrid analysis enabling digital support of the analog simulation is possible.

(*iii*) *Experimental measurements*. SR has been observed in a wide variety of experiments involving electronic circuits, chemical reactions, semiconductor devices, nonlinear optical systems, magnetic systems, and superconducting quantum interference devices (SQUID). The general instruction for experimental procedures can be hardly recommended. They are always developed individually respecting specific character of every research activity. Anyway, be aware that many experiments do not serve for validation of theoretical results. Indeed, the strategy is often opposite. The purpose of the experiment is an initial recognition of the basic principle while the theoretical approach should verify subsequently its validity. It is very frequently observed particularly in neurophysiological experiments related with SR, see monograph [12] and papers [39–43] and others. Three popular examples of this type performed should be named: the mechanoreceptor cells of crayfish, the sensory hair cells of cricket, human visual perception. Another "inverse" experiments (preceding any theoretical modeling) can be seen in a wind tunnel. Here, the divergence instability of the prismatic bar in a cross flow has been observed in the view of SR without any previous theoretical background. A number of primary experimental studies are available also in plasma physics, optics, and in other branches, e.g., [44–46].

2.3. Quantification of stochastic resonance

Occurrence of SR is obviously indicated by periodic transition across the potential barrier which is synchronized in the mean with periodicity of the deterministic excitation component. The frequency should be close to that given by Kramers formula, Eq. (1). The phenomenon emerges markedly, when introducing the optimal noise amount under adequate damping level, as it corresponds to **Figures 2** and **3**, otherwise the response is very small. This rather empirical identification is validated by theoretical means outlined above.

Internal character of the signal provided by SR can be inspected in particular cases using some useful parameters and functions:

(*i*) *Signal to noise ratio*. Very frequently used indicator. It is based on the power spectral density (PSD) attributes of the signal $u(t)$. A couple of variants can be found in literature, see, e.g., [21, 23], etc. Usually the ratio of PSD concerning the periodic signal being proportional to the integral of the Dirac function taken in a small neighborhood $\Delta\omega$ of its frequency $\pm\,\Omega$ and the total (PSD) integral at the same interval is considered. Symbolically expressed:

$$SNR(\omega) = PSD(\omega)/S_N, \quad S_N - \text{output background noise} \tag{11}$$

Strengths and shortcomings of the above expression are obvious. Spectral density $PSD(\omega)$ should be continuous and simple, otherwise Eq. (11) does not provide reliable results applicable in a practical analysis. Nevertheless, other variants of this procedure are evident. They can be based on a certain integral evaluation along the frequency axis, but they should be composed for particular cases.

(*ii*) *Residence time distribution and the first excursion probability*. Observing **Figure 2**, the output signal $u(t)$ is a random process. The time of residence in one basin and the jump to the other one can be regarded as a problem of time of the first excursion probability, see, e.g., [21] and many independent authors like [47], etc. Evaluation of individual periods of residence in one basin can serve as indication of SR stability and quality. This parameter gets an important information because the signal $u(t)$ suffers very often from nonintentional jumps within SR periods. Results provided are more reliable as a rule in comparison with (*i*), but the procedure in a particular case is much more laborious.

(*iii*) *Information entropy based indication*. Widely used in communication theories. This indicator is based on Boltzmann's entropy of information, see monograph [48]. Boltzmann's entropy is defined by the expression:

$$I(\varphi) = \int_X p(\mathbf{x}, t) \cdot \lg\ p(\mathbf{x}, t)\ \mathrm{d}\mathbf{x} \tag{12}$$

where $I(\varphi)$ denotes Boltzmann's entropy of probability and $p(x, t)$ is the cross-probability density of the system response. The procedures working with this tool are usually based on maximization of this entropy with auxiliary constraints, which is the governing dynamic system itself. In particular, PDF is written in a form of the multi-dimensional exponential (mostly a polynomial in a homogeneous form) with free coefficients. These coefficients are subsequently determined by means of the extreme searching using a suitable procedure (Fletcher-Powell, artificial neuronal network, etc.).

This procedure is very effective when impuls character of useful signal is considered, see the SR-focused paper by Neiman [49] or generally oriented [26], etc. As a large source of information can doubtlessly serve relevant chapters in monographs [11] or [50]. A significant step forward to characterize the conventional SR by means of information theory tools has been put by [51], where SR in a nonlinear system driven by an aperiodic force has been studied. See also a number of other papers being more or less on the boundary between classical and nonconventional SR definitions, as for instance [52] dealing with SR capacity enhancement in an asymmetric binary channel.

(*iv*) *Statistics of local random processes in individual basins*. Random processes surrounding the mean value when residing in a basin is evaluated. Then random mean square root is evaluated and compared with the amplitude of the jumping process mean value. Rather special method which appears rarely in SR as a separate tool. If applied, it is more or less smoothly integrated with the analytical process. Its application can be observed more in areas working with more general SR definitions concerning the operator structure and driving noise type, see section 6.

(*v*) *Mutual information*. Let us denote $p_{\varphi\psi}(\varphi, \psi)$ the joint PDF of input and output processes $\varphi(t)$, $\psi(t)$. Being based on Shannon's theorem, see [53], mutual information between processes $\varphi(t)$, $\psi(t)$ is defined as the relative entropy between the joint PDF and the product of partial PDFs, see [48] or [54]:

$$I(\varphi, \psi) = \int\limits_{\varphi, \psi} p_{\varphi\psi}(\varphi, \psi) \cdot \lg\left(\frac{p_{\varphi\psi}(\varphi, \psi)}{p_{\varphi}(\varphi) p_{\psi}(\psi)}\right) d\varphi d\psi \tag{13}$$

It seems that the mutual information is the most effective quantification parameter for assessment in suprathreshold stochastic resonance, see [11] and many more. Take a note that Eq. (13) basically represents a significant generalization of the Boltzmann's entropy procedure Eq. (12) with respect to conditional probability referring some intermediate state analogously with Bayesian updating.

3. Engineering Dynamics and stochastic resonance

It seems that Engineering Mechanics is now gradually discovering SR and is looking for areas of SR applicability. Nevertheless, some areas can already show off tangible results. Research activities are mostly the joint projects with physics, fluid mechanics, electronics, and medical disciples. Similarly like in other branches also in Engineering Mechanics, the direct and inverse tasks are investigated. Due to some delay, it can draw upon experience of other disciplines.

Let us outline some relevant areas of Engineering Mechanics where SR provides (or could provide) significant contribution in various points of the research and application. Then, we present briefly a sample problem of aeroelastic stability related with SR.

3.1. Areas in dynamics related with stochastic resonance

Engineering Dynamics of discrete and continuous systems in classical meaning of the term can come into contact with SR roughly in three areas:

http://dx.doi.org/10.5772/intechopen.70178

(*i*) *Nonlinear SDOF, multi degree of freedom* (*MDOF*) *or possibly continuous dynamic systems subdued to combination of periodic and random excitation*. A number of problems arising in flow structure interaction can be tackled using various models of SR type, e.g., slender structures in a cross flow, soft large roofs, high speed channels with streaming fluids, and propagating solitary waves, etc. Some more examples can be found among other systems with significant Duffing type nonlinearity with meta-stable point of origin, even those more complicated nonlinear system (Van der Pol, Rayleigh, etc.) can exhibit SR effects. They emerge usually after entering into a post-critical regime stabilized by certain nonlinear forces. A sample problem of aeroelastic stability will be shortly looked through.

(*ii*) *Experimental measurements of weak signals below threshold limit*. Subthreshold signal sensing, recording, and filtering is rather a cross discipline widespread nearly everywhere.

Signal sensing and subsequent data processing is a wide area pervading all scientific and engineering disciplines. Hence, relevant problems attracted many researchers all the time. The aim has always been to speed up digitizing frequency, increase resolution and reliability, and to diminish as much as possible differences between input and output processes.

It has been recognized in the past that a weak signal being below the threshold limit of a sensor, can be boosted by adding white noise to the useful signal, see **Figure 4**. For details, see [11]. The sum of both signals can overcome the threshold limit and hence to be detectable by the sensor. Then, random component is filtered out to effectively detect original, previously undetectable signal. Many general studies and special-oriented variants have been performed to detect subthreshold signals using a driving random signal, let us name a few of them [24, 55–58] following various attributes of SR employment in weak signal recognition and reliable recording.

The qualitative jump forward in this strategy brought the suprathreshold stochastic resonance (SSR). The phenomenon of SSR has been discovered by N. Stocks. The first paper informing about SSR is the review paper [8] published in 1999. As the primary source can serve [59], which appeared 1 year later and authored solely by Stocks. Since then, many articles have been published about SSR. Probably, the most comprehensive explanation can be found in the monograph by McDonnell et al. [11].

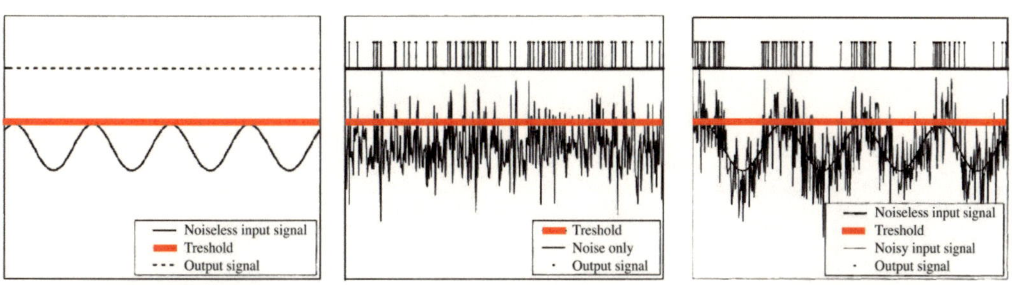

(a) Subthreshold periodic input signal

(b) Noise only

(c) Subthreshold but noisy periodic input signal

Figure 4. Experimental measurements of weak signals below threshold limit, see [11].

(*iii*) *Biomechanics*. Very wide domain gathering experts of many areas constituted interdisciplinary teams worldwide. Domains like heart dynamics, blood streaming, muscle system functionality, and vocal folds are followed. However, predominantly problems of the human skeleton are tackled, see for instance [42, 60]. Here, substantial attention is paid to SR in which noise enhances the response of a nonlinear system to weak signals in various biological sensory systems. In the same time, it has been recognized that adding low magnitude periodic vibration greatly enhances the bone formation in response to loading, which is definitely an excellent contribution of SR for the osteogenic processes. An outcome of these activities are among others the therapies of the whole-body through vibration training on a chair rising in elderly individuals [61, 62]. Very sophisticated stochastic analysis of discrete data sets provided by measuring records has been performed in order to bring an exact evidence of the meaningful healing procedure.

Let us take a note beyond limit of this study. Biomechanics is not far from various medical branches, where a wide range of modern special implants based on the SR principle is successfully used. In particular cochlear stimulators, oftalmological adaptors, pacemakers and others, see for instance [4] or [17] where also many additional references can be found.

3.2. Sample problem of aeroelastic stability

With reference to wind tunnel observations in a wind channel, it seems that SR is promising as a theoretical model inherent for several aeroelastic post-critical effects arising at a prismatic beam in a cross flow. Dealing with relevant projects, these post-critical effects should be carefully investigated in order to eliminate any danger of the bridge deck collapse due to aeroelastic effects. In particular, the divergence or buffeting of a bridge deck can be modeled as a post-critical process of the SR type at an SDOF or two degree of freedom (TDOF) system, see **Figure 5**. For details, see [63]. In **Figure 5(a)**, we can see outline of the TDOF system investigated. **Figure 5(b)** exhibits the stability diagram itself. White or dark fields indicate stable or instable zones, respectively. The stability limits are plotted in the plane of heaving and pitching eigen frequencies ω_u^2 and ω_φ^2. **Figure 5(c)** shows value of the flutter frequency Ω^2 with respect to position on the parabola with axes x_1, x_2 in **Figure 5(b)**.

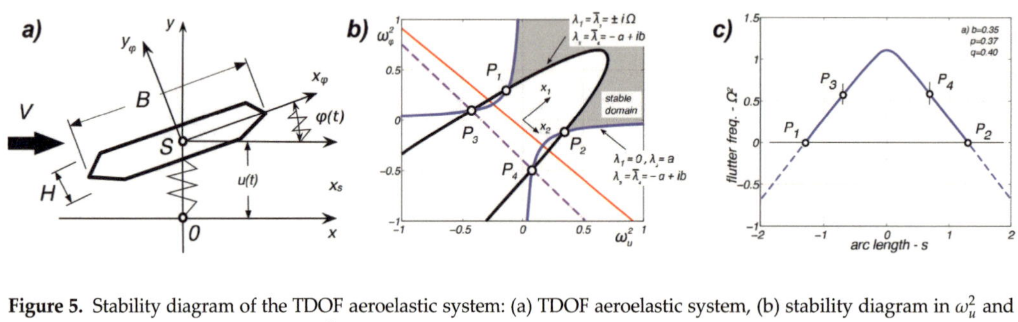

Figure 5. Stability diagram of the TDOF aeroelastic system: (a) TDOF aeroelastic system, (b) stability diagram in ω_u^2 and ω_φ^2 coordinates, (c) flutter frequency Ω^2 as function of a position on the parabolic part of stability limits.

Paper [64] is referred for details and further references. Anyway, let us revisit Eqs. (2–4) for basic mathematical model. Three theoretical solution ways have been followed. FP equation together with boundary conditions is written out in Eqs. (6, 7).

(*i*) *Semi-analytical solution of FP equation*. Galerkin type procedure has been applied in order to respect non-self-adjointness of the FP operator, see [25]. With respect to the linearity of the FP equation, the basic periodicity of the PDF should be equivalent with the frequency of the deterministic excitation component Ω and its integer multiples. See formulation Eqs. (2–4) together with Eqs. (5–7).

Therefore, the series can be written in the following form:

$$p(u,v,t) = p_o(u,v)\sum_{j=0}^{J} q_j(u,v)\cdot\exp(\mathrm{i}j\Omega t) \tag{14}$$

where Ω is the harmonic excitation frequency. The series Eq. (14) represents an approach of a weak solution of FP equation, which repeats in the period $T = 2\pi/\Omega$. It gives a true picture of solution within one period, but cannot express any influence of initial conditions.

In Eq. (14), $p_o(u, v)$ means the solution of FPE Eq. (5) for $P_0 = 0$, it means that the deterministic part of excitation vanishes and the external excitation is limited to random component only. The solution is time independent (solution of the Boltzmann type). For details, see, e.g., [26–28], and other papers and monographs, see also **Figure 6** for symmetric and nonsymmetric potentials V:

$$p_o(u,v) = D\cdot\exp\left(-\frac{2\omega_b}{\sigma^2}H(u,v)\right). \tag{15}$$

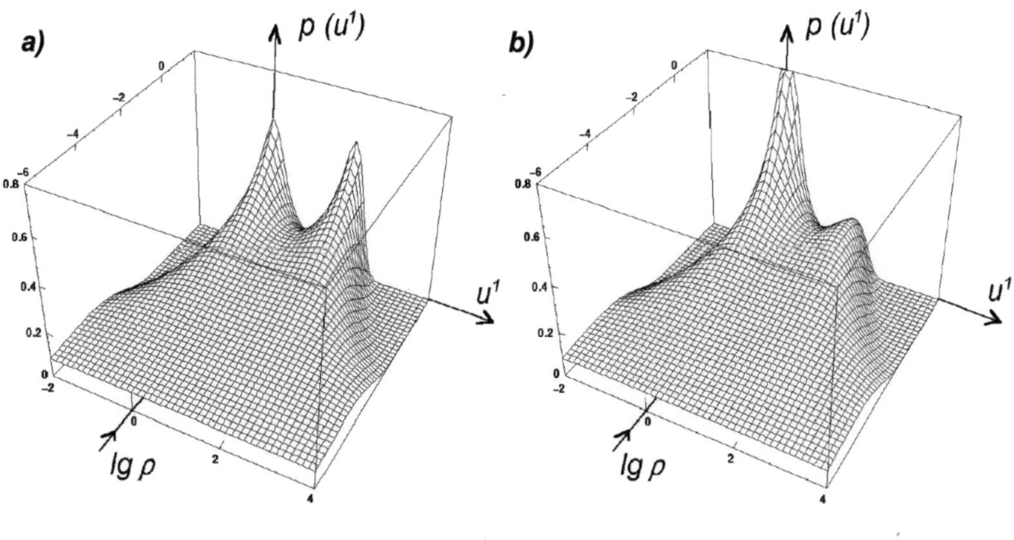

Figure 6. Response PDF of the system excited by white noise only: (a) Symmetric potential; (b) Nonsymmetric potential.

In the above expression, D is the normalizing constant, and $H(u, v)$ represents the Hamiltonian function of the basic system. In particular:

$$H(u,v) = \frac{1}{2}v^2 + V(u) = \frac{1}{2}v^2 - \frac{1}{2}\omega_o^2 u^2 + \frac{1}{4}\gamma^4 u^4 \tag{16}$$

The unknown functions $q_j(u, v)$ in Eq. (14) can be searched for using the generalized method of stochastic moments as it can be found, in [23]. For additional details, see [29]. Using the Galerkin approach, the expression (14) is substituted into Eq. (5) and the whole equation is subsequently multiplied by the testing functions $\alpha(u, v)$.

The testing functions $\alpha(u, v)$ and unknown functions $q_j(u, v)$ are assumed to have a following advantageous form:

$$\alpha(u,v) = \alpha_{r,s}(u,v) = u^r \cdot H_s(\beta v) \; ; \quad r = 0, \ldots, R \; ; \; s = 0, \ldots, S \tag{17}$$

$$q_j(u,v) = \sum_{k,\, l=0}^{R,\, S} q_{j,\,kl} u^k \cdot H_l(\beta v) \tag{18}$$

where $H_s(\beta v)$ are l'Hermite polynomials and $\beta = \sqrt{\omega_b/\sigma^2}$. After applying the mathematical mean operator with respect to probability density function $p_o(u, v)$, see Eq. (15), and employing orthogonality of l'Hermite polynomials, the linear algebraic system for unknown coefficients $q_{j;k,l}$ arises ($q_o(u, v) = 1,\; q_{-1}(u, v) \equiv 0$):

$$2\beta(\mathrm{i}j\Omega + 2\omega_b s)\mathbf{A}\mathbf{q}_{j,s} - 2(s+1)\mathbf{C}\mathbf{q}_{j,s+1} + \mathbf{B}\mathbf{q}_{j,s-1} = 2\beta^2 P_o \mathbf{A}\mathbf{q}_{j-1,s-1} \tag{19}$$

where $\mathbf{q}_{j,s} = [q_{j,0s}, q_{j,1s}, \ldots, q_{j,Rs}]^T$—column vector ($R+1$ components) and $\mathbf{A}, \mathbf{B}, \mathbf{C} \in \mathbb{R}^{(R+1)\times(R+1)}$— square arrays containing moments:

$$A_{r,k} = \int_{-\infty}^{\infty} u^{r+k}\Phi(u)du \; ; \quad B_{r,k} = \int_{-\infty}^{\infty} ku^{r+k-1}\Phi(u)du \; ; \quad C_{r,k} = \int_{-\infty}^{\infty} ru^{r+k-1}\Phi(u)du$$
$$\Phi(u) = \exp\left(\beta\omega_o^2 u^2 - \frac{1}{2}\gamma^4 u^4\right) \tag{20}$$

Function $\Phi(u)$ is symmetric with respect to zero and therefore $A_{r,k} = 0$ for odd $r + k$, while $B_{r,k}$, $C_{r,k}$ vanish for even $r + k$.

For each j, the three-term recurrence formula Eq. (19) forms an algebraic system of size $(S + 1)(R + 1) \times (S + 1)(R + 1)$ for all unknown coefficients $q_{j,rs}$. The block diagonal of the system matrix consists from scaled regular matrices **A**, see Eq. (20), and thus it is invertible.

The resulting probability density function varies in time with periodicity, which corresponds to the frequency of external loading Ω. The individual peaks alternate but the lower peak never vanishes completely, see **Figure 7**. The computed joint probability density is shown in the **Figure 7(a)**, the corresponding curve for the displacement variable u (section for $v = 0$) is in the **Figure 7(b)**. The solid line shows the computed time-dependent probability density for $t = 30$, the dashed line corresponds to the stationary solution of the Boltzmann type $p_o(u, v)$, see Eq. (15).

http://dx.doi.org/10.5772/intechopen.70178

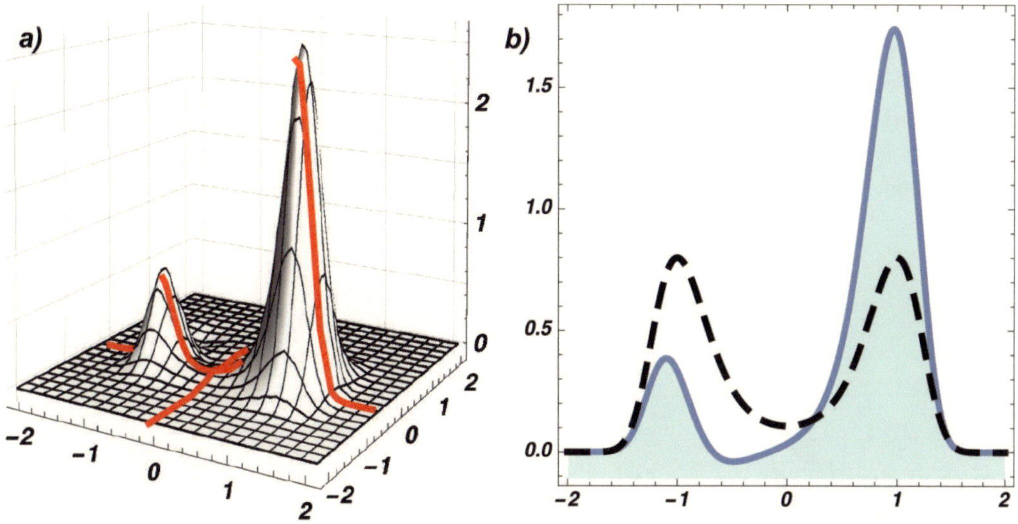

Figure 7. (a) PDF according to relation Eq. (14) for $t = 30$; (b) the corresponding cross section for $v = 0$ during the transition period starting initial condition of the Dirac type (solid line with filling) and stationary solution of the Boltzmann type (dashed).

(*ii*) *Solution of FP equation using FEM*. Solution procedure is based on the approximate solution of Eq. (5) in the Galerin-Petrov meaning on the piecewise smoothly bounded domain $\Psi \in u \times v$, in $\mathbb{R}^d$, $d = 2$. The initial conditions at $t = 0s$ for PDF are considered in a form of very pointed Gaussian distribution function with an initial system position at the point $u_0 = 0$, $v_0 = 0$. For a small values of standard deviation, it approaches to the Dirac function as it is primarily requested. The system of ordinary differential equations emerge with global matrices **M**, **K**(t) and vector of PD values **p**(t) in nodes of the mesh:

$$\mathbf{M}\dot{\mathbf{p}}(t) = \mathbf{K}(t)\mathbf{p}(t). \tag{21}$$

The matrix **K**(t) is time-dependent due to the periodic perturbation entering the drift term of FP equation, and in the result, the solution oscillates periodically between the potential wells. In the regime of SR, the switchings are in phase with the external periodic signal $P(t)$, and the mean residence time is closest to half of the signal period $2\pi/\Omega$.

Some results of numerical analysis are depicted in **Figure 8**. Comparison of those with semi-analytic results plotted in **Figure 7** shows a perfect coincidence.

(*iii*) *Stochastic simulation*. Differential system Eq. (2) has been repeatedly solved numerically respecting its stochastic character, see [35], with the same parameter setting as used before. The white noise was simulated as a finite sum of harmonic functions with uniformly distributed random frequencies $\omega_i \in (0, \omega_{max}\rangle$ ($\omega_{max} = 10\ rad.\ s^{-1}$) and phases $\varphi_i \in (0, 2\pi\rangle$:

$$\xi(t) = \sqrt{2}\sigma \sum_{i=1}^{N} \cos\left(\omega_i t + \varphi_i\right) \tag{22}$$

The results of the SR analysis are illustrated in **Figure 9**, which presents the signal to noise ratio —**Figure 9(a)** as the function of the noise intensity expressed by $2\sigma^2 = \kappa_{vv}$, and the results (Fourier spectra) of the stochastic simulations using the basic system Eq. (2) and **Figure 9(b)**. In the individual spectral lines, it can be seen in the influence of rising the white noise intensity, which acts together with a harmonic force onto the system. For a very low level of the noise, the harmonic component is hardly able to overcome the interwell barrier, and therefore, only

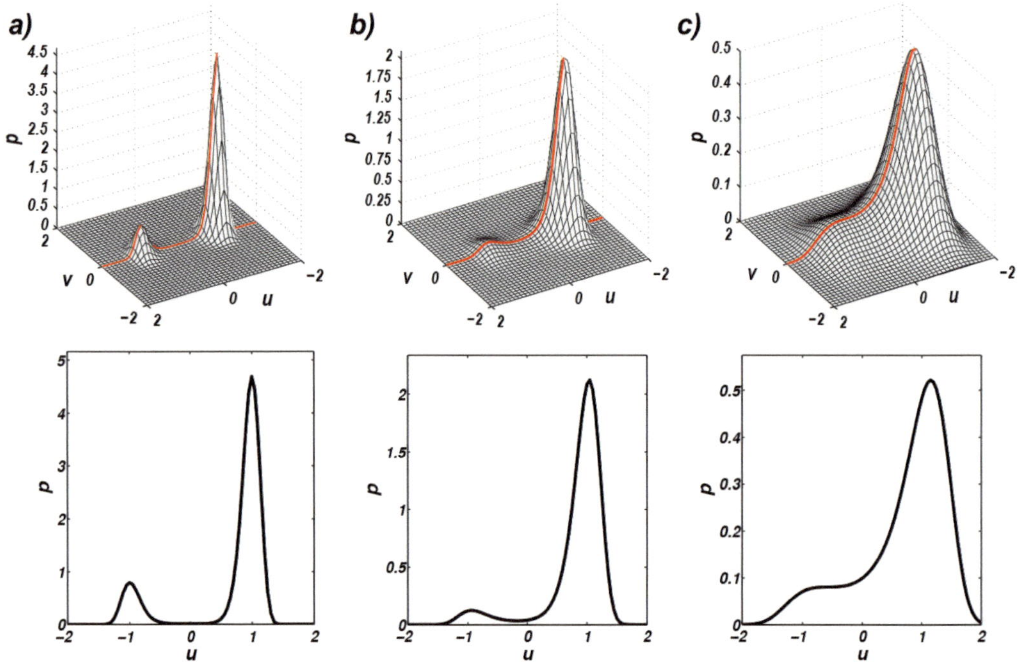

Figure 8. Axonometric and sectional display of the PDF at the highest value of probability of residing in selected potential well: (a) $\kappa_{vv} = 0.10$; (b) $\kappa_{vv} = 0.25$–stochastic resonance; (c) $\kappa_{vv} = 1.0$; the lower pictures are vertical cross-sections of surfaces in the upper row for $v = 0$, see highlighted curves in red.

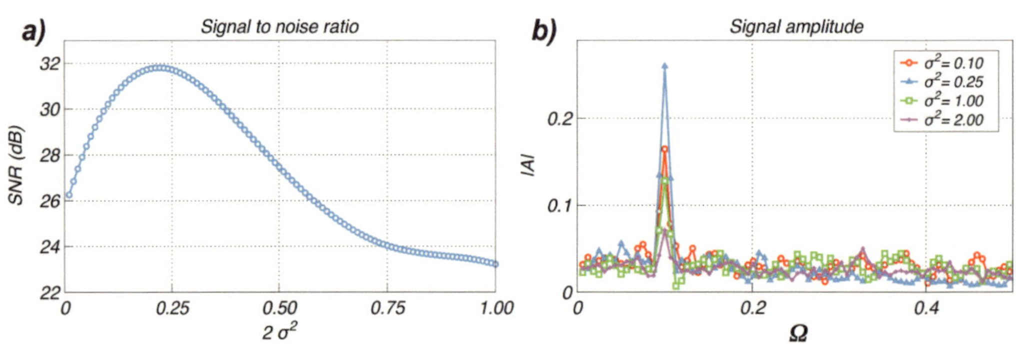

Figure 9. Results of stochastic simulation: (a) the signal to noise ratio as the function of various noise intensity (σ^2) due to SR; (b) Fourier spectra of the response obtained by numerical solution.

seldom irregular jumps between stable points occur, as it has been already demonstrated in **Figure 2**.

In local regimes, the system response is relatively small and nearly linear. Optimal ratio of the noise intensity (σ_0^2), and the amplitude of the harmonic force results for its certain frequency in the system response containing visible spectral peaks (amplification) corresponding with the frequency of the external harmonic modulation. The single peak (in the case of colored noise more peaks may appear) and thus the "optimal" noise strength can be identified.

4. Energy harvesting

A number of sources of harvestable ambient energy exist, including waste heat, vibration, electromagnetic waves, wind, flowing water, solar energy, human motion, and others. They can serve for powering remote wireless sensors, controllers, stimulators in a number of technological and biological applications, without any battery or wiring complements. Therefore, energy harvesting (EH) has emerged as a discipline with the goal of fabricating devices that can generate electrical power by exploiting ambient waste energy, for instance see [65]—ambient vibration, [66]—thermo gradient [67]. Basically following ideas are used: piezoelectric layered parts, magnetic levitation, magneto-rheologic hydraulic elements, ball screw systems, impact systems, and other principles. A pioneering work highlighting theoretical aspects of EH and challenging other authors is [68]. The adequate model follows from SDOF bistable system:

$$\ddot{u} + 2\omega_b \dot{u} + \omega_0^2 u \left(1 - \frac{l}{\sqrt{u^2 + d^2}}\right) = P_0 \cos \Omega t + h\xi(t) \tag{23}$$

where l, d are dimension characterizing von Mieses truss—remembers the system in **Figure 1a**. A couple of modified equations are also used in order to facilitate the insight into the system. The most frequent is the relevant Duffing system with negative linear part of stiffness.

4.1. Small scale energy production and measuring system feeding

A few electro-mechanical principles are used for this purpose. Typically, a cantilevered beam with a piezoelectric strip is used to transform vibrational energy into electrical energy through damping, see **Figure 10**. This figure has been taken over from [69], where many details and systematic background can be found. For small displacements of the beam, peak power generation in the mechanism will occur when the natural frequency of the beam is tuned to the peak of the vibration noise spectrum. Briefly speaking, SR despite being counter-intuitive phenomenon proved to be effective to enhance vibrational EH by adding periodic forcing to a vibration excited energy harvesting. A review of EH suitable piezoelectric materials together with adequate shaping and comprehensive experience in practice can be found in [9]. The most frequent applications cover human stimulation feeding, measuring and transducer system feeding, traffic control feeding, and many other devices with consumption approximately less than $1.0W$.

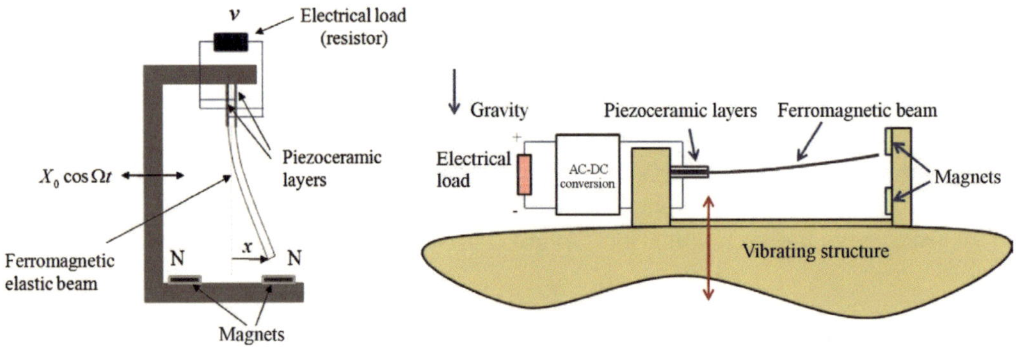

Figure 10. Small scale energy production for capture local feeding, see [69].

4.2. Large scale application and vibration damping

While the first generation of EH devices has been intended for the low power consumption, subsequently an idea of SR-assisted EH application in large scale systems appeared. These systems usually combine auxiliary power production and vibration suppressing in large scale engineering systems and suppose to work with energy approx $1 - 100kW$. Energy is gained from vehicles and transport means operation, vibration of civil and mechanical engineering systems, and other resources.

Comprehensive review of the contemporary knowledge regarding EH in large scale facilities is presented in papers [70] and [71]. Relevant principles are based again on EH assisted by SR phenomena. Possibilities and practical aspects of vibration damping using SR support EH are widely discussed in engineering oriented journals, see **Figure 11**. A number of other facilities is based or supported using this principle. Let us name a few: floating floor, railroad track vertical deflection, vehicle suspension, ocean energy harvesting, and many others.

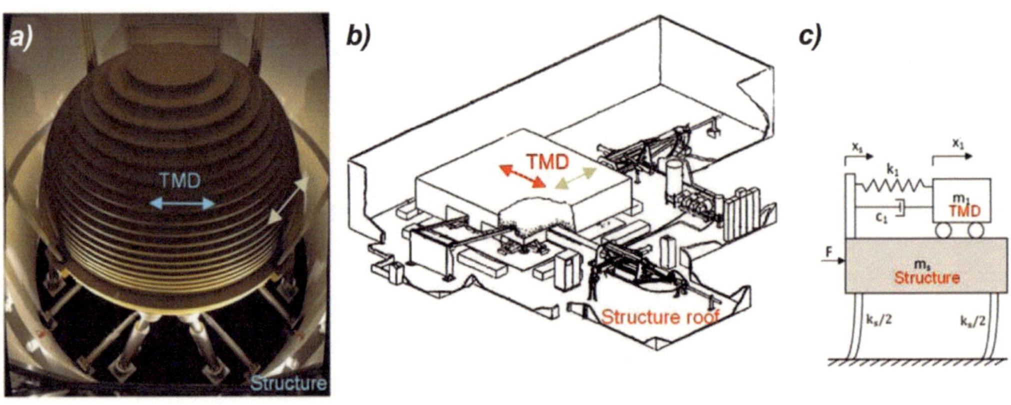

Figure 11. Large scale energy supply of the active TMD, see [70].

5. Climatology

The position of SR in climatology is specific in comparison with other disciplines. It is worthy to be highlighted in the separate section, although apparently it is a bit far from engineering mechanics. The reason is that researchers in climatology demonstrated the first systematic genuine SR in contemporary meaning. This concept was introduced in 1981–1982 by C. Nicolis, see [2, 3], dealing with the problem of climatic changes during the Quaternary. Approximately in the same time appeared papers by Benzi at al. dealing with similar topics [72, 73] preferring a bit more theoretical aspects of the SR phenomena. So that, this pioneering step came from the apparently exotic context of the Earth's climate evolution of the periodic recurrence of Earth's ice ages. For some summary of the starting period and physical motivation analyzing the physical essentials of climatological changes in view of SR, see Scholarpedia [74], other encyclopedia co-authored by C. and G. Nicolis and also a couple of review articles, e.g., [4, 5].

5.1. Physical motivation

It has been known that the climatic system possesses a very pronounced internal variability. A striking illustration is provided by the last glaciation which reached its peak some 18,000 years ago, leading to mean global temperatures of some degrees lower than the present ones and a total ice volume more than twice its present value.

Going further back in the past, it is realized that glaciation has covered, in an intermittent fashion, much of the Quaternary era. Statistical data analysis shows that the glacial/interglacial transitions that have marked the last hundred thousand years display an average periodicity of 10,000 years, indeed. To this process is superimposed a considerable, random looking variability of Sun flux. The conventional explanation was that variations in the eccentricity of Earth's orbital path occurred with a period of about 10^5 years. So that, the energy flux Q impacting the Earth can be characterized as follows:

$$Q = Q_0(1 + \varepsilon \cdot \sin \omega t), \tag{24}$$

where $\varepsilon \approx 0.001$, $\omega \approx 2\pi/10^5 years^{-1}$. This process caused the year average temperature to shift dramatically and produces the ice volume changes on the Earth, which randomly oscillates between limits $30 - 60 \times 10^6 km^3$, see **Figure 12a**.

However, it sounds strange, since the only known time scale in this range is that of the changes in time of the eccentricity of the Earth's orbit around the sun, as a result of the perturbing action of the other solar system bodies. This perturbation modifies the total amount of solar energy received by the Earth but the magnitude of this astronomical effect is exceedingly small, about 0.1%, see above. So that, the measured variation in the eccentricity had a relatively small amplitude compared to the dramatic temperature change. Therefore a question arose, whether one can identify in the Earth-atmosphere-cryosphere system any mechanism capable of enhancing its sensitivity to such small external time-dependent forcing.

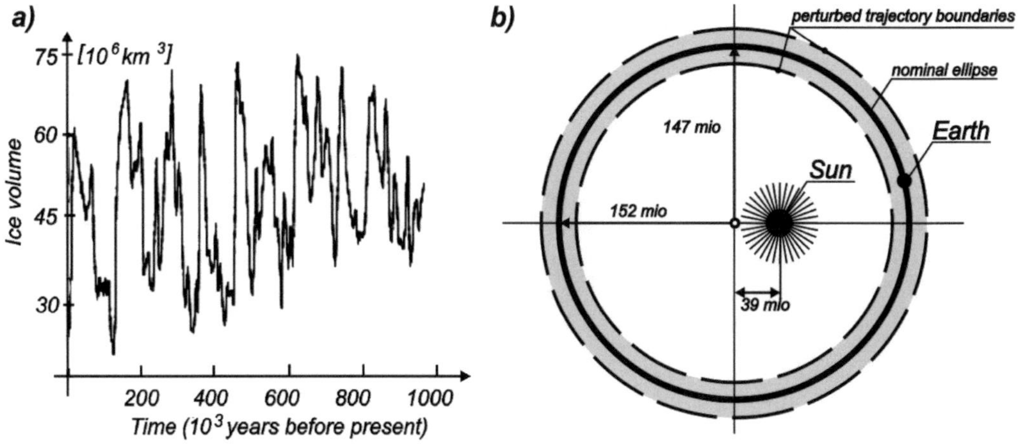

Figure 12. (a) Ice volume on the Earth surface in the past, see [5]. (b) Earth trajectory around the sun.

The search of a response to this question led to the concept of SR, which has been developed out of an effort to understand how the Earth's climate oscillates periodically between two relatively stable global temperature states, one "normal" and the other an "ice age" state. In other words, a theoretical explanation has been elaborated to show that the temperature change due to the weak eccentricity oscillation and added stochastic variation due to the unpredictable energy output of the sun (known as the solar constant) could cause the temperature to move in a nonlinear fashion between two stable dynamic states. Specifically, glaciation cycles are viewed as transitions between glacial and inter-glacial states that are somehow managing to capture the periodicity of the astronomical signal, even though they are actually made possible by the environmental noise rather than by the signal itself. Note that also dynamics of the Earth as a deformable body should be taken into account, see [75], as an indirect source of periodic processes involved.

5.2. Mathematical modeling

The orbit of Earth around the sun is not exactly elliptical, as it is commonly reported. The shape of its trajectory is complex following a form of a spiral. This trajectory is stable within a basin having a form of a closed strip. Its width is approximately 10^7 km, see **Figure 12b**, and exhibits a character of deterministic chaotic attractor. Earth trajectory takes place within the shadow area, see **Figure 12b**. The Lyapunov exponent mostly oscillates nearby 0.

The basic setting of SR in climatology started with SDOF nondynamic system subjected to a stochastic excitation and weak harmonic forcing. It corresponds formally to the Langevin equation of the first order in the form with suppressed inertia term due to high damping (adiabatic approach), compare with Eqs. (26) or (27), Section 6.1:

$$\begin{gathered}\frac{\mathrm{d}^2 u}{\mathrm{d}t^2}+\frac{\partial V(u)}{\partial u}=\eta(t)+Q_0\left(1+\varepsilon\exp(\mathrm{i}\omega t)\right)\\ \dot{\eta}+a\cdot\eta=\xi(t),\quad\text{or}\quad\ddot{\eta}+a\dot{\eta}+b\eta=\xi(t)\end{gathered}\tag{25}$$

where it has been denoted: $V(u)$—conventional symmetric quartic potential, $\eta(t)$—exponentially correlated random process, $\xi(t)$—α stable white noise.

Potential $V(u)$ is considered usually in conventional symmetric quartic form, but also various nonsymmetric variants are regarded in order to respect specific anomalous situations, see also sections 2 and 3. Compare Eq. (25) with FitzHugh-Nagumo equations, see [76, 77]. The contemporary research uses more sophisticated models respecting the space distribution. However, the basic mechanism concerning the time coordinate following Eq. (25) is kept.

Further research in 1990s has been focused to abrupt glacial climatic changes. It has been conducted in view to SR phenomenon related with these changes. Results appeared successively during last 2 decades, see, e.g., [5, 78], and later [79, 80] reflecting furthermore specific attributes of the chaotic dynamics. On the basis of SR, many more studies have been published dealing with general and specific themes. See, e.g., [81] discussing SR in the North Atlantic and a large series of articles by Ditlevsens (senior and junior), e.g., [82] dealing with the rapid climate shifts observed in the glacial climate.

Take a note that the statistical properties of relevant processes are adequately characterized by α-stable processes and so they are widely used in this discipline. For theoretical background see, e.g., monographs [83, 84] and some problem specific papers, see subsection 6.2.

6. Alternative operators and driving processes

The most common SR definition is based on the Duffing equation with the negative linear part of stiffness being excited by an appropriate combination of a harmonic and Gaussian white noise signals. However, it came to light that a few different definitions of SR are possible being based on an alternative differential system or using other driving noise than the Gaussian one. It revealed that many cases can be treated much more effectively than under classical definitions. Application of this background is very wide, and it can be concluded that starting investigation of a particular problem a suitable definition should be carefully selected. So that they can be actually found everywhere in physics, life, and social disciplines.

6.1. Alternative differential operators

Despite of classical definitions of SR, some nonconventional inherent settings appeared together with excellent applications in general theory, nano-scale systems, neurophysiology, etc. Using the linear response theory, some alternative types of SR turned out. For details, see the original papers by Dykman [19, 85, 86], Luchinsky [7, 8], and other authors. They identified SR existence in quite different systems from those commonly studied to date, which are typical by a static double-well potential and being excited by a force equal to the sum of periodic and driving stochastic components.

(*i*) *SR in a monostable system*. The SR can be observed in a monostable nonlinear Duffing oscillator being driven by additive Gaussian white noise $\xi(t)$ of intensity σ. Let us assume the nonlinear mass-unity SDOF oscillator:

$$\ddot{u} + 2\omega_b \dot{u} + \frac{\partial V(u)}{\partial u} = \xi(t) + P_0 \exp(\mathrm{i}\Omega t), \quad V(u) = \frac{\omega_0^2}{2} u^2 + \frac{\gamma^4}{4} u^4 + Bu,$$
$$\omega_b \ll 1, \quad \mathcal{E}\{\xi(t)\} = 0, \quad \mathcal{E}\{\xi(t)\xi(t')\} = 4\omega_b \sigma^2 \cdot \delta(t - t'). \tag{26}$$

Note that the potential $V(u)$ posses the positive quadratic part and therefore the derivative $\partial V(u)/\partial u$ (providing the stiffness force in mechanical system) is a monotonous function. Therefore, the system is monostable unlike conventional systems exhibiting SR. Moreover, the system Eq. (26) is nonsymmetric due to linear term in the potential. It can be understood as a constant external force pre-stressing the system, see **Figure 13(a)**.

The first variant $|B| \leq 0.43$: The eigen frequency is rising monotonously with increasing energy (or the square of response amplitude). In absence of periodic force and under small noise intensity σ, the peak of the response variance, spanning around the eigen frequency $\omega_0(E)$ in an excitation level E, has the width which is approximately given by ω_b, see, e.g., [21] or [23] (in other word Lorenzian peak). That small periodic force inserted on the right side of Eq. (26) will be amplified significantly and therefore SR emerges. The most considerable increment corresponds to the frequency $\Omega = \omega_0(E)$.

The second variant $|B| > 0.43$: The eigen frequency is no more monotonous and exhibits a minimum for a certain $E > 0$. Without periodic force, the system response is given by a narrow spectral density with a maximum at the frequency ω_m lying in the point $\mathrm{d}\omega_0(E)/\mathrm{d}E = 0$. The ω_b is very small, and therefore, in this point the extremely sharp variance of width approximately $\omega_b^{1/2}$ arises and increases nearly exponentially with rising σ. So that for Ω close to ω_m, the SR phenomenon can be expected. It comes to light that the second variant leads to more significant SR phenomenon.

(ii) SR in a bistable system with periodically modulated noise. Potential of the system is similar to the classical version, in particular its quadratic part is negative and hence the system is bistable again. Linear part of the potential is retained. Damping is high (system is over-damped) and therefore the inertia term can be neglected. The system behavior is modeled is follows:

$$\dot{u} + \frac{\partial V(u)}{\partial u} = f(t) \equiv \xi(t)\left(\frac{1}{2} P_0 \exp(\mathrm{i}\Omega t) + 1\right), \quad V(u) = -\frac{\omega_0^2}{2} u^2 + \frac{\gamma^4}{4} u^4 + Bu, \tag{27}$$

Unlike Eq. (26), a harmonically modulated white noise is applied on the right side. Parameter B characterizes again the asymmetry of the potential. For $-2/(3\sqrt{3}) < B < 2/(3\sqrt{3})$, the potential possesses two minima. Simple manipulation gets the intensity of the driving force, see **Figure 13(b)**. As the amplitude P_0 is considered small, its square can be neglected. So the real part reads:

$$\mathcal{E}\{f(t)f(t')\} = 2\sigma^2 \delta(t - t')(1 + P_0 \cos(\Omega t)) \tag{28}$$

and we can see that the intensity of the driving force is periodic. Herewith the phenomenon of SR type emerges.

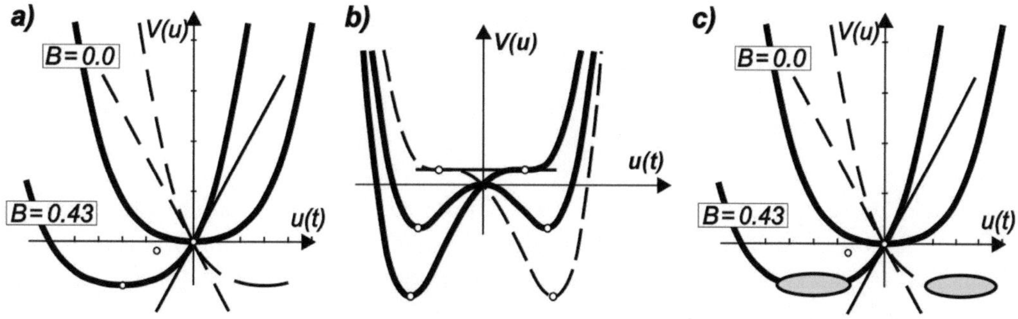

Figure 13. Alternative operators: (a) monostable system; (b) bistable system with periodically modulated noise; (c) system with coexisting periodic attractors.

(iii) SR in a system with coexisting periodic attractors. The third form of nonconventional SR is entirely different form of bistability. The SR concept can be based on coexisting stable states having the form of periodic or chaotic attractors, if there are any. The coexisting attractors are not static, but periodic. Theoretical analysis of these more involved situations draw on the existence (for relevant systems) of generalized potentials, not necessarily analytic in the state variables, possessing local minima on the corresponding attractors. For simplicity, the case where the period of vibration for each of the two attractors is the same can be considered and, consequently, it can be assumed that they correspond to two different stable states of forced vibration induced by an external periodic field driving the system, see **Figure 13(c)**. This interesting approach has been proposed in [87] where chaotic SR is studied to enhance attractors reconstruction using an appropriated random additional noise.

The under-damped nonlinear oscillator to be considered provides a well-known simple, but nontrivial, example of a system that behaves in just this way; its bistability under periodic, nearly resonant driving has been investigated in the context of nonlinear optics and in experiments on a confined relativistic electron excited by cyclotron resonant radiation. The particular model we treat, the nearly-resonantly-driven, under-damped, single-well Duffing oscillator with additive noise, which serves as an archetype for the study of fluctuation phenomena associated with coexisting periodic attractors, is described by

$$\begin{gathered}\ddot{u} + 2\omega_b \dot{u} + \frac{\partial V(u)}{\partial u} = \xi(t) + P_0 \exp(\mathrm{i}\Omega t), \quad V(u) = \frac{\omega_0^2}{2} u^2 + \frac{\gamma^4}{4} u^4 + Bu, \\ \omega_b \ll 1, \quad \mathcal{E}\{\xi(t)\} = 0, \quad \mathcal{E}\{\xi(t)\xi(t')\} = 4\omega_b \sigma^2 \cdot \delta(t - t').\end{gathered} \tag{29}$$

The appearance of new types of SR in systems far from the conventional static double-well potential shows that SR is a very general phenomenon. In other words, there are many physical situations where noise can be used to increase the response of a system to periodic driving. The effect is not confined to systems with coexisting static stable states. Correspondingly, SR may be more widespread in nature, and potentially of wider relevance in science and technology, than has hitherto been appreciated.

(*iv*) *Logistic map*. Let us note that each of differential operators above can be formulated in term of its discretized variant. Then the whole stochastic differential system can be rewritten in form of a logistic map:

$$\mathbf{u}_{i+1} = (\mathbf{u}_i, \mathbf{u}_{i-1}, \ldots, t_i, t_{i-1}, \ldots) \tag{30}$$

where $\mathbf{u}_i$ is the state vector of the system in *i*-th point. This scheme includes an explicit time point to indicate that additive excitation (deterministic, stochastic in time) is acting. This discretized version is widely used if the immediate stochastic simulation is foreseen. Anyway, a care should be taken and Ito system is to be formulated the first respecting principles of manipulation with stochastic processes, see, e.g., [20–22]. These operations related with SR are very close to optimal (suboptimal) filtering and other stochastic data treatment. They can provide valuable contribution to SR application especially in numerical processing. This concerns particularly one-pass filtering where evaluation processes with SR algorithms are very close. For details see, for instance, [23] and other monographs, where even more general models than Eq. (30) are formulated.

6.2. NonGaussian driving noise

Although Gaussian random noise is mostly used as driving component, there approved well also other than Gaussian processes. This finding results from the inherent nature of a number of processes originally characterized by different PDF.

(*i*) *α-stable processes*. A number of papers deal with α-stable processes in the role of SR generator. For comprehensive acquainting with α –stable and other useful nonGaussian processes monographs [83, 84] are recommended, see **Figure 14**. Indeed, α-stable processes are suitable for use in nondynamical application, see, e.g., [88]. Authors thoroughly analyzed specific attributes of this class of problems and show doubtless advantages of α-stable instead Gaussian processes in certain nondynamical cases. They obtained these conclusions by means of theoretical and experimental procedures with white and arbitrarily colored noise. Further contribution being neurophysiology motivated are papers [40, 41]. These large

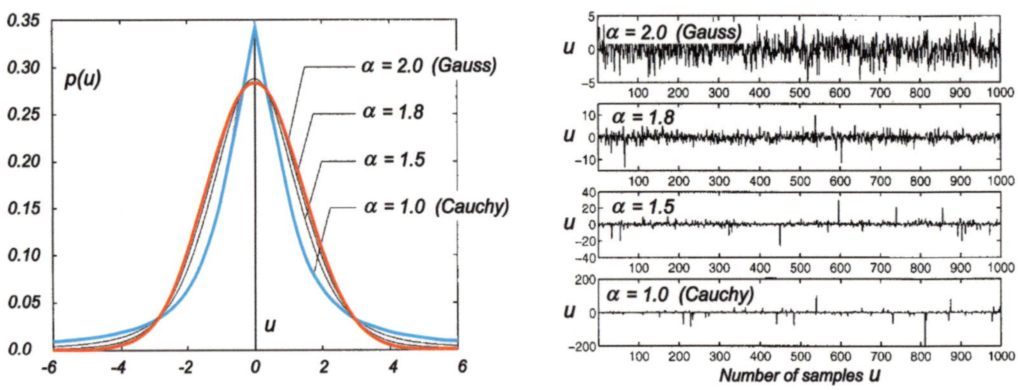

Figure 14. α-stable, Gaussian, and Cauchy processes.

studies treat problems of robust SR and adaptive SR in noisy neurons based on mutual information assessment.

(ii) Impuls chains–Poisson driven processes. Concerning nonGaussian driving noise, the impuls chains and various Poisson driven processes can be used in special cases [46]. A couple of authors investigated the basis of signal detection and adaptation in impulsive driving noise in framework of plasma physic, see [89].

(iii) Colored noise. Employment of colored noise is studied in review and particular problem focused articles. This noise being used intentionally has an effect, which is close to window filtering. It can be adjusted suitably to instantaneous needs. If it follows from the frequency limited "white noise" (finite correlation times), then influence of this low pass filter should be examined. The role of such physically realistic noise is studied for exponentially correlated Gaussian noise with constant intensity, see, e.g., [46, 90], etc. In principle, in over-damped dynamics (first order equation of SDOF system), the role of colored noise generally results in a reduction of SR efficiency. In contrast, finite inertia effects (second order equation of SDOF system), induced by moderate damping, tends to increase SR system response.

(iv) High frequency deterministic signal. Interesting idea is to use a high frequency deterministic signal in a meaning of a driving noise instead a random noise. For this reason, the final phenomenon is called vibrational resonance (VR), see [91]. This phenomenon analogous with SR occurs when the excitation frequency is well separated from the forcing frequency of the potential well. This setting approved very well when machine vibration is treated. Machine vibration is never truly stochastic, this provides a mechanism to link stochastic resonance to real mechanical devices, such as those used for vibrational energy harvesting, see section 3 or 4 referring among others about vibration damping.

6.3. Some other nonconventional settings

Let us briefly remark some specific SR settings. They are valuable not only for the area where they usually have been evolved, but serve as a possible inspiration for the whole SR community. Although for the full understanding, the adequate papers should be studied, have a look at some of them:

(i) Useful signal has the impulsive or rectangular form. Driving random signal is still Gaussian white noise. Amplification and distortion of a periodic rectangular driving signal by a noisy bistable system has been studied in [92]. Impulsive signals emerging in plasma physics are thoroughly reported in a series of publications by Nurujjaman et al. see [89]. Anyway, these papers attract attention also beyond plasma physics being interesting from general methodological points of view.

(ii) SR in systems exhibiting chaos. Dynamical systems in the regime of deterministic chaos evolve under certain conditions through a sequence of intermittent jumps between two preferred regions of phase space and without the intervention of a driving noise. Such systems, which give rise to multi-modal probability distributions, display an enhanced sensitivity to external periodic forcings through a stochastic resonance-like mechanism, see, e.g., [87]. For

further reading about chaotic response of deterministic systems, see monographs e.g., [93] or [94].

Let us include to this paragraph also reference to the adaptive SR, see, e.g., [41]. This approach seems to be promising as it makes possible to change parameters of the system dynamically during signal transmission in noisy neurons ambiance. The main goal of that concept consists in the fact that fuzzy and other adaptive systems can learn to induce SR based only on samples from the process. Application in other fields like control of electro-hydraulic testing equipment or smart control of vibration damping are obvious.

(iii) Slowly varying parameters. In many systems, the dynamics in the absence of both noise and forcing is controlled by a number of parameters λ_i describing the constraints acting from the external world. Ordinarily these parameters are assumed to remain constant, but there are situations where this strategy constitutes an oversimplification (gradual switching on/off a device, man-biosphere-climate interactions, etc.). In the absence of external periodic forcing, the simultaneous action of noise and of a slow variation of λ_i in the form of a ramp may lead to freezing of the system in a preferred state by practically quenching the transitions across the barrier. The interaction between SR and the action of the ramp provides an alternative method for the control of the transition rates by allowing the system to perform (transiently) a certain number of transitions (depending on the forcing frequency and the noise strength) prior to quenching.

7. Conclusion

The chapter tried to indicate the essence of SR. This is for the first view counter-intuitive phenomenon brings a large impact on physical, biological, and engineering systems. It is clear that SR is generic enough to be observable in a large variety of systems. The SR emerges in all scales, we can imagine. It governs the processes from nuclear fusion in the sun to the intra-atomic structures on the level of quantum mechanics. Amazing results of the basic research have been achieved and excellent industrial programs have been launched being based on many variants of SR. This concept of SR enabled to obtain an insight and exact description of many effects in macro and micro (nano) world and to fight successfully against various nondesirable phenomena in engineering. It resulted in many actually nonreplaceable products of signal sensing and processing, medical instruments, and treatment procedures. Many SR-inspired neurophysiological implants represent cornerstones at the field.

The SR can be perceived as a natural phenomenon ruling inside of certain dynamic systems. In such a case, it can act either positively as for instance to help stabilize the dynamic system and therefore, to improve the system reliability or oppositely it can affect the system negatively, e.g., as a strong periodic exciting force, which is necessary to be avoided. The second view of SR understanding is considered in active synthesis and manipulation with the noise. Addition of appropriate dose of (mostly) random noise onto the useful signal provides a significant increase of sensitivity and reliability of the equipment and enlarge its ability of data sensing, processing, and possibly their usage in a feedback. The same is valid concerning an increase of information transfer capacity and reliability.

The chapter outlines a short history of SR. An overview of SR utilization in various disciplines in physical, life, and social sciences is briefly looked through. Some possibilities of modeling in dynamics using SR strategy are indicated. Mathematical treatment and the most popular solution methods of investigation are pointed out including semi-analytic, numerical, simulation based and experimental approaches. Nevertheless, aspects related with Engineering Dynamics make intentionally a core of the chapter. Also the section dealing with energy harvesting has been highlighted as it shares many joint attributes with dynamics itself.

The phenomenon of SR in whatever variant is worthy to be employed in Engineering Dynamics having a large potential of specific basic research as well as of engineering applications. Industrial aerodynamics seems to be promising wide branch where several effects of stability loss could be explained as effects related with SR. This approach approved to describe the divergence stability loss in the nonlinear formulation of a slender beam post-critical behavior in a cross flow. Additional problems are waiting for similar type of theoretical description and subsequent experimental verification. The same probably emerge at area of panel flutter, various variants of buffeting, etc. This strategy could enable to formulate new ideas for development of nonconventional measures for vibration damping. Another area of SR application prove to be problems of vehicle dynamic stability and its post-critical behavior. Similarly like in aeroelasticity the results obtained can be used for development of new generation of vibration quenching devices of both passive and actively controlled types.

It should be highlighted that adequate experiments will be absolutely necessary. However, they should be newly proposed and performed properly, as they will differ in many ways from conventional experiments. On the other hand, a lot of inspiration at both theoretical as well as experimental fields can be taken from solid state physics and energy harvesting area.

Let us be aware that SR is a challenging discipline for Engineering Dynamics offering a large variety of possibilities of new developments at theoretical as well as experimental platform. It could significantly enhance the top areas of nonlinear and stochastic dynamics closely related with Computational Mechanics, which is very advanced and widely used in comparison with other fields of numerical analysis. It provides strong support to Engineering Dynamics, which stands on the threshold to enter the field of research and application of SR.

Acknowledgements

The kind support of the Czech Science Foundation project No. 15-01035S and of the RVO 68378297 institutional support are gratefully acknowledged.

Author details

Jiří Náprstek* and Cyril Fischer

*Address all correspondence to: naprstek@itam.cas.cz

Institute of Theoretical and Applied Mechanics ASCR, Prague, Czech Republic

References

[1] Kramers HA. Brownian motion in a field of force and the diffusion model of chemical reactions. Physica. 1940;**VII**(4):284-304

[2] Nicolis C. Solar variability and stochastic effects on climate. Solar Physics. 1981;**74**:473-478

[3] Nicolis C. Stochastic aspects of climatic transitions-response to a periodic forcing. Tellus. 1982;**34**:1-9

[4] Gammaitoni L, Hänggi P, Jung P, Marchesoni F. Stochastic resonance. Reviews of Modern Physics. 1998;**70**(1):223-287

[5] Nicolis G, Nicolis C, McKernan D. Stochastic resonance in chaotic dynamics. Journal of Statistical Physics. 1993;**70**(1):125-140

[6] Wellens T, Shatokhin V, Buchleitner A. Stochastic resonance. Reports on Progress in Physics. 2004;**67**(1):45-48

[7] Luchinsky DG, Mannella R, McClintock PVE, Stocks NG. Stochastic resonance in electrical circuits I. Conventional stochastic resonance. IEEE Transactions on Circuits and Systems II. 1999;**46**(9):1205-1214

[8] Luchinsky DG, Mannella R, McClintock PVE, Stocks NG. Stochastic resonance in electrical circuits II. Nonconventional stochastic resonance. IEEE Transactions on Circuits and Systems II. 1999;**46**(9):1215-1224

[9] Anton SR, Sodano HA. A review of power harvesting using piezoelectric materials (2003–2006). Smart Materials and Structures. 2007;**16**(3):R1

[10] Moss F, Ward L, Sannita W. Stochastic resonance and sensory information processing: A tutorial and review of application. Clinical Neurophysiology. 2004;**115**:267-281

[11] McDonnell MD, Stock NG, Pearce CEM, Abbott D. Stochastic Resonance: From Suprathreshold Stochastic Resonance to Stochastic Signal Quantization. Cambridge, New York: Cambridge University Press; 2008

[12] Tuckwell HC. Introduction to Theoretical Neurobiology. Vol. 2. Cambridge: Cambridge University Press; 1988

[13] Moss F. Stochastic resonance: From the ice ages to the Monkey's ear. In: Weiss GH, editor. Contemporary Problems in Statistical Physics. Philadelphia: SIAM; 1994. pp. 205-253

[14] Berglund N, Gentz B. Noise-induced Phenomena in Slow-Fast Dynamical Systems. Berlin, Heidelberg: Springer; 2006

[15] Anishchenko VS, Astakhov VV, Neiman AB, Vadivasova TE, Schimansky-Geier L. Non-Linear Dynamics of Chaotic and Stochastic Systems. Berlin, Heidelberg: Springer; 2003

[16] Weisstein EW. Stochastic Resonance. From MathWorld—A Wolfram Web Resource; [Accessed: 05-05-2017]. Available from: http://mathworld.wolfram.com/StochasticResonance.html

[17] Náprstek J. Stochastic resonance—Challenges to engineering dynamics. In: Topping BHV, Kruis J, editors. Computational Technology Reviews. Vol. 12. Scotland: Civil-Comp Ltd.; 2015. pp. 53-101

[18] Inchiosa ME, Bulsara AR. Signal detection statistics of stochastic resonators. Physical Review E. 1996;**53**(3):2021-2024

[19] Dykman MI, et al. Resonant subharmonic absorption and second-harmonic generation by a fluctuating non-linear oscillator. Physical Review E. 1996;**54**(3):2366-2377

[20] Wong E, Zakai M. On the relation between ordinary and stochastic equations. International Journal of Engineering Sciences. 1965;**3**(2):213-229

[21] Lin YK, Cai GQ. Probabilistic Structural Dynamics—Advanced Theory and Applications. NewYork: McGraw-Hill; 1995

[22] Náprstek J. Real and Markov processes in stochastic systems. In: Dobiáš I, editor. Proceedings of the Dynamics of Machines. Prague: IT ACSR; 2003. pp. 127-134

[23] Pugachev VS, Sinitsyn IN. Stochastic Differential Systems—Analysis and Filtering. Chichester: J. Willey; 1987

[24] Kang YM, Xu JX, Xie Y. Observing stochastic resonance in an underdamped bistable Duffing oscillator by the method of moments. Physical Review E. 2003;**68**(3):036123

[25] Mikhlin SG. Variational Methods in Mathematical Physics (in Russian). Moscow: Nauka; 1970

[26] Náprstek J. Application of the maximum entropy principle to the analysis of non-stationary response of SDOF/MDOF systems. In: Půst L, Peterka F, editors. Proceedings of the 2nd European Non-Linear Oscillations Conference—EUROMECH. Prague: IT ASCR; 1996. pp. 305-308

[27] Cai GQ, Lin YK. On exact stationary solutions of equivalent non-linear stochastic systems. International Journal of Non-Linear Mechanics. 1988;**23**(4):315-325

[28] Zhu WQ, Cai GQ, Lin YK. On exact stationary solutions of stochastically perturbed Hamiltonian systems. Probabilistic Engineering Mechanics. 1990;**5**(2):84-87

[29] Náprstek J, Fischer C, Král R, Pospíšil S. Comparison of numerical and semi-analytical solution of a simple non-linear system in state of the stochastic resonance. In: Papadrakakis M, Papadopoulos V, Plevris V, editors. COMPDYN 2015–5th ECCOMAS Thematic Conference on Computational Methods in Structural Dynamics and Earthquake Engineering. Athens: National Technical University of Athens; 2015. pp. 1971-1982

[30] Grasman J, van Herwaarden OA. Asymptotic Methods for the Fokker-Planck Equation and the Exit Problem in Application. Berlin, New York: Springer; 1999

[31] Náprstek J, Král R. Finite element method analysis of Fokker-Planck equation in stationary and evolutionary versions. Journal of Advances in Engineering Software. 2014;**72**:28-38

[32] Spencer BF, Bergman LA. On the numerical solution of the Fokker-Planck equation for non-linear stochastic systems. Non-Linear Dynamics. 1993;**4**:357-372

[33] Bergman LA, Spencer BF, Wojtkiewicz SF, Johnson EA. Robust numerical solution of the Fokker-Planck equation for second order dynamical system under parametric and external white noise excitation. In: Langford W, Kliemann W, Namachchivaya NS, editors. Proceedings of Non-Linear Dynamics and Stochastic Mechanics. Providence: American Mathematical Society; 1996. pp. 23-37

[34] Náprstek J, Král R. Theoretical background and implementation of the finite element method for multi-dimensional Fokker–Planck equation analysis. Journal of Advances in Engineering Software. 2017; in print. https://doi.org/10.1016/j.advengsoft.2017.02.013

[35] Kloeden PE, Platen E. Numerical Solution of Stochastic Differential Equations. Berlin, Heidelberg: Springer; 1992

[36] Ott E. Chaos in Dynamical Systems. Cambridge: Cambridge University Press; 1993

[37] Gang H, Qing GR, Gong DC, Weng XD. Comparison of analog simulation of stochastic resonance with adiabatic theory. Physical Review A. 1991;**44**(10):6414-6420

[38] Gammaitoni L, Marchesoni F, Menichella-Saetta E, Santucci S. Stochastic Resonance in Bistable Systems. Physical Review Letters. 1989;**62**(4):349-352

[39] Ohka M, Beceren K, Jin T, Chami A, Yussof HB, Miyaoka T. Experiments on stochastic resonance toward human mimetic tactile data processing. International of Journal of Social Robotics. 2012;**4**(1):65-75

[40] Kosko B, Mitaim S. Robust stochastic resonance: Signal detection and adaptation in impulsive noise. Physical Review E. 2001;**64**(5):051110

[41] Mitaim S, Kosko B. Adaptive stochastic resonance. Proceedings of the IEEE. 1998;**86**(11):21 52-2183

[42] Tanaka SM, Alam I, Turner CH. Stochastic resonance in osteogenic response to mechanical loading. The FASEB Journal. 2003;**17**(2):313-314

[43] McDonnell MD, Ward LM. The benefits of noise in neural systems: bridging theory and experiment. Nature Reviews Neuroscience. 2011;**12**:415-426

[44] Dinklage A, Wilke C, Klinger T. Spatio-temporal response of stochastic resonance in an excitable discharge plasma. Physics of Plasmas. 1999;**6**(8):2968-2971

[45] McNamara B, Wiesenfeld K, Roy R. Observation of stochastic resonance in a ring laser. Physical Review Letters. 1988;**60**(25):2626-2629

[46] Gingl Z, Kiss LB, Moss F. Non-dynamical stochastic resonance: Theory and experiments with white and arbitrarily coloured noise. Europhysics Letters. 1995;**29**:191-196

[47] Naess A, Gaidai O, Karpa O. Estimation of extreme values by the average conditional exceedance rate method. Journal of Probability and Statistics. 2013;**2013**. Article ID 797014

[48] Cover TM, Thomas JA. Elements of Information Theory. New York: Wiley; 1991

[49] Neiman A, Shulgin B, Anishchenko V, Ebeling W, Schimansky-Geier L, Freund J. Dynamical entropies applied to stochastic resonance. Physical Review Letters. 1996;**76**:4299-4302

[50] Strong SP, Koberle R, van Steveninck RR, Bialek W. Entropy and information in neural spike trains. Physical Review Letters. 1998;**80**(1):197-200

[51] Gang H, De-chun G, Xiao-dong W, Chun-yuan Y, Guang-rong Q, Rong L. Stochastic resonance in a nonlinear system driven by an aperiodic force. Physical Review A. 1992 Sep;**46**:3250-3254

[52] Chapeau-Blondeau F. Noise-enhanced capacity via stochastic resonance in an asymmetric binary channel. Physical Review E. 1997 Feb;**55**:2016-2019

[53] Shannon CE. Communication in the presence of noise. Proceedings of the IRE. 1949;**37**:10-21

[54] Munakata T, Sato A, Hada T. Stochastic resonance in a simple threshold system from a static mutual information point of view. Journal of Physical Society Japan. 2005;**74**:2094-2098

[55] Luo ACJ, Gu K, Han RPS. Resonant-separatrix webs in stochastic layers of the twin-well Duffing oscillator. Non-Linear Dynamics. 1999;**19**(1):37-48

[56] Almog R, Zaitsev S, Shtempluck O, Buks E. Signal amplification in a nanomechanical Duffing resonator via stochastic resonance. Applied Physics Letters. 2007;**90**(1):013508

[57] Wu X, Guo W, Cai W, Shao X, Pan Z. A method based on stochastic resonance for the detection of weak analytical signal. Talanta. 2003;**61**(6):863-869

[58] Zhi-Hui L, Yong-Gang L, Sheng-Bo F. Stochastic resonance of cascaded bistable Duffing system. Acta Physica Sinica. 2013;**62**(7):070503

[59] Stocks NG. Suprathreshold stochastic resonance in multilevel threshold systems. Physical Review Letters. 2000;**84**(11):2310-2313

[60] Tanaka SM, Li J, Duncan RL, Burr DB, Turner CH. Mechanical stimulation with broad frequency vibration promotes differentiation of osteoblasts in 3D culture. Journal of Bone and Mineral Research. 2001;**16**:S481

[61] Rogan S, Hilfiker R, Schmid S, Radlinger L. Stochastic resonance whole-body vibration training for chair rising performance on untrained elderly: A pilot study. Archives of Gerontology and Geriatrics. 2012;**55**(2):468-473

[62] Collins A, Blackburn JT, Olcott C, Yu B, Weinhold P. The impact of stochastic resonance electrical stimulation and knee sleeve on impulsive loading and muscle co-contraction during gait in knee osteoarthritis. Clinical Biomechanics. 2011;**26**(8):853-858

[63] Náprstek J, Pospíšil S. Response types and general stability conditions of linear aeroelastic system with two degrees-of-freedom. Journal of Wind Engineering and Industrial Aerodynamics. 2012;**111**:1-13

[64] Náprstek J. Combined analytical and numerical approaches in dynamic stability analysis of engineering systems. Journal of Sound and Vibration. 2015;**338**:2-41

[65] Stephen NG. On energy harvesting from ambient vibration. Journal of Sound and Vibration. 2006;**293**(1–1):409-425

[66] Sodano H, Simmers G, Dereux R, Inman D. Recharging batteries using energy harvested from thermal gradients. Journal of Intelligent Material Systems and Structures. 2007;**18**(1):3-10

[67] Mann BP, Sims ND. Energy harvesting from the non-linear oscillations of magnetic levitation. Journal of Sound and Vibration. 2009;**319**:515-530

[68] McInnes CR, Gorman DG, Cartmell MP. Enhanced vibrational energy harvesting using non-linear stochastic resonance. Journal of Sound and Vibration. 2008;**318**(4–4):655-662

[69] Erturk A, Inman DJ. Broadband piezoelectric power generation on high-energy orbits of the bistable Duffing oscillator with electromechanical coupling. Journal of Sound and Vibration. 2011;**330**(10):2339-2353

[70] Zuo L, Tang X. Large-scale vibration energy harvesting. Journal of Intelligent Material Systems and Structures. 2013;**24**(11):1405-1430

[71] Wang Y, Inman DJ. A survey of control strategies for simultaneous vibration suppression and energy harvesting via piezoceramics. Journal of Intelligent Material Systems and Structures. 2012;**23**:2021-2037

[72] Benzi R, Sutera A, Vulpiani A. The mechanism of stochastic resonance. Journal of Physics A. 1981;**14**:453-457

[73] Benzi R, Parisi G, Sutera A, Vulpiani A. Stochastic resonance in climatic change. Tellus. 1982;**34**:10-16

[74] Rouvas-Nicolis C, Nicolis G. Stochastic resonance. Scholarpedia;**2**(11):1474

[75] Forte AM, Mitrovica JX. A resonance in the Earth's obliquity and precession over the past 20 Myr driven by mantle convection. Nature. 1997;**390**:676-680

[76] FitzHugh R. Mathematical models of excitation and propagation in nerve. In: Schwan HP, editor. Biological Engineering. New York: McGraw-Hill Book Co.; 1969. pp. 1-85

[77] Nagumo J, Arimoto S, Yoshizawa S. An active pulse transmission line simulating nerve axon. Proceedings of the IRE. 1962;**50**:2061-2070

[78] Nicolis C. Long-term climatic transitions and stochastic resonance. Journal of Statistical Physics. 1993;**70**:3-14

[79] Ganopolski A, Rahmstorf S. Abrupt glacial climatic changes due to stochastic resonance. Physical Review Letters. 2002 Jan;**88**:038501

[80] Alley RB, Marotzke J, Nordhaus WD, Overpeck JT, Peteet DM, Pielke RA, et al. Abrupt Climate Change. Science. 2003;**299**(5615):2005-2010

[81] Alley RB, Anandakrishnan S, Jung P. Stochastic resonance in the North Atlantic. Paleoceanography. 2001;**16**:190-198

[82] Ditlevsen PD, Ditlevsen OD. On the stochastic nature of the rapid climate shifts during the last ice age. Journal of Climate. 2009;**22**:446-457

[83] Nikias CL, Shao M. Signal Processing with Alpha-Stable Distributions and Applications. New York: Wiley; 1995

[84] Grigoriu M. Applied Non-Gaussian Processes. Englewood Cliff, NJ: Prentice Hall; 1995

[85] Dykman MI, Luchinsky DG, Mannella R, McClintock PVE, Stein ND, Stocks NG. Nonconventional stochastic resonance. Journal of Statistical Physics. 1993;**70**(1):479-499

[86] Dykman MI, Horita T, Ross J. Statistical distribution and stochastic resonance in a periodically driven chemical system. Journal of Chemical Physics. 1995;**103**(3):966-972

[87] Castro R, Sauer T. Chaotic stochastic resonance: Noise-enhanced reconstruction of attractors. Physical Review Letters. 1997;**79**(6):1030-1033

[88] Szczepaniec K, Dybiec B. Non-Gaussian, non-dynamical stochastic resonance. Europhysical Journal B. 2013;**86**(11). paper 468

[89] Nurujjaman M, Iyengar ANS, Parmananda P. Emergence of the stochastic resonance in glow discharge plasma. In: Mago VK, Ananthapadmanabhan PV, Patil DS, Das AK, editors. 23rd National Symposium on Plasma Science and Technology (PLASMA-2008). Vol. 208Journal of Physics Conference Series; 2010. Article-Number 012084

[90] Hänggi P, Inchiosa ME, Fogliatti D, Bulsara AR. Non-linear stochastic resonance: The saga of anomalous output-input gain. Physical Review E. 2000;**62**(5):6155-6163

[91] Landa PS, McClintock PVE. Vibrational resonance. Journal of Physics A: Mathematical and General. 2000;**33**(45):L433

[92] Morillo M, nez JGO. Amplification and distortion of a periodic rectangular driving signal by a noisy bistable system. Physical Review E. 1995;**51**(2):999-1003

[93] Lichtenberg AJ, Lieberman MA. Regular and Chaotic Dynamics. New York: Springer; 1992

[94] Tondl A, Ruijgrok T, Verhulst F, Nabergoj R. Autoparametric Resonance in Mechanical Systems. Cambridge: Cambridge University Press; 2000

Chapter 4

Chaotic, Stochastic Resonance, and Anti-Resonance Phenomena in Optics

Vladimir L. Kalashnikov

Additional information is available at the end of the chapter

http://dx.doi.org/10.5772/intechopen.70737

Abstract

Existence of different, frequently incommensurate scales is a common phenomenon in nature. Interactions between processes characterized by different scales can result in a multitude of emergent phenomena when a system cannot be described as a scale-separated hierarchy of underlying processes but presents a substantially new entity with qualitatively new properties and behavior. Striking examples are life, fractals, and chaos. Here, we shall demonstrate the quite nontrivial phenomena: chaotic and stochastic resonances and anti-resonance on examples of laser systems. The phenomena of resonant stochastization (stochastic anti-resonance), self-ordering (stochastic resonance), and resonant chaotization of coherent structures (dissipative solitons) are considered on the examples of mode-locked lasers and Raman fiber amplifiers. Despite a well-known effect of noise suppression and global regularization of dynamics due to the resonant interaction of noise and regular external periodic perturbation, here we report about the reverse situation when the regular and noise-like perturbations result in the emergent phenomena ranging from the coherent structure formation to the fine-grained chaotic/ noisy dynamics. We guess that the nonlinear optical systems can be considered in this context as an ideal test-bed for "metaphorical modeling" in the area of deterministic and stochastic dynamics of resonance systems.

Keywords: chaotic and stochastic resonance/anti-resonance, soliton-emergence phenomena, resonant soliton–linear wave interaction, noise-assisted coherence, "metaphorical" optical modeling, resonance vector mode-locking

1. Introduction

Is a noise so destructive? This question is not only philosophical because it is directly addressable. We live in a noisy environment, and who knows, would such environment be extremely constructive, namely constructive? Why not? For instance, the growth of initial quantum gravitational perturbations gives birth to our Universe as a quite-ordered structure, and our brains are very flowing but constructive, isn't that so? The key point is a resonant interaction of noise with a *nonlinear* system [1, 2]. The resonance phenomena in nonlinear systems are fraught with counterintuitive consequences. Noise can enhance a system's internal coherence [*stochastic resonance,* (SR)] or damage it [*stochastic anti-resonance,* (SAR)]. Both effects are resonantly sensible to the system parameters that allow naming both phenomena as *a resonance* with taking into account a principal difference between the *linear* and *nonlinear* systems far from an equilibrium state [3].

The notion of SR occurred unexpectedly from the studies regarding the long-term climatic changes (i.e., the ice ages) when the short-term (1-year scaled) climatic noise enhances resonantly an incommensurable weak variation (~10^5 years) of the Earth ecliptic [4, 5]. The excellent surveys expose a further progress forwarding this direction [3, 6–11]. A development of the SR ideology in the fields of neuroscience, biology, and information processes was especially exciting. A noise-induced resonant enhancement of neural sensibility, adaptivity, and learning capability was demonstrated and analyzed [10, 12–18].

The classical theory of SR was based on the resonant transitions in noisy bi-stable nonlinear systems [19–21]. Further studies revealed that both SR and SAR cover an extremely broad range of phenomena including escape from the metastable state, threshold "firing" dynamics, dynamics assisted by deterministic chaos, regularization induced by coherent periodic or continuous structures without a noise assistance, etc [3, 22, 23]. Therefore, the terms of SR and SAR can be misleading in some respects, and it is better to speak about a broad range of phenomena in the nonlinear systems far from equilibrium, which is caused by the resonant-like interaction between processes with incommensurable characteristic scales [3, 24].

As a classical illustration of SR, one may consider the so-called FitzHugh-Nagumo (FN) model (e.g., see [3] and references ibidem), which describes a noise excitable evolution in a very simple two-dimensional form:

$$\epsilon\frac{dx}{dt}=f(x)-y,\quad \frac{dy}{dt}=\gamma x-\beta y-s(t)+\sqrt{2D}\zeta(t), \tag{1}$$

where a potential function is defined as $f(t)=x-ax^3$, typically ϵ defines a ratio of evolutional scales between x and y variables, γ is a coupling parameter, β is a friction coefficient, and $s(t)$ is a periodic external force ($s(t)=\alpha\cos[\omega t]$, usually). The last term in Eq. (1) describes a Wiener stochastic process with volatility $2D$. The stochastic Eq. (1) is treated in the Stratonovich's sense. Evolution of dynamical variables in the absence of noise and periodic modulation is shown in **Figure 1**, which demonstrates a relaxation to local minimum of potential.

Separated effects of small harmonic modulation and noise are shown in **Figure 2**. One can see that they have a "perturbative" character and induce the small oscillations/fluctuations around the potential local minimum.

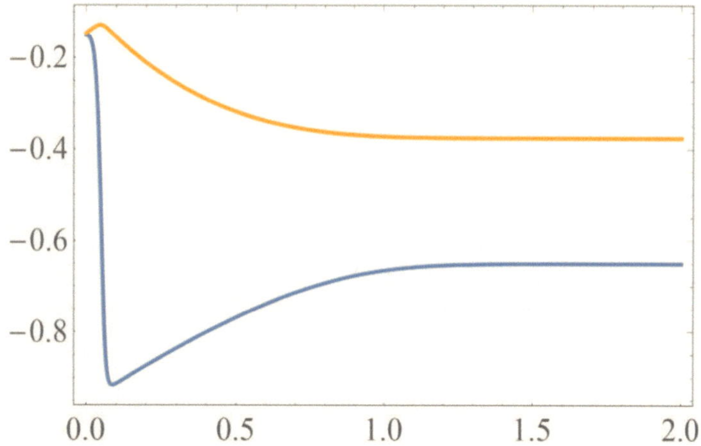

Figure 1. Evolution of $x(t)$ (the lower curve) and $y(t)$ (the upper curve) in the absence of stochastic and modulation terms in Eq. (1) (i.e., $(D, \alpha)=0$) within the range of $t\in[0, 2]$. $\epsilon=0.01$, $a=1$, $b=0.6$, $\beta=1$, $\gamma=1.5$.

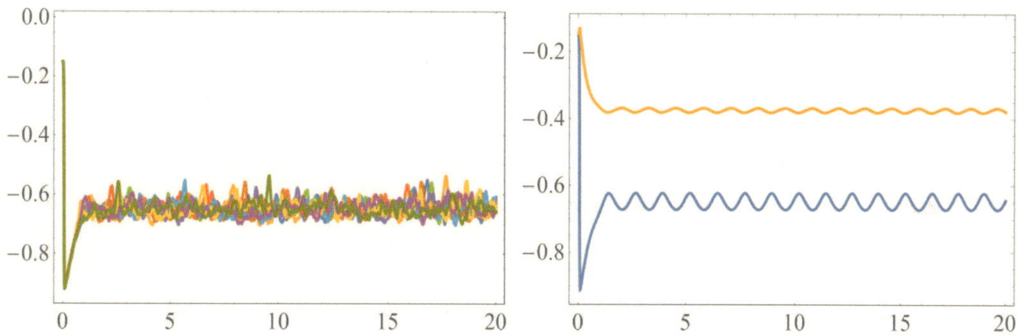

Figure 2. Left: Evolution of $x(t)$ (the lower curve) and $y(t)$ (the upper curve) in the absence of stochastic and presence of modulation terms in Eq. (1); $\alpha=0.05$ and $\omega=5$. Right: Ten stochastic trajectories for $x(t)$ in the absence of harmonic modulation; $D=0.01$. Other parameters correspond to **Figure 1**.

However, the situation changes drastically under the common action of noise and external modulation (**Figure 3**). Extremal and almost regular spikes appear at a frequency, which is lower than the modulational one and incommensurable with the noise scale. One may consider this example based on the FN-model as an impressive and quite simple illustration of SR.

At this moment, there is a huge amount of work concerning the SR and SAR as well as their variations and modifications. We refer a reader to the above-cited books and surveys (the reference list is not exhaustive, of course). A spectacular demonstration of SR in a ring dye laser [25] gave impetus to an intensive exploration of this field. Therefore, we intend to discuss some aspects of SR, SAR, coherent resonant, and multi-scale phenomena regarding laser optics and solitonics.

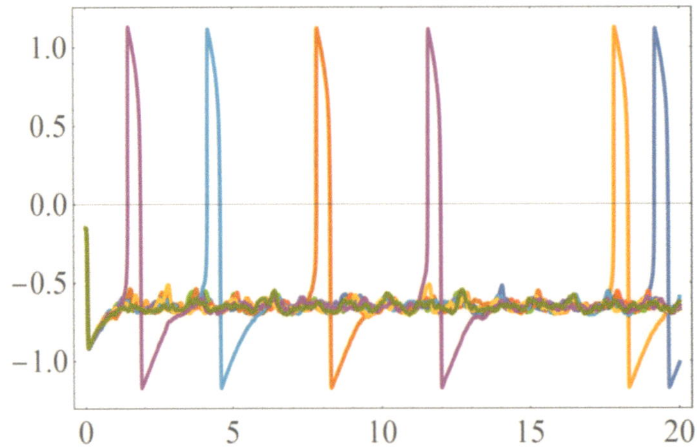

Figure 3. SR appearing under joined action of factors illustrated in **Figure 2**.

This paper is organized as follows. In the next section, we expose the SAR phenomenon in a Raman fiber amplifier. Then, the chaotic resonance of dissipative soliton with linear waves will be considered. Further, the SR and SAR as well as multi-scale resonant phenomena in mode-locked lasers will be exposed. Finally, the resonance vector mode-locking will be described in a nutshell.

2. SAR in a Raman fiber amplifier

A Raman amplifier can be considered as a test-bed for the study of SAR and multi-scale dynamics due to a comparative simplicity and realizability and, simultaneously, high practical significance. The latter is defined by the fact that Raman amplification provides an efficient tool for optical telecommunication lines with frequency multiplexing (for details see [26]). In such lines, there are very different scales: a length corresponding to the width of pulse carrying information (~10 – 100 mm), commensurable lengths of polarization beats and inherent stochastic distortions of a fiber (~10 – 100 m), attenuation length (~10 km), nonlinear and dispersion lengths (>100 km), and overall propagation length (>10^7 m) [27].

The Raman amplification is sensitive to the relative polarization of gain and signal—a gain is maximum for copolarized pump and signal but minimum for their mutually transverse polarizations. Since beat rates for signal (b_s) and pump (b_p) differ, it causes a periodical modulation of the Raman gain with fiber length [26]. Simultaneously, the polarization properties (birefringence) of fiber are sensitive to the inevitable *stochastic* breakdowns of the fiber cylindrical symmetry [27]. Thus, one has all necessary prerequisites for the manifestation of SR and SAR phenomena.

The extended vector theory of the stimulated Raman scattering with taking into account the random birefringence is presented in [26–28]. The system of stochastic differential equations describing an evolution of copropagating pump and signal states of polarization (SOP) under the action of random fiber birefringence can be written in the following form [27]:

$$\frac{d\vec{S}}{dz}=\frac{g_R}{2}\left(\left|\vec{P}\right|\vec{S}+\left|\vec{S}\right|\vec{P}\right)-\alpha_s\vec{S}+\beta\begin{pmatrix}S_2\\-S_1\\0\end{pmatrix}+2b_s\begin{pmatrix}0\\-S_3\\S_2\end{pmatrix},\tag{2a}$$

$$\frac{d\vec{P}}{dz}=-\frac{\omega_p}{\omega_s}\frac{g_R}{2}\left(\left|\vec{P}\right|\vec{S}+\left|\vec{S}\right|\vec{P}\right)-\alpha_p\vec{P}+\beta\begin{pmatrix}P_2\\-P_1\\0\end{pmatrix}+2b_p\begin{pmatrix}0\\-P_3\\P_2\end{pmatrix},\tag{2b}$$

where $\vec{S}=S_0\vec{s}$ and $\vec{P}=P_0\vec{p}$ are the projections of signal and pump powers with the corresponding unit vectors $\vec{s}$ and $\vec{p}$ ($S_0=\left|\vec{S}\right|, P_0=\left|\vec{P}\right|$) in the Stokes representation. The Raman gain coefficient is g_R, and the pump/signal frequencies are $\omega_{p,s}$, respectively. The attenuation constants for the pump and signal are $\alpha_{p,s}$. The most interesting parameters are $b_{p,s}=2\pi/L_{p,s}$ ($L_{p,s}$ are the pump/signal beat lengths, respectively) and the Wiener *stochastic* term with the zero drift and the volatility $\sigma^2=1/L_c$: $\langle\beta(z),\beta(z')\rangle=\sigma^2\delta(z-z')$ (L_c is a correlation length of the stochastic material birefringence).

The variation of L_c (correlation length defining a noise "strength") relatively $L_{p,s}$ (periods of the deterministic polarization beatings) causes a transition between the different regimes.

i. A strong polarization pump/signal coupling (**Figure 4**, left) corresponds to a case of $L_{p,s}\gg L_c$ when a noise is too "fine-grained" and cannot distort nonlinear trapping of signal by pump. As a result, the mutual polarizations of pump and signal are highly correlated, and the signal fluctuations are small (≈1% in the case under consideration; see **Figure 5**, left).

ii. When $L_{p,s}$ approach L_c (i.e., relative strength of noise increases), the signal and pump decouple (**Figure 4**, middle), and the signal evolution becomes extremely noisy (**Figure 5**, middle).

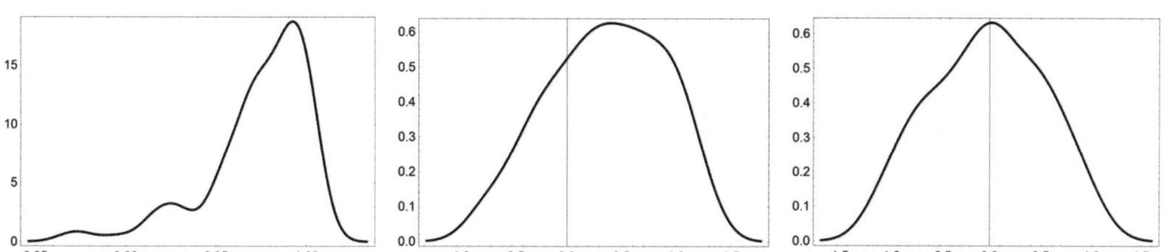

Figure 4. PDF of the normalized output signal-pump scalar product $\vec{S}\vec{P}/\left|\vec{S}\right|\left|\vec{P}\right|$ with lowering beat lengths $L_{p,s}$ [29]. L_c=100 m, L_s=1 km (left), 150 m (middle), 30 m (right); L_p=1.55L_s/1.465 (an Er-doped fiber), the propagation length L=5 km. The input powers of pump and signal are 1 and 0.01 W, respectively.

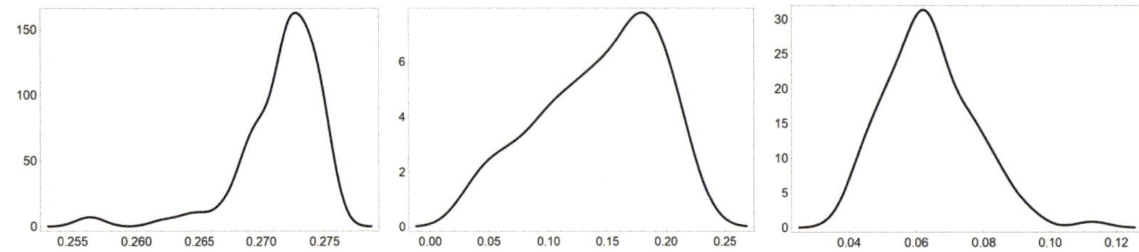

Figure 5. PDF of the output signal power $\left|\vec{S}\right|^2$ (in Watts) with lowering beat lengths $L_{p,s}$ as in **Figure 4** [29].

iii. Further decrease of $L_{p,s}$ relatively L_c causes almost complete decoupling of pump and signal (so-called diffusion limit; **Figure 4**, right) when a fiber behaves like an isotropic medium. Noise plays important but diminishing role (**Figure 5**, right).

Such a resonant-like enhancement of irregularity that depends on the relative strengths of noise and regular oscillations is an example of SAR. **Figure 6** is a spectacular illustration of this phenomenon based on the model of Eqs. (2a) and (2b) [30]. We can see here the resonant enhancement of a Raman gain standard deviation defined as $\sigma_G = \sqrt{\overline{\left|\vec{S}(L)\right|^2} / \left|\overline{\vec{S}(L)}\right|^2 - 1}$ in dependence on a fiber length L and a polarization mode-dispersion parameter $D_p = 2\lambda_s\sqrt{L_c}/cL_s$ defining relative contribution of stochastic and deterministic polarization effects (λ_s and c are the signal wavelength and the speed of light, respectively).

The phenomenon of SAR can be explained as an escape from a metastable state corresponding to pump-signal pulling with an effective potential barrier ΔU and an "intra-well relaxation time"

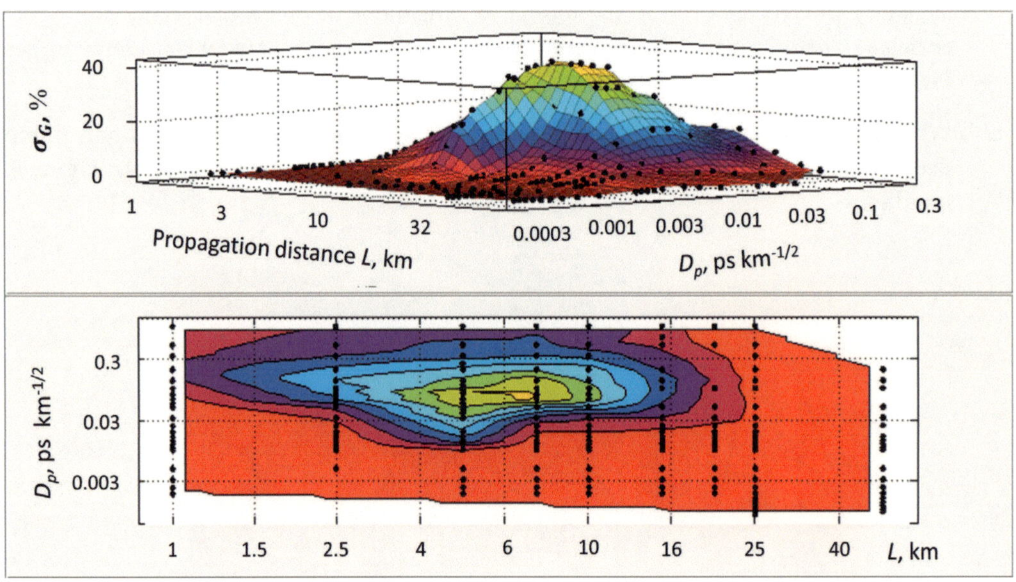

Figure 6. Relative standard deviation of the maximum Raman gain coefficient illustrating the SAR in a fiber Raman amplifier [30].

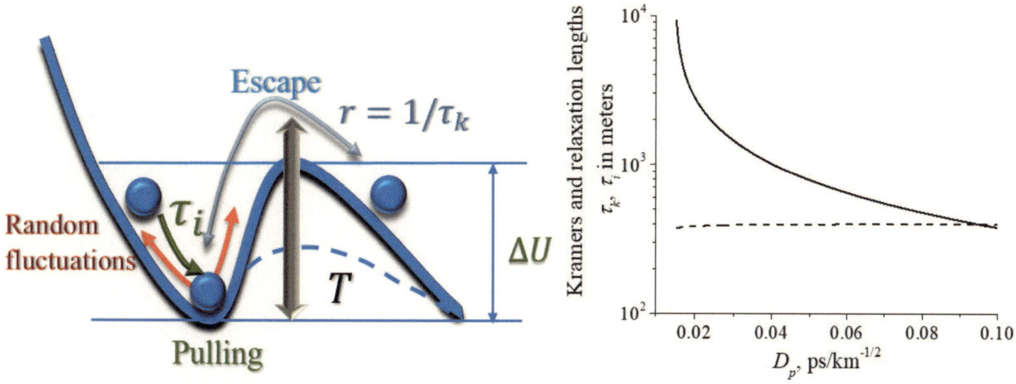

Figure 7. Left: Noise causes an escape from metastable (polarization pulling) state through a potential barrier ΔU controlled by periodic modulation T induced by regular polarization beating. Right: Dependence of Kramers length τ_k (solid curve) and intra-well relaxation length τ_i (dashed curve) on the polarization mode-dispersion parameter D_p [27, 31].

(or length in our case) τ_i (**Figure 7**, left) [27, 31]. The random fluctuations can cause an escape from this metastable state with an escape rate $r \propto \exp\left(-\frac{\Delta U}{D}\right) = 1/\tau_k$ defined by the so-called *Kramers time* (*length*) τ_k (D is an effective "temperature" defining a noise strength) [6, 32]. The periodic (T) modulation of potential barrier caused by the polarization beatings can enlarge this effective temperature and, thereby, increase the escape rate (**Figure 7**, right) [27, 31].

Thus, a Raman fiber amplifier can be considered as a simple and practically valuable test-bed for a demonstration of SAR that is a phenomenon of noise-induced escape from the metastable state. Practical control of this phenomenon is especially important for the development of modern high-speed optical communication lines that promise to exceed the limits of existing broadband information infrastructure.

3. Chaotic resonance between a dissipative soliton and linear waves

Dissipative soliton (DS) is a well-localized structure self-emergent in dissipative systems. Such structures appear in different areas ranging from physics to biology, medicine, and even economy and sociology [33, 34]. The simplest equation regarding the DS modeling is the so-called generalized complex nonlinear Ginzburg-Landau equation (CNGLE) [33–35]:

$$\begin{aligned}\frac{\partial A(z,t)}{\partial z} &= \left\{-\sigma + \alpha\frac{\partial^2}{\partial t^2} + \left[\kappa\left(1-\zeta|A(z,t)|^2\right)|A(z,t)|^2\right]\right\}A(z,t) + i\left\{\frac{\beta_2}{2}\frac{\partial^2}{\partial t^2} - \gamma|A(z,t)|^2\right\}A(z,t)\\ &\quad + \frac{\beta_3}{6}\frac{\partial^3}{\partial t^3}A(z,t) + s(t),\end{aligned} \tag{3}$$

where a field of amplitude A, slowly-varying with a local time t, propagates along a coordinate z under action of dissipative (first braces) and nondissipative (second braces) factors. Then,

σ-parameter corresponds to energy-dependent net-loss, α-parameter defines spectral losses, κ- and ζ-parameters describe effective nonlinear gain and its depletion, respectively. Nondissipative factors are self-phase modulation (SPM, γ), group-delay dispersion (GDD, β_2) with low-order correction to the latter (third-order dispersion or TOD, β_3). $s(t)$ describes a complex white noise.

The general-form of DS solutions of Eq. (3) is unknown, and the extensive numerical simulations are required to investigate the complexity of DS dynamics. However, there are some very simple considerations based on resonance/balance relations, which allow understanding some basic properties of DS.

Indeed, a steady-state solution of Eq. (3) has a form $A(t, z) = E(t)\exp(-iqz)$, where the soliton wave number q is related to the carrier-envelope offset [36], which results from nonlinear phase shift caused by SPM: $q = \gamma P_0$ (P_0 is a DS peak power) [37]. The dispersion relation for linear waves is $k(\omega) = \beta_2\omega^2/2$. Hence, to be stable (i.e., nonradiating), the DS spectrum has to be truncated at the frequencies $\pm\Delta$: $k(\pm\Delta) = q$, where $\Delta = \sqrt{2\gamma P_0/\beta_2}$ (**Figure 8**) [35].

Simultaneously, the spectral loss ~$\alpha\Delta^2$ has to be compensated by the nonlinear gain ~κP_0. This condition plus the resonant condition give the rough stability criterion for DS:

$$\alpha\gamma/\kappa\beta_2 \leq 1/2, \tag{4}$$

which interrelates dissipative and nondissipative factors contributing to DS formation (more precise analysis can be found in [38]).

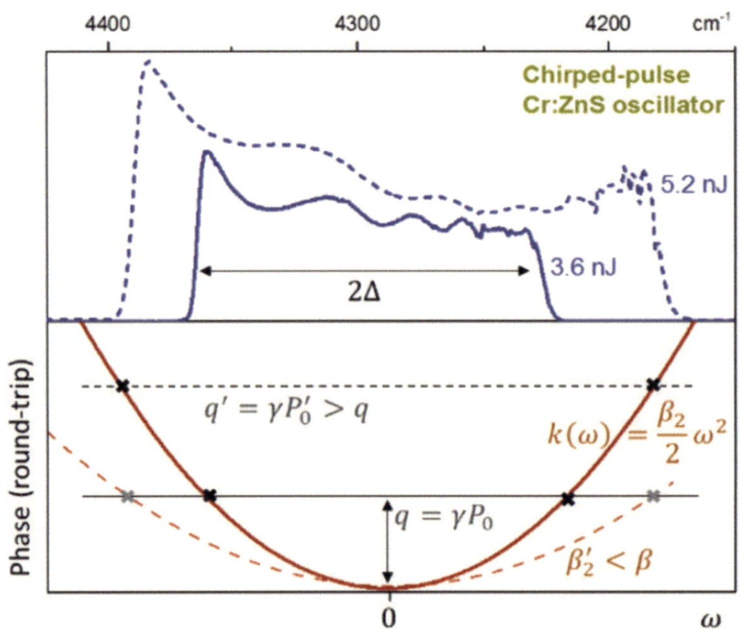

Figure 8. *Resonance condition* (black and gray crosses at the bottom panel) for the DS defines the spectrum width 2Δ. Changing the power and/or the dispersion (solid and dashed lines in the bottom panel) controls the spectrum width. Lines in the top panel show the experimental spectra corresponding to different energies [35].

TOD modifies the dispersion condition for a linear wave so that the resonant condition becomes: $q = \frac{\beta_2 \omega^2}{2} + \frac{\beta_3 \omega^3}{6}$ (black/red online/curve in **Figure 9c**). It can cause an appearance of additional resonant frequency which proximity to DS spectrum (i.e., to one of the other resonant frequencies) can initiate chaotic dynamics [39, 40]. This conjecture was confirmed in [35] both experimentally and numerically.

Figure 9 demonstrates an example of such chaotization obtained from numerical simulations of Eq. (3) (for details, see [35]). The Wigner function (time-frequency diagram, **Figure 9a**) consists of strongly distorted DS-part near 2.3 μm (dark-red region online) and long dispersive tale in spectral domain around 2.4 μm (yellow – light blue region online), which co-propagates with DS and, as an analysis shows, collides with it in time domain. As a result, the DS spectrum becomes modulated chaotically (**Figure 9b**), but the averaged spectrum looks quite smooth with the characteristic shape of "*Boa constrictor* digesting an elephant" (**Figure 9d**) [35]. As was mentioned, these phenomena can be explained as a nonlinear resonance of three-coupled oscillators [41, 42] when the TOD-induced resonant point (DW in **Figure 9c**) approaches one of the other two (R_2, in our case).

The last statement can be confirmed by a reconstruction of phase space corresponding to the chaotic dynamics in **Figure 9**. Such a reconstruction is based on the standard lag-delayed procedure when one tries to reconstruct an N-dimensional phase space from time-series data $V(t)$ by the means of following discretization: $[V(t), V(t+L), V(t+2L), \ldots, V(t+(N-1)L)]$, where L is a time-lag [43]. As a rule, an appropriative time-lag is defined from the first zero of autocorrelation function of time-series (peak powers in our case). The corresponding reconstruction is shown in **Figure 10** [35]. One can see, that the chaotic trajectory of $P_0(t)$ is

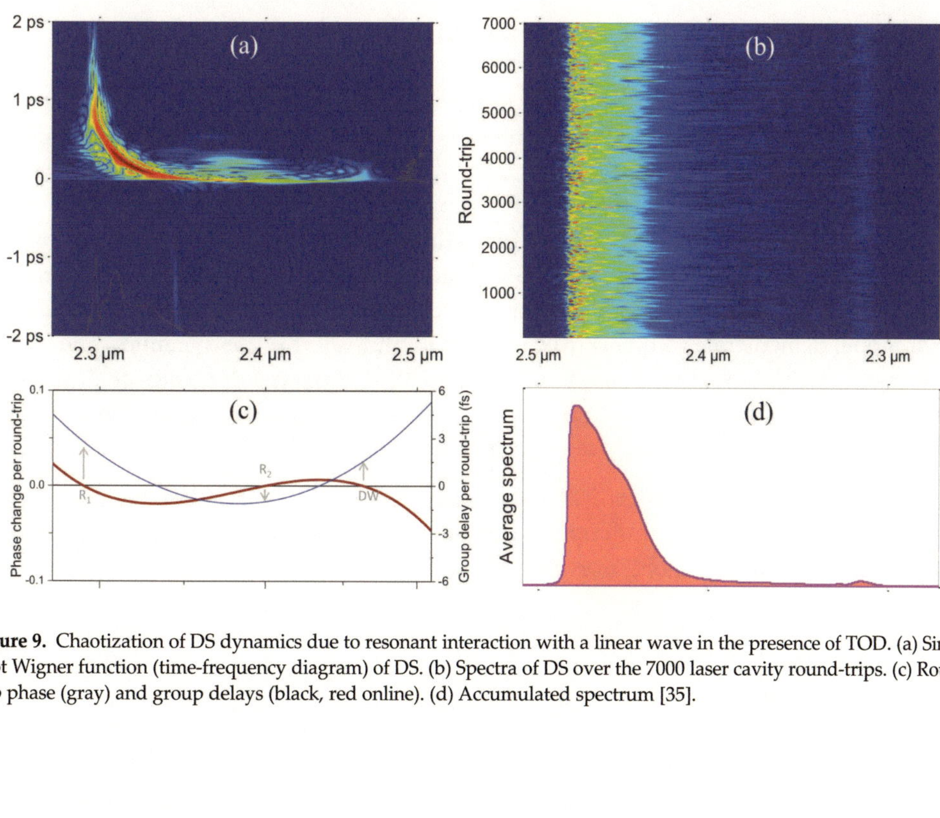

Figure 9. Chaotization of DS dynamics due to resonant interaction with a linear wave in the presence of TOD. (a) Single-shot Wigner function (time-frequency diagram) of DS. (b) Spectra of DS over the 7000 laser cavity round-trips. (c) Round-trip phase (gray) and group delays (black, red online). (d) Accumulated spectrum [35].

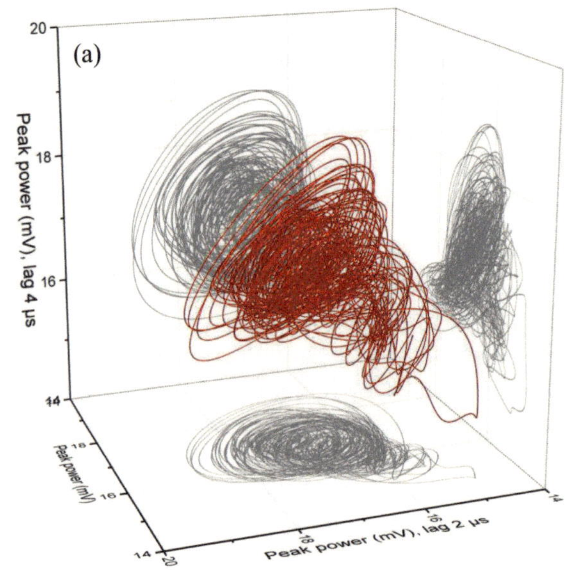

Figure 10. Phase space reconstructed from the experimental DS peak power set [35].

completely embedded in the three-dimensional manifold, and the attracting manifold has a typical toroidal shape. Both facts validate a model of nonlinear resonance of three coupled oscillators.

4. Stochastic resonance and anti-resonance in mode-locked lasers

A laser, as a device locking electromagnetic waves, possesses a discrete set of longitudinal modes, i.e., set of standing waves, which interacts irregularly due to random mutual phases. Locking of a mutual phase between modes, namely mode-locking, results in the generation of a high-intensive ultra-short laser pulse circulating with the repetition rate multiple of the period of laser resonator (e.g., see [44, 45]). In the time domain, the ultra-short pulse formation can be described in the frameworks of the so-called *fluctuation model* [46, 47], which treats a pulse[1] emerging as a process of amplification and selection of noise fluctuations (see **Figure 11**). Such a model demonstrates a crucial role of noise in ultra-short pulse dynamics. The noise is not only a source of pulse formation, but it can also affect the pulse dynamics at all stages. In particular, it is a source of "linear" (dispersive) waves, which can resonantly interact with a pulse and randomize its dynamics (see the previous section).

The typical equation with "distributed" laser parameters is Eq. (3). It describes a multitude of realistic phenomena intrinsic to the pulse dynamics. But in many real-world situations, the "discretized" models are more relevant. For instance, let us consider a system of laser

[1]We use the term "pulse" instead of DS because it is more appropriate to "discretized" systems for which the notion "soliton" can be misleading.

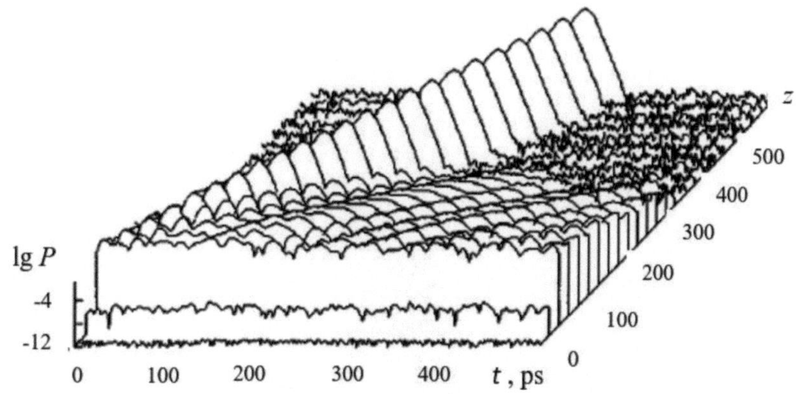

Figure 11. Formation of the ultra-short pulse from initial noise fluctuations (P is a normalized power, t is a local time, z is a cavity round-trip number) [48]. Noise is suppressed on the final stage, but remains on a "vacuum level.".

resonantly coupled with an external resonator. In the dispersion-less case, the field evolution in a laser can be described as follows [49]:

$$\begin{aligned} A'(z,t) &= A(z,t)\exp\left[g(t) - i\gamma|A(z,t)|^2\right] + s(t), \\ \frac{\partial g(t)}{\partial t} &= \sigma_{14}\left(g_m - g(t)\right)I_p/h\nu_p - \sigma_{32}g(t)|A(z,t)|^2/h\nu - \frac{g(t)}{T_{31}}. \end{aligned} \tag{5}$$

Here, both z and t are discretised so that z is a cavity transit number, and t is a local time discretised with a step Δt. The coefficient $g(t)$ describes a local gain for a 4-level active medium with the maximal gain g_m for full population inversion: σ_{14}, σ_{32}, and T_{31} are absorption, emission cross-section, and gain relaxation time, respectively. ν_p and ν are pump and generation wavelength, respectively.

$$s(t) = s(t - \Delta t)\exp\left[-\frac{\Delta t}{t_{coh}}\right] + s_0 \exp\left[i\phi(t)\right], \tag{6}$$

is a noise term with the coherence time t_{coh}, a noise level s_0, and a random phase $\varphi(t)$ [50].

Spectral dissipation is provided by a Fabry-Pérot etalon with a group-delay t_f [51]:

$$A''(z,t) = \left(1 - R_f\right)A'(z,t) + R_f A''(z, t - \Delta t), \tag{7}$$

where $R_f = t_f/(t_f + \Delta t)$.

The field A is coupled with the field in external resonator $B(z, t)$ [49]:

$$\begin{aligned} A(z+1,t) &= RA''(z,t) - i\theta T\exp\left[i\pi\phi\right]B(z,t), \\ B(z+1,t) &= -i\theta TA''(z,t) + \theta^2 R\exp\left[i\pi\phi\right]B(z,t), \end{aligned} \tag{8}$$

where R, T, and θ are reflection, transmission, and coupling coefficients, respectively.

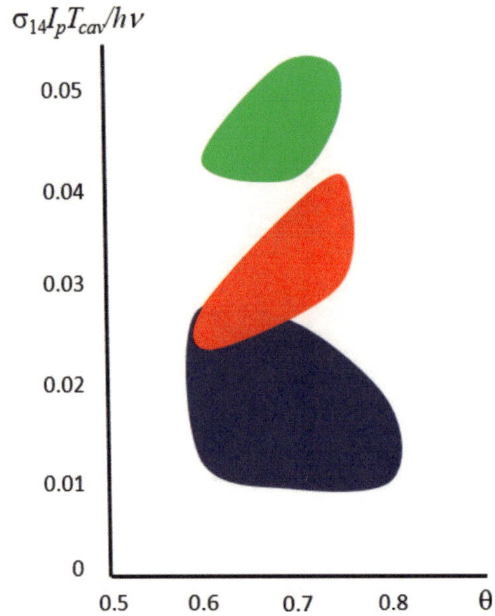

Figure 12. Regions of mode-locking self-start for the model (5–8). $\gamma h\nu/\sigma_{32}T_{cav}=0.01$, $\varphi=0$, $t_f=1$ ps, $t_{cor}=\infty$ (bottom region), 1 μs (middle region), and 1 ps (top region). T_{cav} is a cavity period.

It is interesting to consider the question about disruptiveness of noise for pulse formation. **Figure 12** shows the regions of formation of stable pulses from noise (so-called regions of mode-locking self-start) for different t_{coh}. One can see, that the decrease of noise coherence is destructive from the points of view of the mode-locking regions size and the threshold pump intensity providing mode-locking.

However, the simulations demonstrate that even very low-frequency external modulation (e.g., by moving resonator mirror inducing the Doppler shift of optical wave) can suppress noise (see **Figure 13**) and stabilize dynamics [52, 53]. This phenomenon can be interpreted as a manifestation of resonant interaction of scale-incommensurable processes. Moreover, exactly such a resonance provides mode-locking self-start in the majority of lasers (moving mirror technique [54] or even simple mirror knocking).

In close connection with the phenomenon mentioned above, one has to note that a nonlinear interconnection between the scale-incommensurate processes is a crucial factor defining all considered phenomena. An interesting example closely connected with previous one is a laser mode-locked by external phase modulation [55, 56]. This system can be described by following equation (compare with Eq. (3)):

$$\frac{\partial A(z,t)}{\partial z}=\left\{-\sigma-\delta\frac{\partial}{\partial t}+\alpha\frac{\partial^2}{\partial t^2}-i\gamma|A(z,t)|^2\right\}A(z,t)+i\omega tA(z,t), \quad (9)$$

where $-\sigma$ is a saturated gain, ω is a modulation frequency normalized to modulation depth, and δ is a mismatch between the modulation and cavity periods. In the absence of SPM, the

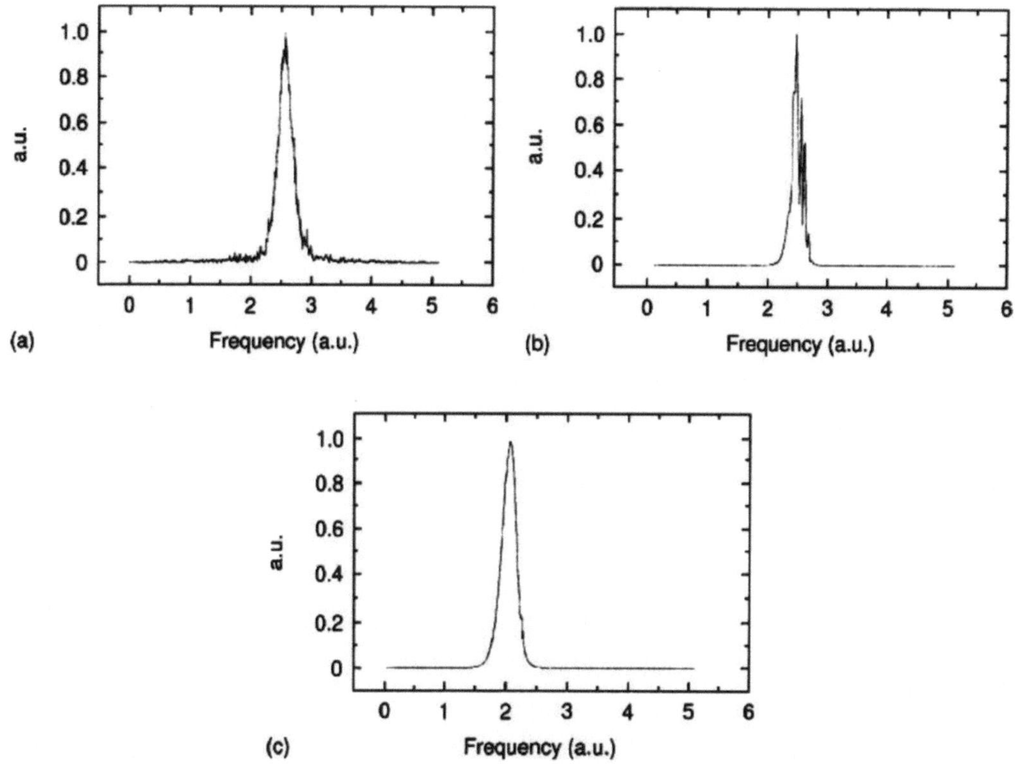

Figure 13. Evolution of noisy pulse with external phase modulation (moving mirror, the modulation frequency is 1 kHz) [53]. (a) Initial noisy pulse; (b) pulse at 1000th cavity transit (noise is sweeping out due to Doppler effect); (c) pulse at 5000th cavity transit (noise is swept out).

pulse width τ is $\tau = 2\sqrt{-\sigma}/|\omega|$. It increases with saturated net-gain $-\sigma$, is closely associated with the modulation frequency ω, and exceeds substantially the minimal value $\sqrt{\alpha}$ defined by spectral dissipation. However, the nonlinearity (namely, SPM) can modify a situation crucially [55, 56]. Firstly, the pulse width decreases (not increases) with a gain that allows generating high energy, and simultaneously, short pulses. Secondly, and it is a first nontrivial fact, pulse width can be extremely short ($\sim 10\,\sqrt{\alpha}$) and reach scales incommensurable with the modulation frequency. Third impressive fact is that the modulation frequency providing stable mode-locking can be extremely small in comparison with the laser cavity period (ω is lower by approximately three orders of magnitude in comparison with a linear case). It seems that this effect is closely related to the above considered noise suppression due to the Doppler shift.

Returning to an effect of noise on the mode-locking self-start illustrated in **Figure 12**, one may consider another interesting manifestation of SR/SAR in mode-locked lasers. External resonator providing mode-locking can be considered as a Fabry-Pérot interferometer resonantly matched with a laser cavity (see above). This interferometer can contain a nonlinear medium, and such a system possesses rich dynamical properties, in particular, it can cause spontaneous formation of ultra-short pulses (mode-locking). Examples of regions of such mode-locking are

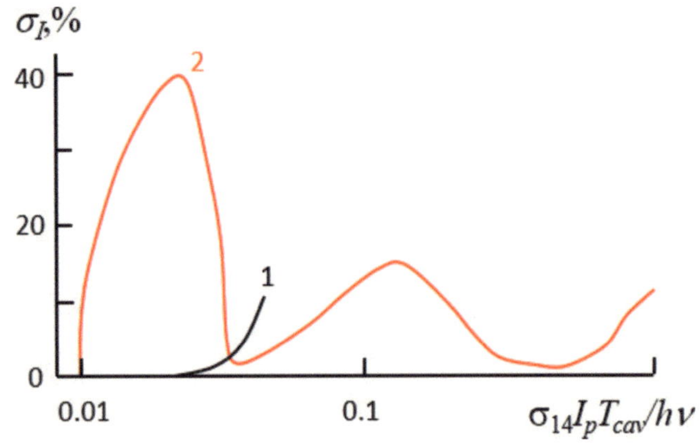

Figure 14. Dependence of mean-square deviation of pulse intensity on the pump in a laser mode-locked by a nonlinear Fabry-Pérot interferometer [53] for the phase mismatch $\varphi = \pi$, reflection coefficients of coupling mirror 0.96 (1), 0.7 (2), and internal transmission of interferometer 0.9 (1), 0.5 (2).

shown in **Figure 12**. It is interesting that such regions are inhomogeneous. **Figure 14** demonstrates the mean-square deviation σ_I of pulse peak intensity inside two such regions in dependence on pump.

One can see, that the pulse is highly stable inside the mode-locking region and destabilizes only on stability border in the case of (1). But in the case of (2), the behavior of σ_I becomes strongly nonmonotonic. Pronounced peaks in the σ_I-dependence is the classical SAR manifestation caused by the excitation of noise with subsequent formation of the pulse satellites whose interaction with main pulse perturbs strongly the latter. The regions of SAR alternate with the regions of regular dynamics. Thus, the mode-locking region can be granulated.

In all examples above considered, a mode-locking resulting in the pulse appearance was caused by either loss self-modulation or external periodical modulation. However, the mode-locking can appear spontaneously due to spontaneous multimode instability (so-called Risken-Nummedal-Graham-Haken effect, RNGH) [57, 58]. However, such self-mode-locking is unstable. Nevertheless, the stable self-mode-locking was obtained in Er-fiber laser due to beatings induced by the difference of intra-laser (fiber + polarization components) birefringence and that induced by polarization hole burning in active medium (Er-doped fiber). That is the so-called *resonance vector mode-locking* [59]. The beatings generate the spectral satellites (sidebands) for each laser mode produced by multimode instability (see **Figure 15**). Adjusting of intra-laser birefringence by polarization controller shifts these sidebands to adjacent modes that cause a resonance between them with subsequent stable mode-locking.

But that is not all. The generated comb of locked modes can excite the transverse acoustic waves in a fiber through electrostriction effect [61]. The resonance between the comb and these waves lock (trap) a pulse in time domain that provides an unprecedented stability of the pulse train. The last is highly required for metrology, high-resolution spectroscopy, etc [62].

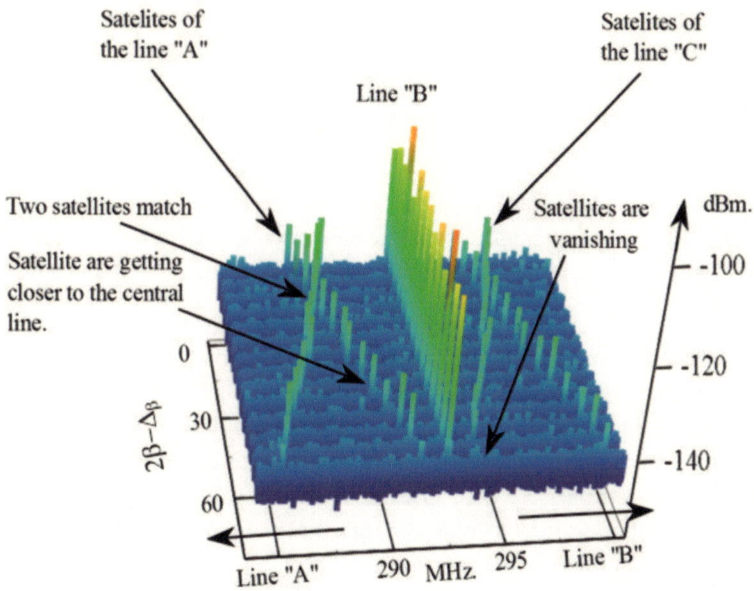

Figure 15. Evolution of spectrum in the vicinity of allocated mode "B" neighboring with modes "A" and "C" [59]. Control parameter is an intra-laser birefringence 2β, Δ_β is a birefringence caused by polarization hole burning in the active fiber [60].

5. Conclusions

The nonlinear resonance phenomena are illustrated as examples of fiber Raman amplifiers and mode-locked lasers. These systems proved their advantage as an ideal test-bed for "metaphoric modeling" of complex nonlinear systems due to comparative simplicity, high-speed statistic gathering, and precise controllability.

We considered the phenomenon of the so-called *stochastic anti-resonance* as examples of a fiber Raman amplifier and a laser mode-locked by resonant coupling with a nonlinear Fabry-Pérot interferometer.

In the first case, the regular polarization beatings between pump and Raman signal are coupled resonantly with the stochastic birefringence caused by material defects (stochastic changes of fiber symmetry). As a result, there is a region of parameters (first of all, so-called polarization mode-dispersion parameter) where the evolution of the state of polarization becomes highly irregular that manifests itself in resonance growth of relative standard deviation of Raman gain. This phenomenon was interpreted as a noise-induced escape from metastable state and quantitatively characterized by an abrupt decrease of the characteristic Kramers length.

In the second case, it is shown a crucial dependence of mode-locking ability on noise correlation time so that the growth of irregularity squeezes the mode-locking region and increases the mode-locking threshold. Nontrivial effect of noise manifests itself inside the region of parameters, where spontaneously born pulse exists. Namely, a monotonic variation of the pump causes the alternation of maximums and minimums of the pulse peak intensity mean-square deviation. Thus, the stochastic anti-resonances exist inside the mode-locking region, which, thereby, has a granular structure.

The interesting example of *scale hierarchy* in a mode-locked laser is demonstrated. The matter is that extremely slow (~1 kHz), external modulation can suppress noise in mode-locked laser through the Doppler effect. This effect is broadly known for experimenters using the resonator mirror knocking for the mode-locking self-start.

Active mode-locked lasers can demonstrate another aspect of *scale hierarchy* in the nonlinear resonance phenomena. The laser phase nonlinearity coupled with the external phase modulation can provide generation of pulses whose widths are not limited by modulation frequency but only by intra-laser spectral dissipation. Moreover, laser mode-locking can be reached at anomalously low (in comparison with laser resonator round-trip) modulation frequencies. One may bring the last effect into correlation with that mentioned in the previous paragraph, but this issue demands a further consideration.

The phenomenon, which is connected closely with the resonance in systems possessing a scale hierarchy, is a so-called *resonance vector mode-locking*. In this case, a spontaneous locking of laser modes emerging as a result of multimode instability (RNGH) is stabilized due to the polarization beating caused by intra-laser birefringence and birefringence induced by polarization hole burning in the active medium. That results in stable self-mode-locking, which is stabilized additionally through a resonant coupling with the acoustic waves excited by the mode-locking itself through the electrostriction effect.

The *dissipative soliton resonance* with linear waves originating from noise can be considered separately in some way. It is shown, that this resonance defines the dissipative soliton characteristics, namely, its spectral width. When the resonant conditions change due to the contribution of higher-order dispersions (third-order in our case), the dissipative soliton can emit a radiation, which interacts with soliton in turn. As a result, the dissipative soliton dynamics becomes chaotic, that can be classified as *chaotic resonance* in terms of nonlinear resonance of three coupled oscillators.

The unified viewpoint on the nonlinear stochastic and chaotic phenomena in the field of laser physics and solitonics remains undeveloped yet. Such a viewpoint would be a part of general thermodynamic and kinetic theory of dissipative systems promising a strong practical impact in different areas ranging from physics to biology, medicine, and sociology.

Acknowledgements

I acknowledge my colleagues Dr. E. Sorokin and Dr. S. Sergeyev for their valuable contribution to the presented research findings. The author acknowledges the support from Austrian Science Fund (FWF project P24916-N27). Part of computational results has been achieved using the Vienna Scientific Cluster (VSC).

http://dx.doi.org/10.5772/intechopen.70737

Author details

Vladimir L. Kalashnikov

Address all correspondence to: vladimir.kalashnikov@tuwien.ac.at

Institute of Photonics, Vienna University of Technology, Vienna, Austria

References

[1] Oono Y. The Nonlinear World. Tokyo: Springer; 2013

[2] Scott AC, editor. The Nonlinear Universe. Berlin: Springer; 2010

[3] Lindner B, Garcia-Ojalvo J, Neiman A, Schimansky-Geier L. Effects of noise in excitable systems. Physics Reports. 2004;**392**:321

[4] Benzi R, Parisi G, Sutera A, Vulpiani A. Stochastic resonance in climatic change. Tellus. 1982;**34**:10

[5] Nicolis C. Long-term climatic transitions and stochastic resonance. Journal of Statistical Physics. 1993;**70**:3

[6] Gammaitoni L, Hänggi P, Jung P, Marchensoni F. Stochastic resonance. Reviews of Modern Physics. 1998;**70**:223

[7] Anishchenko VS, Neiman AB, Moss F, Schimansky-Geier L. Stochastic resonance: noise-enhanced order. Physics – Uspekhi. 1999;**42**:7

[8] Wellens T, Shatokhin V, Buchleitner A. Stochastic resonance. Reports on Progress in Physics. 2004;**67**:45

[9] Anishchenko VS, Astakkov V, Neiman A, Vadivasova T, Schimansky-Geier L. Nonlinear Dynamics of Chaotic and Stochastic Systems. Berlin: Springer; 2007

[10] McDonnell MD, Stocks NG, Pearce CEM, Abbott D. Stochastic Resonance: From Superthreshold Stochastic Resonance to Stochastic Signal Quantization. Cambridge: Cambridge University Press; 2008

[11] Andò B, Gaziani S, editors. Stochastic Resonance: Theory and Applications. Dordrecht: Kluwer; 2000

[12] Longtin A. Stochastic resonance in neuron models. Journal of Statistical Physics. 1993;**70**:309

[13] Hänggi P. Stochastic resonance in biology: how noise can enhance detection of weak signals and help improve biological information processing. Chemphyschem. 2002;**3**:285

[14] Lindner B, Schimansky-Geier L, Longtin A. Maximizing spike train coherence or incoherence in the leaky integrate-and-fire model. Physical Review E. 2002;**66**:031916

[15] Moss F, Ward LM, Sannita WG. Stochastic resonance and sensory information processing: a tutorial and review of application. Clinical Neurophysiology. 2004;**115**:267

[16] McDonell MD, Abbott D. What is stochastic resonance? Definitions, misconceptions, debates, and its relevance to biology. PLoS Computational Biology. 2009;**5**:e1000348

[17] McDonell MD, Word LM. The benefits of noise in neural systems: bridging theory and experiment. Nature Reviews/Neuroscience. 2011;**12**:415

[18] Haykin S, Kosko B, editors. Intelligent Signal Processing. New York: IEEE Press; 2001

[19] Benzi R, Sutera A, Vulpiani A. The mechanism of stochastic resonance. Journal of Physics A. 1981;**14**:L453

[20] Benzi R, Sutera A, Vulpiani A. Stochastic resonance in the Landau-Ginzburg equation. Journal of Physics A. 1985;**18**:2239

[21] McNamara B, Wiesenfeld K. Theory of stochastic resonance. Physical Review A. 1989;**39**: 4854

[22] Hänggi P. Escape from a metastable state. Journal of Statistical Physics. 1986;**42**:105

[23] Nikolis G, Nicolis C, McKernan D. Stochastic resonance in chaotic dynamics. Journal of Statistical Physics. 1993;**70**:125

[24] Landa PS, McClintock PVE. Nonlinear systems with fast and slow motions: changes in the probability distribution for fast motions under the influence of slower ones. Physics Reports. 2013;**532**:1

[25] McNamara B, Wiesenfeld K, Roy R. Observation of Stochastic Resonance in a Ring Laser. Physical Review Letters. 1988;**60**:2626

[26] Headley P, Agrawal G, editors. Raman Amplification in Fiber Optical Communication Systems. Amsterdam: Elsevier; 2005

[27] Kalashnikov VL, Sergeyev SV. In: Skiadas C, editor. The Foundations of Chaos Revisited: From Poincaré to Recent Advancements. Switzerland: Springer; 2016. p. 159-179

[28] Lin Q, Agrawal GP. Vector theory of stimulated Raman scattering and its application to fiber-based Raman amplifiers. Journal of the Optical Society of America B: Optical Physics. 2003;**20**:1616

[29] Kalashnikov VLStochastic Anti-Resonance in a Fibre Raman Amplifier "(Commented Matematica/Matlab Code)"https://www.researchgate.net/publication/308233643_Stochastic_Anti-Resonance_in_a_Fibre_Raman_Amplifier

[30] Kalashnikov V, Sergeyev SV, Ania-Castanón JD, Jacobsen G, Popov S. Stochastic phenomena in a fiber Raman amplifier. Annalen der Physik. 2017;**529**:1600238

[31] Kalashnikov V, Sergeyev SV, Jacobsen G, Popov S. Multi-scale polarisation phenomena. Nature Light: Science & Applications. 2016;**5**:e16011

[32] Kramers H. Brownian motion in a field of force and the diffusion model of chemical reactions. Physica. 1940;**7**:284

[33] Akhmediev NN, Ankiewicz A, editors. Dissipative Solitons. Berlin: Springer; 2005

[34] Akhmediev NN, Ankiewicz A, editors. Dissipative Solitons: From Optics to Biology and Medicine. Berlin: Springer; 2008

[35] Sorokin E, Tolstik N, Kalashnikov VL, Sorokina IT. Chaotic chirped-pulse oscillators. Optics Express. 2013;**21**:29567

[36] Telle HR, Steinmeyer G, Dunlop AE, Stenger J, Sutter DH, Keller U. Carrier-envelope offset phase control: A novel concept for absolute optical frequency measurement and ultrashort pulse generation. Applied Physics B: Lasers and Optics. 1999;**69**:327-332

[37] Podivilov E, Kalashnikov VL. Heavily-chirped solitary pulses in the normal dispersion region: New solutions of the cubic-quintic complex Ginzburg-Landau equation. JETP Letters. 2005;**82**:467-471

[38] Kalashnikov VL. In: Al-Khursan AH, editor. Solid-State Laser. Rijeka: InTech; 2012. p. 145-184

[39] Kalashnikov VL, Fernández A, Apolonski A. High-order dispersion in chirped-pulse oscillators. Optics Express. 2008;**16**:4206

[40] Kalashnikov VL. In: Skiadas CH, Dimotikalis I, Skiadas C, editors. Chaos Theory. Modeling, Simulation and Applications. Singapore: World Scientific; 2011. p. 199-206

[41] Baesens C, Guckenheimer J, Kim S, MacKay RS. Three coupled oscillators: mode-locking, global bifurcations and toroidal chaos. Physica D: Nonlinear Phenomena. 1991;**49**:387

[42] Pazó D, Sánchez E, Matías MA. Transition to high-dimensional chaos through quasiperiodic motion. International Journal of Bifurcation and Chaos. 2001;**11**:2683

[43] Abarbanel HDI, Brown R, Sidorowich JJ, Tsimring LS. Practical method for determining the minimum embedding dimension of a scalar time series. Reviews of Modern Physics. 1993;**65**:1331

[44] Kuizenga DJ, Siegman AE. FM and AM mode locking of the homogeneous laser - Part I: Theory. IEEE Journal of Quantum Electronics. 1970;**QE-6**:694

[45] Smith PW. Mode-locking of lasers. Proceedings of the IEEE. 1970;**58**:1342

[46] Kryukov PG, Letokhov VS. Fluctuation mechanism of ultrashort pulse generation by laser with saturable absorber. IEEE Journal of Quantum Electronics. 1972;**QE-8**:766

[47] Glenn WH. The fluctuation model of a passively mode-locked laser. IEEE Journal of Quantum Electronics. 1975;**QE-11**:8

[48] Kalashnikov VL, Kalosha VP, Mikhailov VP, Poloyko IG. Self-mode locking of four-mirror-cavity solid-state lasers by Kerr self-focusing. Journal of the Optical Society of America B: Optical Physics. 1995;**12**:462

[49] Kalashnikov VL, Kalosha VP, Mikhailov VP, Demchuk MI. Self-starting of cw mode-locked solid-state lasers with a linear external cavity. Optics Communications. 1993;**96**: 249

[50] Fleck JA. Ultrashort-Pulse Generation by Q-Switched Lasers. Physics Review. 1970;**B1**:84

[51] Catherall JM, New GHC. Role of spontaneous emission in the dynamics of mode locking by synchronous pumping. IEEE Journal of Quantum Electronics. 1986;**QE-22**:1593

[52] Blow KJ, Wood D. Mode-locked lasers with nonlinear external cavities. Journal of the Optical Society of America. 1988;**5**:629

[53] Kalashnikov VL, Kalosha VP, Mikhailov VP, Poloyko IG. The general approach to mode-locking mechanism analysis for CW solid-state lasers with a nonlinear Fabry-Pérot interferometer. Optical and Quantum Electronics. 1995;**27**:1061

[54] Liu YM, Sun KW, Prucnal PR, Lyon SA. Simple method to start and maintain self-mode-locking of a Ti:sapphire laser. Optics Letters. 1992;**17**:1219

[55] Kalashnikov VL, Poloiko IG, Kalashnikov VP. Generation of ultrashort pulses in lasers with external frequency modulation. Quantum Electronics. 1998;**28**:264

[56] Kalashnikov VL, Poloiko IG, Mikhailov VP. Phase modulation of radiation of solid-state lasers in the presence of Kerr optical nonlinearity. Optics and Spectroscopy. 1998;**84**:104

[57] Risken H, Nummedal K. Self-Pulsing in Lasers. Journal of Applied Physics. 1968;**39**:4662

[58] Lugiato L, Prati F, Brambila M. Nonlinear Optical Systems. Cambridge: Cambridge University Press; 2015

[59] S.A. Kolpakov, S.V. Sergeyev, Y. Loika, N. Tarasov, V. Kalashnikov, G. Agrawal, Resonance vector mode locking. arXiv:1508.05933 [physics.optics]

[60] Sergeyev SV. Fast and slowly evolving vector solitons in mode-locked fibre lasers. Philosophical Transactions of the Royal Society A. 2014;**372**:20140006

[61] Grudinin AB, Gray S. Passive harmonic mode locking in soliton fiber lasers. Journal of the Optical Society of America B: Optical Physics. 1997;**14**:144

[62] Ye J, Cundiff ST. Femtosecond Optical Frequency Comb Technology. Boston: Springer; 2005

Chapter 5

Optimization of Double-Well Bistable Stochastic Resonance Systems and Its Applications in Cognitive Radio Networks

Di He

Additional information is available at the end of the chapter

http://dx.doi.org/10.5772/intechopen.70517

Abstract

In this chapter, the optimization method of double-well bistable stochastic resonance (SR) system and one of its applications in cognitive radio networks are introduced, especially in the energy detection problem. The chapter will be divided into five sections. Firstly, the conventional double-well bistable stochastic resonance system is introduced with its special properties. Then based on the conventional discrete overdamped double-well bistable SR oscillator, the optimization method and the analyses results are given especially under low signal-to-noise ratio (SNR). In the applications, a novel spectrum sensing approach used in the cognitive radio networks (CRN) based on SR is proposed. The detection probability is also derived theoretically under a constant false-alarm rate (CFAR). Moreover, a cooperative spectrum sensing technique in CRN based on the data fusion of various SR energy detectors is proposed. Finally the whole chapter is summarized.

Keywords: stochastic resonance, optimization, cognitive radio networks, spectrum sensing, energy detection, cooperative spectrum sensing

1. Introduction of conventional double-well bistable stochastic resonance system

In many different dynamic systems, it can be found that the stochastic resonance (SR) is a kind of complex nonlinear phenomenon with many applications [1, 2]. In this kind of dynamic system, it possesses some good performances, while it can help to increase the periodic driving signal power under some special conditions. A lot of researches have demonstrated that the SR system may help to convert some power of the state variable signal in the SR system into the spectral power of the single-frequency driving signal in the SR system [1–3]. So, the SR system has been widely used in many applications, such as the weak target identification, weak signal detection and estimation, and so on [4–6].

In the dynamic SR processing, according to the SR noise power influence to the SR system, it can be found that the improvement effects, which include the signal power amplification and the signal-to-noise ratio (SNR) enhancement, have great relationships between the SR driving sinusoidal signal power and the SR noise power [3].

Mathematically, an SR system in a continuous form can be written as [3]

$$d\mathbf{x}(t)/dt = f[\mathbf{x}(t), r(t)], \tag{1}$$

while in the above equation, $f[\cdot]$ is the dynamic SR system, $\mathbf{x}$(t) is the state vector, and $r(t)$ is the driving signal of the SR system.

In many SR systems, it can be found that the quartic double-well bistable system is a widely used SR system with many researches and discussions, and it has been applied in many applications. It can be expressed as

$$dx(t)/dt = ax(t) - bx^3(t) + k \cdot r(t). \tag{2}$$

In the expression above, $x(t)$ is the state variable of the SR system, parameters a and b determine the properties of the SR system, and the driving parameter k influences the effect of driving signal $r(t)$ seriously. In many studies, $r(t)$ is set as a single-frequency sinusoidal signal, which is also influenced by some additive noise $n(t)$, which is

$$r(t) = \varepsilon \sin\omega_s t + n(t), \tag{3}$$

while in the above equation, the parameters ε and ω_s are the corresponding signal amplitude and signal angular frequency of the driving signal; $n(t)$ is the additive noise. To simplify the analyses, $n(t)$ can always be supposed to obey the Gaussian distribution, which possesses mean 0 and variance σ_n^2. So, the SNR of the driving signal $r(t)$ (or the SNR of the input SR system) can be expressed by

$$SNR_i = \varepsilon^2/2\sigma_n^2. \tag{4}$$

According to the linear response theory of SR system [3], the output of the SR system state variable $x(t)$ can be expressed as a sum of two components as

$$x(t) = \varepsilon_o \sin(\omega_s t + \varphi_o) + n_o(t), \tag{5}$$

where ε_o is amplitude of the output signal, φ_o is the phase of the output signal at the input frequency point ω_s, and $n_o(t)$ is the additive noise in the output signal.

Based on the above assumptions, when $\omega_s \to 0$ or even $\omega_s = 0$, the output SNR of the SR system may be estimated by [2]

$$SNR_o \approx \left[\frac{\sqrt{2}\, a\, \varepsilon^2 c^2}{k^3 \sigma_n^4} e^{-\frac{2U_0}{k^2\sigma_n^2}}\right] \cdot \left[1 - \left(\frac{4 a^2 \varepsilon^2 c^2}{\pi^2 k^3 \sigma_n^4} e^{-\frac{4U_0}{k^2\sigma_n^2}}\right) \Big/ \left(\frac{2 a^2}{\pi^2} e^{-\frac{4U_0}{k^2\sigma_n^2}} + \omega_s^2\right)\right]^{-1}, \tag{6}$$

where $c = \sqrt{a/b}$ and $U_0 = a^2/4b$ are constants corresponding to the selection of parameters a and b in (2). It can also be found in (6) that in many real applications, the parameter k is the only parameter which can be adjusted, and also it cannot influence the parameter SNR_i in (4), so it is a very important factor which can determine the SR phenomenon [2].

2. Optimization of double-well bistable stochastic resonance system

2.1. System optimization and performance analyses

As described in last section, to make the SR system more applicable to the weak target identification or detection problems, we investigate a kind of optimization method to the quartic double-well bistable SR system in (2), and the target is to guarantee the enhancement of the signal SNR and also reach a maximal output SNR at the same time.

Although the result in (6) is based on the assumption $\omega_s \to 0$, when under some conditions that $\omega_s \to 0$ cannot be guaranteed, some traditional down-conversion methods can be applied if the frequency of the sinusoidal signal cannot fulfill $\omega_s \to 0$. Without loss of generality, an additive SR noise $n_1(t)$ is also introduced into the SR system, which possesses mean 0 and variance 1; then the quartic double-well bistable system in (2) can be rewritten as

$$\begin{aligned} dx(t)/dt &= ax(t) - bx^3(t) + k_1 \cdot r(t)/\|r(t)\| + k_2\, n_1(t) \\ &= ax(t) - bx^3(t) + k_1\, \varepsilon \sin\omega_s t/\|r(t)\| + k_1\, n(t)/\|r(t)\| + k_2\, n_1(t), \end{aligned} \tag{7}$$

where k_1 and k_2 are the positive driving parameters corresponding to $r(t)$ and $n_1(t)$, respectively. $r(t)$ is normalized by $\|r(t)\|$ to simplify the analyses, which is defined by

$$\|r(t)\| \overset{def}{=} \lim_{N\to\infty} \frac{1}{N} \sum_{t=1}^{N} r^2(t) = \frac{1}{2}\varepsilon^2 + \sigma_n^2, \tag{8}$$

where N is the sampling number. And when SNR_i is small enough, we have

$$\hat{\sigma}_n^2 \approx \|r(t)\|. \tag{9}$$

Based on the analyses in Ref. [4], if we want to reach an optimal result, it requires that the SR noise should be symmetric, and then $n_1(t)$ can also be chosen as a kind of noise signal with Gaussian distribution. So, (7) can be rewritten as

$$dx(t)/dt = ax(t) - bx^3(t) + k_3\, \varepsilon \sin\omega_s t + k_4\, n_{SR}(t), \tag{10}$$

where the parameters are defined by

$$k_3 \stackrel{\text{def}}{=} k_1 / \|r(t)\|, \tag{11}$$

$$k_4 \stackrel{\text{def}}{=} \sqrt{k_1^2 \cdot \hat{\sigma}_n^2 / \|r(t)\|^2 + k_2^2}, \tag{12}$$

and $n_{SR}(t)$ in (10) is a Gaussian noise with mean 0 and variance 1.

With the assumptions above, the SNR_o in (6) can be rewritten as

$$SNR_o \approx \left[\sqrt{2}\, a k_3^2 \varepsilon^2 c^2 k_4^{-4} e^{-\frac{2U_0}{k_4^2}}\right] \cdot \left[1 - 2 k_3^2 \varepsilon^2 c^2 k_4^{-4}\right]^{-1} = \frac{\sqrt{2}\, a k_3^2 \varepsilon^2 c^2}{k_4^4 - 2 k_3^2 \varepsilon^2 c^2} e^{-\frac{2U_0}{k_4^2}}. \tag{13}$$

Firstly, to ensure the SNR improvement effect of the SR system, it requires $SNR_o > SNR_i$, so

$$\frac{\sqrt{2}\, a k_3^2 \varepsilon^2 c^2}{k_4^4 - 2 k_3^2 \varepsilon^2 c^2} e^{-\frac{2U_0}{k_4^2}} > \frac{\varepsilon^2}{2 \sigma_n^2}. \tag{14}$$

And when SNR_i is low enough, (14) can be simplified to

$$k_3^2 > k_4^4 e^{\frac{2U_0}{k_4^2}} / (2\sqrt{2}\, U_0 \sigma_n^2). \tag{15}$$

When U_0 and σ_n^2 are fixed, it is obvious that $k_4 = \sqrt{U_0}$ will lead to the maximal value of the right side expression of (15). So when we have:

$$k_3^2 > U_0 e^2 / (2\sqrt{2}\, \hat{\sigma}_n^2), \tag{16}$$

the SNR enhancement can be achieved.

What is more, to reach the maximum output SNR of the system, we can set up an optimization problem, where we suppose (13) as the corresponding objective function and let k_1 be fixed, and then we let

$$\partial SNR_o / \partial k_4^2 = 0. \tag{17}$$

And the result can be changed to

$$k_4^6 - U_0 k_4^4 + 2 U_0 k_3^2 \varepsilon^2 c^2 = 0, \tag{18}$$

or k_4 is the solution of the above equation.

By calculating the discriminant Δ of (18), we have

$$\Delta = U_0^2 k_3^4 \varepsilon^4 c^4 - \frac{2}{27} U_0^4 k_3^2 \varepsilon^2 c^2 = \frac{U_0^2 a k_3^2 \varepsilon^2 c^2}{216 b^2} (216 k_3^2 \varepsilon^2 b - a^3). \tag{19}$$

Then the optimization result can be decided by the power or the amplitude of the driving sinusoidal signal. It can be found that k_3 should also satisfy (16), so we can choose a reasonable k_3 big enough to fulfill $\Delta > 0$ and guarantee that the optimization result of k_4 can be achieved. When substituting the optimal values of k_3 and k_4 into (11) and (12), the optimal driving parameters k_1 and k_2 can finally be achieved.

2.2. Computer simulations

To testify the effectiveness of the above proposed optimization method, we give out a testifying example and carry out corresponding computer simulation results based on the analyses in the last section.

To simplify the simulations, a single-frequency sinusoidal signal corrupted with additive white Gaussian noise (AWGN) is assumed as the signal r(t), and the amplitude and angular frequency of the signal are chosen as $\varepsilon = 1$, $\omega_s = 0.01$, respectively. The sampling number is $N = 1 \times 10^5$; and the parameters are chosen as $a = 1$ and $b = 1$ in the SR system.

In the following computer simulations, the maximum likelihood estimate (MLE) method [7] is applied to estimate the amplitude of the signal as

$$\hat{\varepsilon} = \sqrt{\hat{\alpha}_1^2 + \hat{\alpha}_2^2}, \quad (20)$$

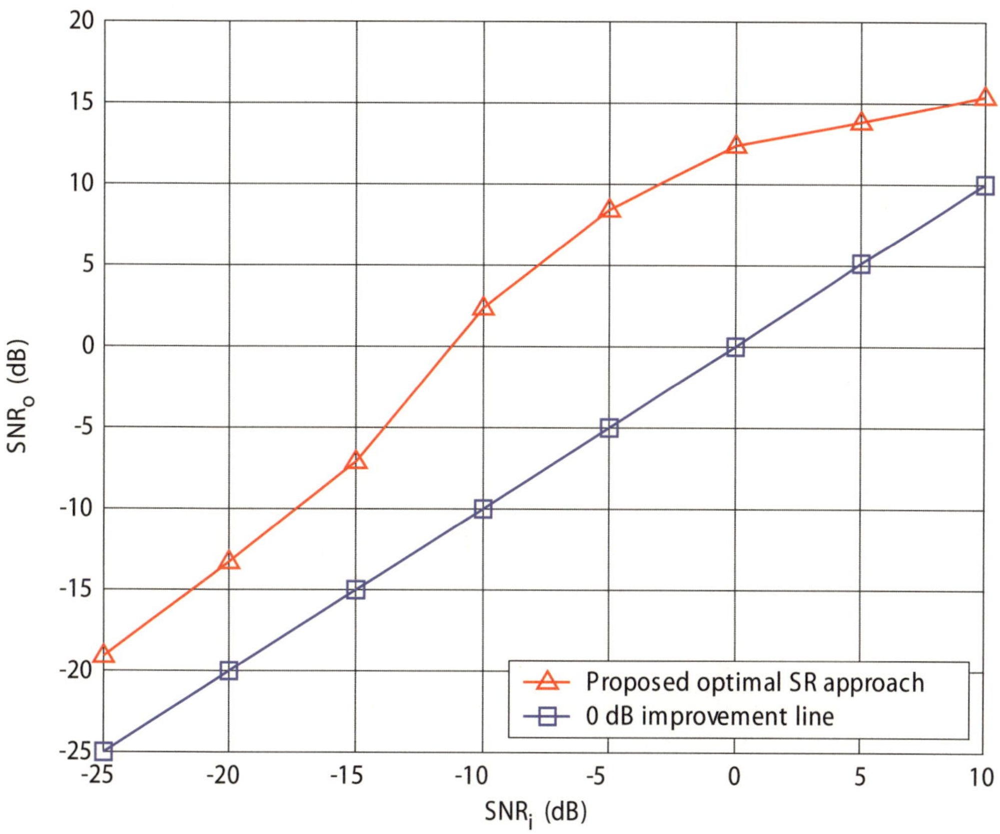

Figure 1. SNR_o vs. SNR_i by using the proposed optimal SR approach.

where we have

$$\hat{\alpha}_1 = \frac{2}{N}\sum_{t=0}^{N-1} r(t)\cos\omega_s t,\ \hat{\alpha}_2 = \frac{2}{N}\sum_{t=0}^{N-1} r(t)\sin\omega_s t. \quad (21)$$

Eq. (7) is changed to the following difference equation for simulations [8]:

$$x(t+1) = x(t) + \Delta t \cdot \left[ax(t) - bx^3(t) + k_1 \cdot r(t)/\|r(t)\| + k_2 n_1(t)\right],\ (t = 0, 1, \cdots, N-2), \quad (22)$$

where the parameter Δt is chosen as 0.0195.

Figure 1 gives the SNR_o vs. SNR_i comparison performance through the proposed method, while the range of SNR_i is between −25 dB and 10 dB. From the result, it can be discovered that the SNR of r(t) has been enhanced especially under low SNR, for example, $SNR_i < 0$ dB.

Figure 2 shows a result regarding to the SNR enhancement through the proposed optimal SR method. The SNR_i also changes from −25 dB to 10 dB. It can be found that the SNR enhancement can also be reached even under low SNR.

Figure 3 shows the SNR_o performance when the parameters k_1 and k_2 are adjustable under $SNR_i = -25$ dB. It can be discovered clearly that a maximal SNR_o can be reached with some certain optimal k_1 and k_2 values. **Figure 4** shows the performance of SNR_o vs. k_2 under optimal k_1 under the condition $SNR_i = -25$ dB. **Figures 5** and **6** give the same computer simulation results under the condition $SNR_i = -20$ dB, and they also verify the reliability of the proposed method.

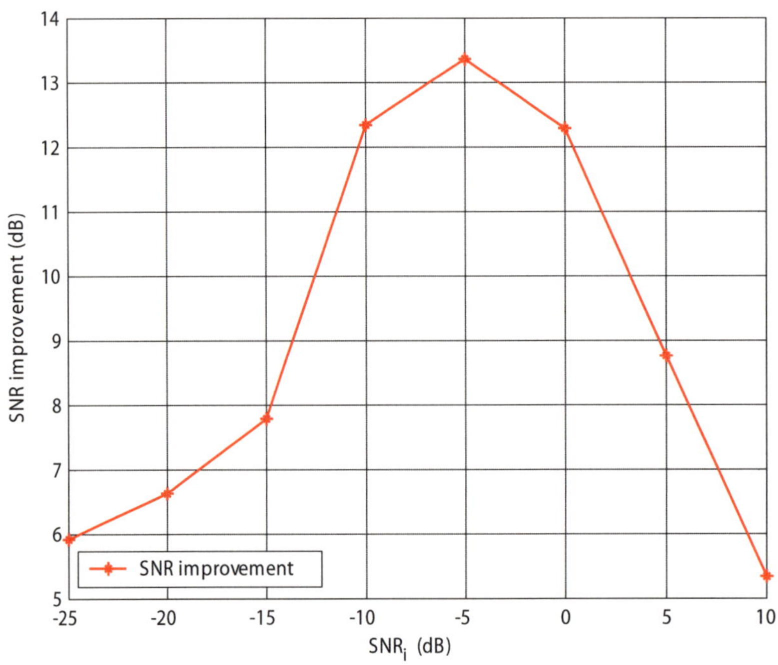

Figure 2. SNR improvement by using the proposed optimal SR approach.

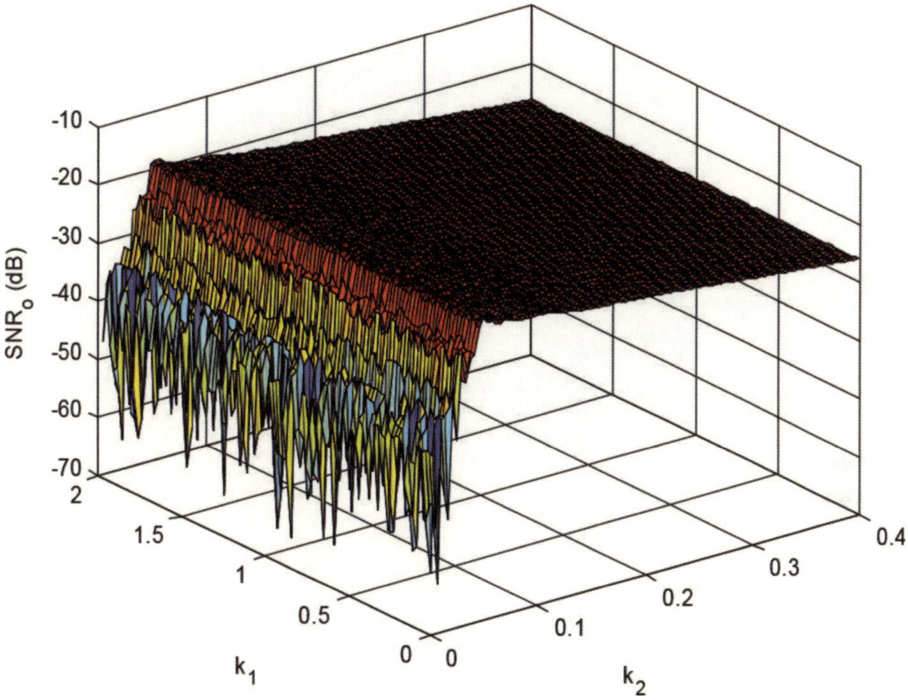

Figure 3. Performance of SNR_o under $SNR_i = -25$ dB when k_1 and k_2 are adjustable.

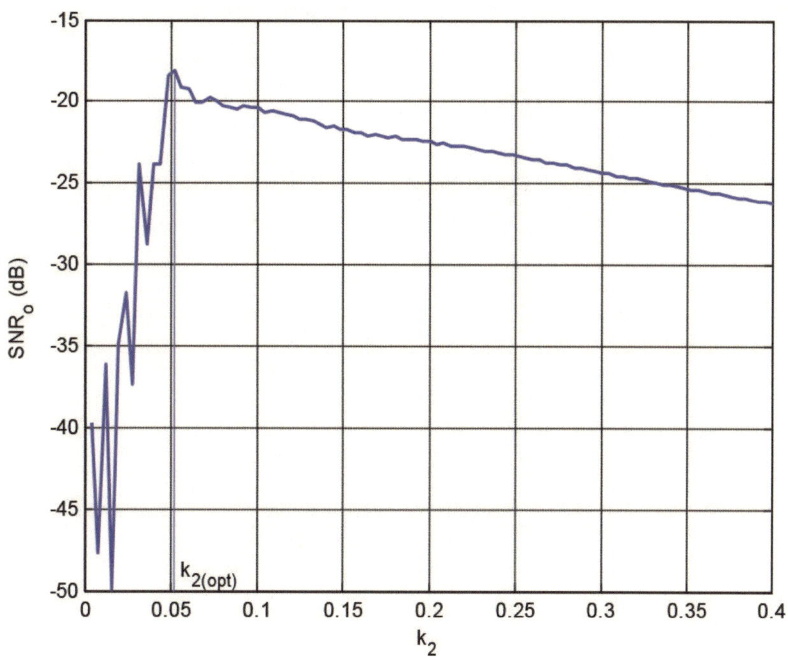

Figure 4. SNR_o vs. k_2 under optimal k_1 and $SNR_i = -25$ dB.

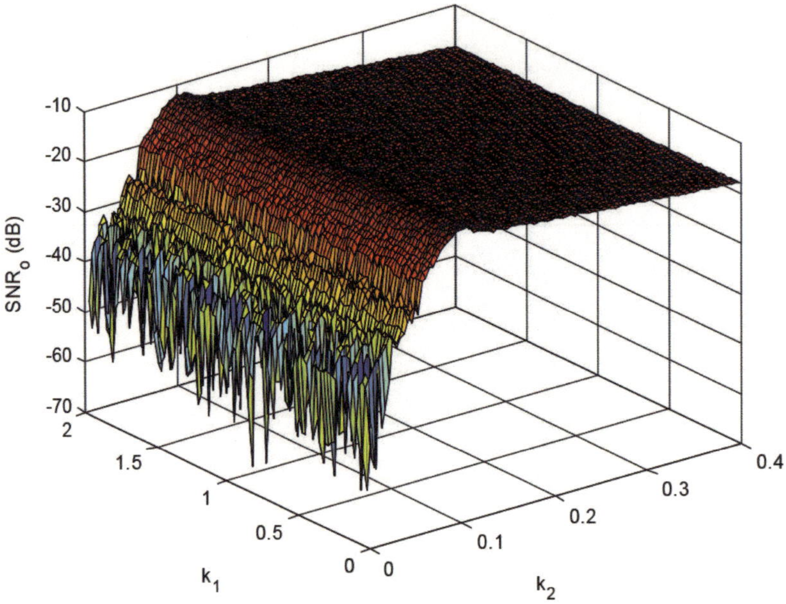

Figure 5. Performance of SNR_o under SNR_i = −20 dB when k_1 and k_2 are adjustable.

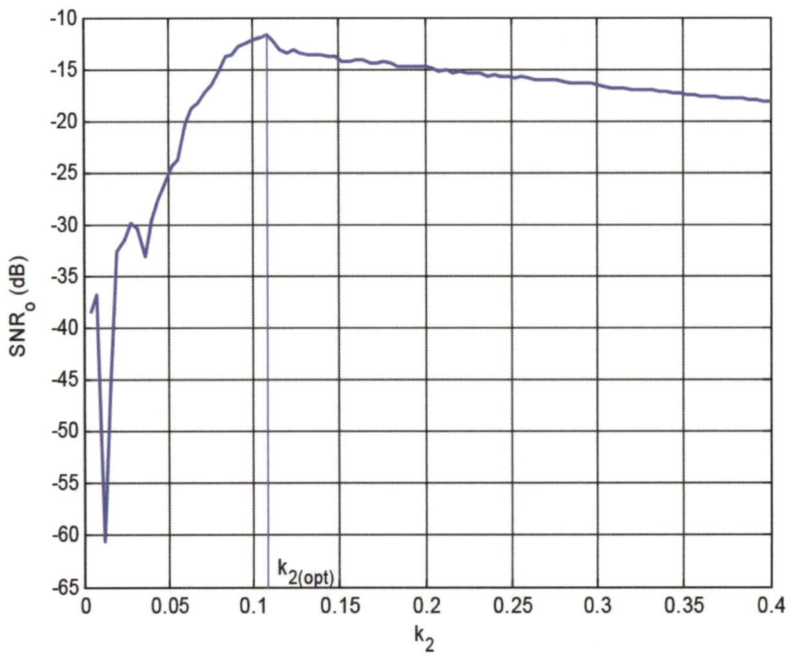

Figure 6. SNR_o vs. k_2 under optimal k_1 and SNR_i = −20 dB.

3. Applications in the energy detection problem in cognitive radio networks

3.1. Energy detection problem in cognitive radio networks

In the past years, the research works in the area of cognitive radio (CR) network have been widely reported with fast progress. A lot of novel research developments make the research topics in the related areas more and more attractive [9]. As is known, in CR, the spectrum sensing approaches play an important role in CR network because it helps the secondary or opportunistic users (SUs) to detect the existence of the primary users (PUs) and define whether they can transmit the information or not [10]. Without loss of generality, we suppose that only the overlay mode in CR networks is considered in this discussion. The main target of spectrum sensing is to define the presence of PUs under some unpredictable noisy wireless communications conditions. So when the PUs are detected to be absent, the SUs are permitted to use the spectrum holes on an opportunistic basis which are occupied by PUs before, so that it can enhance the spectrum utility significantly [11]. In other words, the spectrum sensing can be regarded as a base of CR networks seriously.

In the literatures, many approaches have been proposed to ensure the performance of spectrum sensing and minimize the interference to some other users, including the PUs [9]. With the previous studies, it is found that the energy detection is a very general spectrum sensing method which does not need any prior knowledge of PUs; and based on the traditional Neyman-Pearson criterion [7], the spectrum sensing problem can be converted to a detection problem as the following two hypotheses:

$$\begin{aligned} H_0&: r(t) = n(t), \quad (t = 0,1,\ldots,N-1) \\ H_1&: r(t) = h \cdot s(t) + n(t), \quad (t = 0,1,\ldots,N-1), \end{aligned} \tag{23}$$

where $r(t)$ is the signal at the receiver, $s(t)$ is the PU signal, and it is assumed that $s(t)$ obeys the distribution with mean 0 and variance σ_s^2 and h is the channel fading factor between the transmitter (PU) and the receiver (SU). In the wireless communications applications, it can always be assumed that it has Rayleigh distribution with the second-order moment $E[h^2] = m_h^2$ independent to PU, and $n(t)$ is the additive noise independent to $s(t)$ and h. Simultaneously, sometimes the co-channel interference or multiuser interference of the PU signal can also be regarded as another additive part of $n(t)$. So to simplify the analyses, we suppose that h is predictable or can be estimated properly at the receiver and $n(t)$ also obeys the additive white Gaussian noise (AWGN) distribution with mean 0 and variance σ_n^2.

For the traditional Neyman-Pearson detection, the assumption or decision H_1 can be made when the likelihood ratio exceeds a certain threshold γ, as follows:

$$L(\mathbf{r}) = p(\mathbf{r}; H_1)/p(\mathbf{r}; H_0) > \gamma, \tag{24}$$

where $\mathbf{r} = [r(0), r(1), \ldots, r(N-1)]^T$ is the receiving signal vector and $p(\mathbf{r}; H_0)$ and $p(\mathbf{r}; H_1)$ represent the probability density functions (PDFs) of the receiving signal vector $\mathbf{r}$ under H_0 and H_1, respectively, while $L(\mathbf{r})$ is the likelihood ratio to be calculated.

Based on the analyses above, under two different hypotheses, the receiving signal $\mathbf{r}$ obeys Gaussian distribution with different variances, which can be expressed by

$$\mathbf{r} \sim N(\mathbf{0}, \sigma_n^2 \mathbf{I}) \quad \text{under } H_0, \tag{25}$$

$$\mathbf{r} \sim N(\mathbf{0}, (m_h^2 \sigma_s^2 + \sigma_n^2)\mathbf{I}) \quad \text{under } H_1. \tag{26}$$

Thus, H_1 is decided when

$$T(\mathbf{r}) = \sum_{t=0}^{N-1} r^2(t) > \left[2 \ln \gamma - N \ln\left(\frac{\sigma_n^2}{m_h^2 \sigma_s^2 + \sigma_n^2}\right)\right] \left[\sigma_n^2 (m_h^2 \sigma_s^2 + \sigma_n^2)\right] m_h^{-2} \sigma_s^{-2} = \gamma_{ED}, \tag{27}$$

where $T(\mathbf{r})$ is the statistic of the traditional energy detector and γ_{ED} is the threshold to satisfy $P_{fa} = \alpha$ for a given CFAR α. Because the Neyman-Pearson detector calculates the energy of the receiving signal $r(t)$, it is also called an energy detector.

In the following, the corresponding false alarm rate $P_{fa(ED)}$ and the detection probability $P_{d(ED)}$ of the above energy detector can be given as

$$P_{fa(ED)} = \Pr\{T(\mathbf{r}) > \gamma_{ED}; H_0\} = \Pr\left\{\frac{T(\mathbf{r})}{\sigma_n^2} > \frac{\gamma_{ED}}{\sigma_n^2}; H_0\right\} = Q_{\chi_N^2}\left(\frac{\gamma_{ED}}{\sigma_n^2}\right), \tag{28}$$

$$P_{d(ED)} = \Pr\{T(\mathbf{r}) > \gamma_{ED}; H_1\} = \Pr\left\{\frac{T(\mathbf{r})}{m_h^2 \sigma_s^2 + \sigma_n^2} > \frac{\gamma_{ED}}{m_h^2 \sigma_s^2 + \sigma_n^2}; H_1\right\} = Q_{\chi_N^2}\left(\frac{\gamma_{ED}}{m_h^2 \sigma_s^2 + \sigma_n^2}\right), \tag{29}$$

while $Q_{\chi_N^2}(\cdot)$ is the right-tail probability function with N degrees of freedom.

It can be found in the researches that this kind of energy detection method could perform well under high SNR. But its performance degrades seriously when SNR is reduced, especially when SNR < −10 dB. For example, the value of detection probability P_d under $N = 10^3$ and $P_{fa(ED)} = 0.1$ will decrease from 0.795 to 0.283 when the SNR changes from −10 dB to −15 dB, which may be a very general case in CR networks [12].

3.2. SR-based spectrum sensing approach

In this subsection, we propose a novel spectrum sensing method with the combination of traditional energy detector and the SR processing. First, let the receiving signal pass the SR system, and the amplified signal can be observed at the output of the SR system. Then the amplified signal goes through the conventional energy detector to get the final spectrum sensing decision.

In this proposed scheme based on SR, first, we set the normalized signal of $r(t)$ in (23), say $r_0(t)$, as the input of an SR system $f[\cdot]$; then we have

$$\dot{\mathbf{x}}(t) = f[\mathbf{x}(t), r_0(t) + n_0(t)], \tag{30}$$

$$r_0(t) = r(t)/\sqrt{\text{var}[r(t)]}, \quad (t = 0, 1, \ldots, N-1) \tag{31}$$

where x(t) is still the SR system status vector and $n_0(t)$ is the SR noise with mean 0 and variance $\sigma^2_{n_0}$, so $r_0(t) + n_0(t)$ can be taken as the drive signal of the SR system.

Based on the SR linear response theory [3], the status vector of SR system can be divided into two independent additive parts, say

$$\mathbf{x}(t) = \mathbf{s}_{SR}(t) + \mathbf{n}_{SR}(t), \tag{32}$$

where $\mathbf{s}_{SR}(t)$ is the system response signal corresponding to the normalized PU signal $h{\cdot}s(t)/\sqrt{\text{var}[r(t)]}$ and $\mathbf{n}_{SR}(t)$ is the system response signal corresponding to the noise signal $n(t)/\sqrt{\text{var}[r(t)]} + n_0(t)$. It can be found that the additive channel noise $n(t)$ also plays a part role of SR noise.

From the above analyses, to reach a maximal SNR_o, the optimal variance of the introduced SR noise $\sigma^2_{n_0(\text{opt})}$ can be calculated according to the derivations in the last section.

3.3. Experimental and comparison results

In the following, we present some experimental and comparison outcomes. In the computer simulations, the discrete overdamped bistable oscillator in (22) is used as the dynamic SR system model.

As is known QPSK and QAM are the mostly used modulation methods [13, 14] in the broadcasting systems. So in the computer simulations thereafter, a QPSK signal as the PU signal together with a co-channel interference QPSK signal with AWGN through the Rayleigh fading channel is utilized as the driving signal of the SR system, which can be expressed by

$$r(t) = h \cdot [A_P \cdot \sin(\omega_P t + \varphi_P) + A_M \cdot \sin(\omega_M t + \varphi_M)] + n(t), \tag{33}$$

where h is the Rayleigh channel gain with mean 1; A_P, ω_P, and φ_P are the amplitude, angular frequency, and phase of the PU sinusoidal carrier signal; A_M, ω_M, and φ_M are the amplitude, angular frequency, and phase of the multiuser interference sinusoidal carrier signal, respectively. Here, $\varphi_P, \varphi_M \in \{\pi/4, 3\pi/4, 5\pi/4, 7\pi/4\}$ in QPSK. In this case, the input SNR can be calculated by [15]

$$SNR_i = \frac{1}{2} m_h^2 A_P^2 \Big/ \left(\frac{1}{2} m_h^2 A_M^2 + \sigma_n^2\right). \tag{34}$$

In the following simulations, we choose $\omega_P = 0.04\pi$, $\omega_M = 0.2\pi$, and $\sigma_n^2 = 1$. So the optimal variance of the introduced white Gaussian SR noise with mean 0 can be calculated through the analyses in the last section, and we can get $\sigma^2_{n_0(\text{opt})} = 1 - k^2$ which requires $k \leq 1$.

Figures 7 and **8** give the performance comparison results of the receiver operating characteristic (ROC) plots between the traditional energy detector and the proposed SR-based energy detector under the conditions SNR = −15 dB and SNR = −20 dB. The total sampling number is N = 10^3. In the figures, both the theoretical results and the computer simulation results of the above two methods are given, and the theoretical results of detection probability of the proposed method are calculated based on (29). It can be discovered that the detection probabilities of the proposed approach are higher than the energy detector, especially under low SNR as SNR < −10 dB, which is a good performance to the real applications; and it can also be discovered that even under SNR = −20 dB which is also very common in CR networks, the proposed detection method can still perform better than the energy detection method with a significant detection probability enhancement.

Besides the ROC curve performance comparison, the results of the detection probability versus SNR under CFAR are also presented. **Figures 9** and **10** give the performance comparison results between the proposed detection method and the conventional energy detection method under P_{fa} = 0.05 and P_{fa} = 0.1, respectively. The total sampling number is still selected as N = 10^3. In the following simulations, the input SNR changes from −20 dB to 0 dB. And both the theoretical analyses results and the computer simulation results are given in **Figures 9** and **10**. It is obvious that the detection probability of the proposed SR-based method can be improved, especially under low SNR of SNR < −10 dB, and also it can be discovered that a 5 dB SNR enhancement can be achieved. Based on the simulation results, the main problems of the conventional energy detection method can be solved.

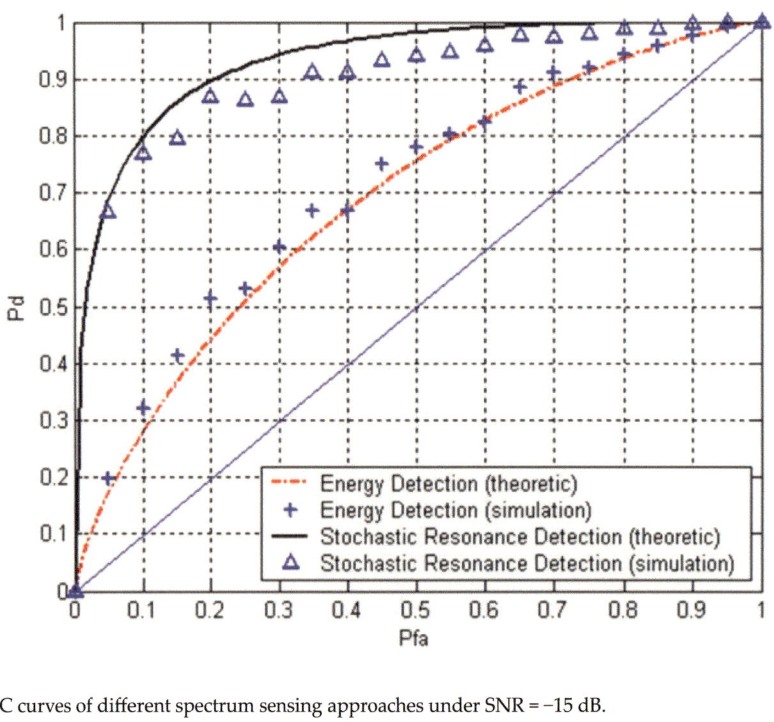

Figure 7. ROC curves of different spectrum sensing approaches under SNR = −15 dB.

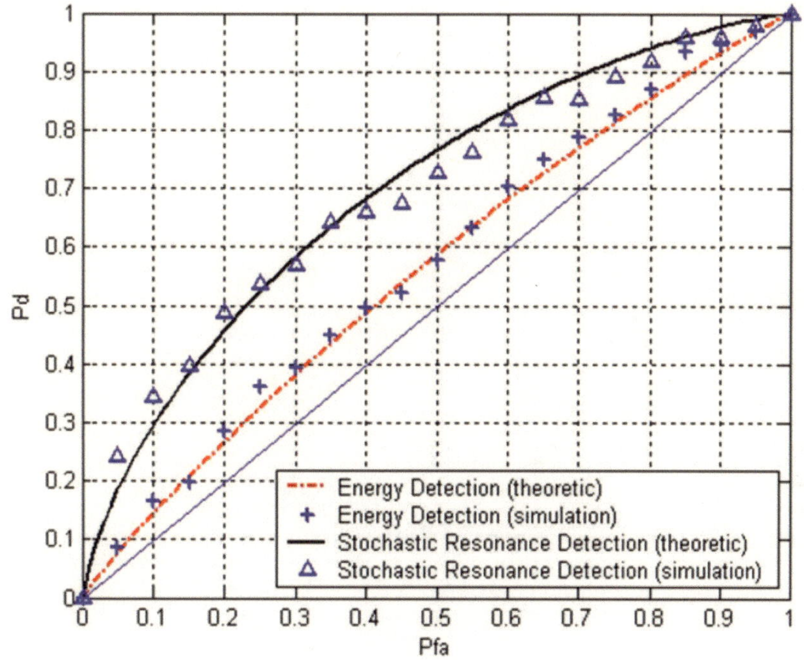

Figure 8. ROC curves of different spectrum sensing approaches under SNR = −20 dB.

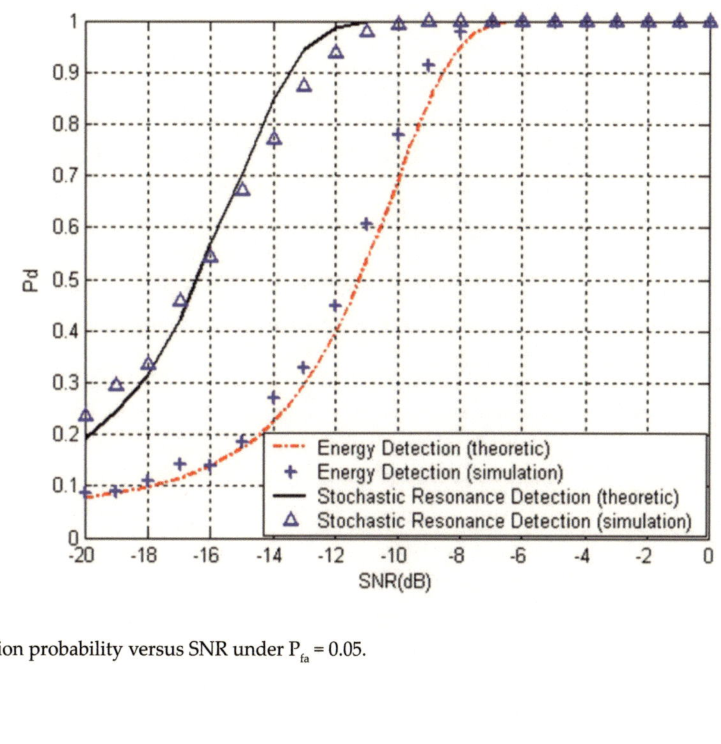

Figure 9. Detection probability versus SNR under P_{fa} = 0.05.

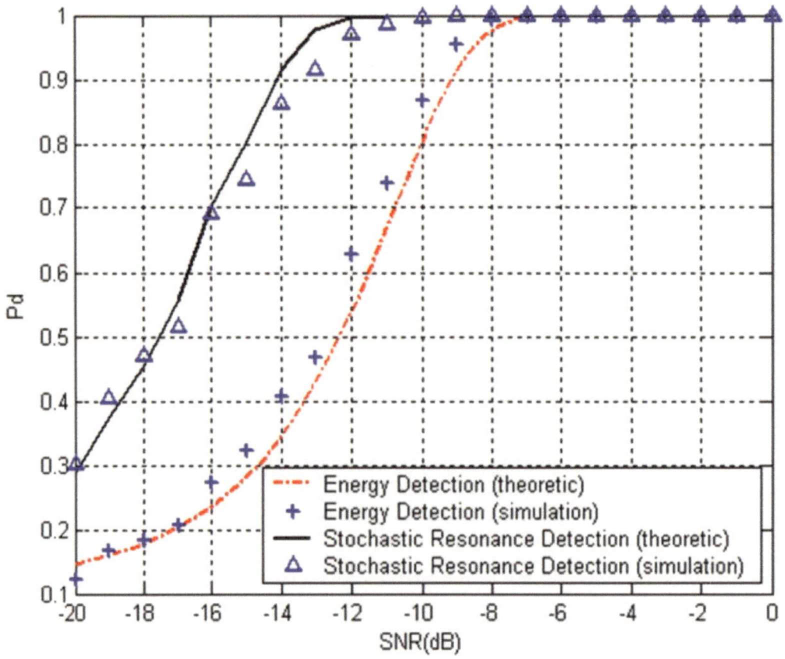

Figure 10. Detection probability versus SNR under $P_{fa} = 0.1$.

4. Application of cooperative stochastic resonance in the energy detection problem in cognitive radio networks

4.1. Cooperative SR-based spectrum sensing approach

In these years' studies, the spectrum sensing techniques in physical layer can be divided into two classes: noncooperative sensing techniques and cooperative sensing techniques. Recently it has become a new direction by introducing some cooperation methods into the spectrum sensing or PU signal detection procedure with the cooperation of different secondary user (SU) sensing results [16, 17]. So based on the results in the last two sections, here we introduce the chaotic stochastic resonance (CSR) system to improve the spectrum sensing performance especially under low SNR circumstances.

At the first step, we could randomly select K SUs to carry out the spectrum sensing process independently to the communication channel. To realize the data fusion, we carry out two kinds of sensing methods at each SU at the same time. One is the traditional energy detection method with the statistics A(x), and another is the CSR energy detector with the statistics B(x). While in the real applications, it is also very difficult to determine what kind of SR noise will be best or optimal, so we choose different types of CSR noise as the CSR noise signal candidate $\beta_0(t)$ in each of the CSR systems, but the CSR systems in each SUs are all the same. To get some certain fusion result, some commonly used noise types in the wireless communication systems can be used, for example, AWGN signal, lognormal distribution noise,

Weibull distribution noise, etc. Here we list the noise-type candidates as $\{\beta_1(t), \beta_2(t),\ldots, \beta_K(t)\}$. When some certain CSR system f [.] and the noise types are fixed, the corresponding optimal parameters with these noise-type candidates can also be calculated.

Theoretically and without loss of generality, it can be assumed that all the receiving signal at different SU obey the same distribution, so (23) is suitable for each SU. To simplify the analyses thereafter, we suppose that h = 1. So we have

$$\begin{cases} H_0: \; E[A_1(\mathbf{x})] = E[A_2(\mathbf{x})] = \cdots = E[A_K(\mathbf{x})] = E[A(\mathbf{x})] = \sigma_n^2, \\ H_1: \; E[A_1(\mathbf{x})] = E[A_2(\mathbf{x})] = \cdots = E[A_K(\mathbf{x})] = E[A(\mathbf{x})] = \sigma_s^2 + \sigma_n^2, \end{cases} \tag{35}$$

while $A_1(x)$, $A_2(x)$, ..., and $A_K(x)$ are the statistics of SUs 1, 2, ..., K, respectively.

In the data fusion processing, we introduce the traditional Bayesian fusion method to realize the cooperative spectrum sensing. Simultaneously, if the same traditional energy detection method and the same threshold γ_{ED} are used at each SU detector, the expectation result $E[A^{1,2,\cdots,K}(\mathbf{x})]$ of the Bayesian fusion can be written as

$$\begin{aligned} E[A^{1,2,\cdots,K}(\mathbf{x})] &= \frac{\prod_{k=1}^{K} E\left[A_k(\mathbf{x}) \,\middle|\, A_k(\mathbf{x}) = \frac{1}{N}\sum_{t=1}^{N} x_k^2(t)\right]}{\prod_{k=1}^{K} E\left[A_k(\mathbf{x}) \,\middle|\, A_k(\mathbf{x}) = \frac{1}{N-1}\sum_{t=1}^{N-1} x_k^2(t)\right]} \cdot E\left[A^{1,2,\cdots,K}(\mathbf{x}) \,\middle|\, A_k(\mathbf{x}) \right. \\ &= \left. \frac{1}{N-1}\sum_{t=1}^{N-1} x_k^2(t)\right] = E[A(\mathbf{x})], \end{aligned} \tag{36}$$

where $x_k(t)$ is the output of the kth SU's CSR system.

Let the receiving signal r(t) goes through the dynamic CSR system with different CSR noise $\beta_1(t), \beta_2(t),\ldots, \beta_K(t)$, and we can denote the output of each SU's CSR energy detector to be $B_1(x)$, $B_2(x)$, ..., $B_K(x)$, respectively. Introducing the conventional Bayesian fusion method to fuse all K SUs' statistical results $\{A_1(x), A_2(x), \ldots, A_K(x)\}$, $\{B_1(x), B_2(x), \ldots, B_K(x)\}$, and $A_{1,2\ldots,K}(x)$, then the following Theorem 1 [18] could verify the effectiveness of the proposed cooperative spectrum sensing method.

Theorem 1. The cooperative spectrum sensing approach proposed by using the Bayesian fusion to all K SUs' statistics $\{A_1(x), A_2(x), \ldots, A_K(x)\}$, $\{B_1(x), B_2(x), \ldots, B_K(x)\}$, and $A_{1,2,\ldots,K}(x)$ shown in **Figure 1** can improve the sensing performance of conventional energy detection method.

Proof: Please refer Theorem 1 in Ref. [18] for details.

4.2. Computer simulation results

In the following, some computer simulations are carried out to certify the correctness of the proposed method. Here, a QPSK signal is selected as the PU signal, that is

$$s(t) = A_P \sin(\omega_P t + \varphi_P), \tag{37}$$

where A_P, ω_P, and φ_P are the amplitude, angular frequency, and phase of the PU signal and $\varphi_P \in \{\pm\pi/4, \pm 3\pi/4\}$ in QPSK. In the following simulations, we set $A_P = 5$ and $\omega_P = 0.02\pi$.

Also in the simulations, a kind of conventional discrete overdamped bistable oscillator is utilized as the CSR system, that is [19]

$$x_i(t+1) = \left[g \cdot x_i(t) - x_i^3(t)\right] e^{-x_i^2(t)/h} + d \cdot r(t) + \beta_i(t). \tag{38}$$

In the equation above, $x_i(t)$ is the state variable and g and h are the corresponding parameters which determine the performance of the system seriously. In the simulations, we choose $g = 2.85$ and $h = 10$. d is the driving parameter of the CSR system.

The additive channel noise $n(t)$ is supposed to be composed by a sinusoidal interference signal and an AWGN signal in the computer simulations as

$$n(t) = n_0(t) + \varepsilon \cdot \sin\omega_\varepsilon t, \tag{39}$$

while $n_0(t)$ is the AWGN signal, and the amplitude and angular frequency of the sinusoidal signal are set as $\varepsilon = 0.1$ and $\omega_\varepsilon = 0.8\pi$.

Simultaneously, we choose the following types of CSR noise: uniform distribution noise, Weibull distribution noise, and lognormal distribution noise. While the uniform distribution noise is evenly distributed within the range [−1,+1].

The pdf of the Weibull distribution noise is

$$g(x; u, v) = \mathrm{uv}^{-u} x^{u-1} e^{-(x/v)^u}, \tag{40}$$

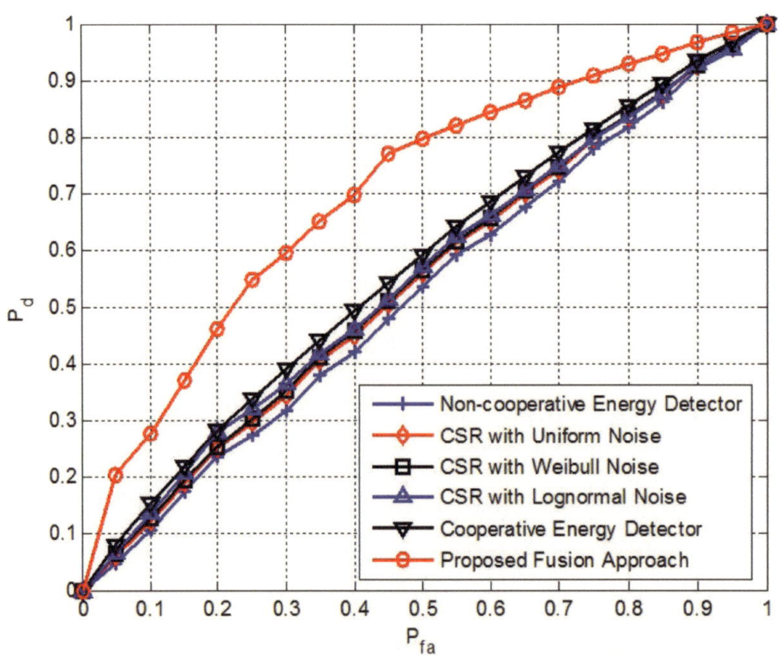

Figure 11. ROC curves of different spectrum sensing approach under SNR = −20 dB.

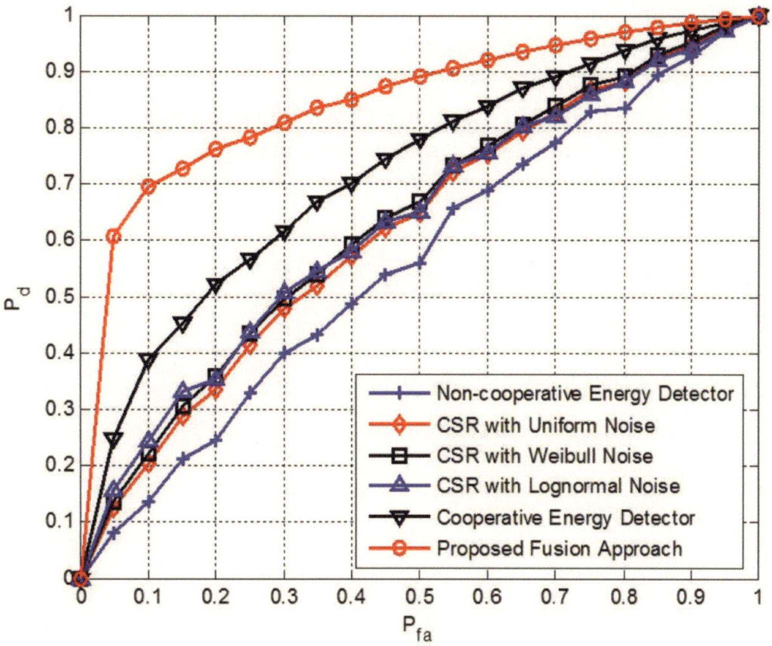

Figure 12. ROC curves of different spectrum sensing approach under SNR = −15 dB.

where $u = 2$ and $v = 1$. The pdf of the Lognormal distribution noise is

$$g(x; \mu, \sigma) = e^{-(\ln x-\mu)^2/2\sigma^2} / (x\sigma\sqrt{2\pi}), \tag{41}$$

where the parameters of are fixed as $\mu = 1$ and $\sigma = 1$.

In the computer simulations, the total sampling number is $N = 10^6$, and the Bayesian fusion process is performed under the CSR energy detection spectrum sensing driven by these three various kinds of noises, respectively.

Both **Figures 11** and **12** give the ROC curves of different spectrum sensing results under SNR = −20 dB and −15 dB, respectively. It can be found obviously that the proposed cooperative approach can achieve some better performance than the conventional noncooperative spectrum sensing methods.

5. Summary

In this chapter, some conventional double-well bistable SR systems are introduced first. Then based on the conventional discrete overdamped double-well bistable SR oscillator, the optimization method and the corresponding analyses results are given especially under low SNR circumstances. Besides, a novel spectrum sensing approach used in CRN based on SR is proposed. And a cooperative spectrum sensing technique in CRN based on the data fusion technique is also proposed. The last section summarizes the whole chapter.

The optimization approach introduced is especially applicable under low SNR, which are familiar in the wireless communications. In the applications, the performance analyses and computer simulations show that the effectiveness of the proposed spectrum sensing approach is better than the traditional energy detection methods, and this methodology can be extended to some other problems with the same two-hypothesis decisions.

Author details

Di He

Address all correspondence to: dihe@sjtu.edu.cn

Shanghai Jiao Tong University, Shanghai, China

References

[1] Mitaim S, Kosko B. Adaptive stochastic resonance. Proceedings of the IEEE. 1998 Nov;**86**(11):2152-2183

[2] McNamara B, Wilesenfeld K. Theory of stochastic resonance. Physical Review A. 1989 May;**39**(9):4854-4869

[3] Anishchenko VS, Astakhov VV, Neiman AB, Vadivasova TE, Schimansky-Geier L. Nonlinear Dynamics of Chaotic and Stochastic Systems: Tutorial and Modern Developments. Berlin Heidelberg: Spring-Verlag; 2002

[4] Chen H, Varshney PK, Kay SM, Michels JH. Theory of the stochastic resonance effect in signal detection: Part I—Fixed detectors. IEEE Transactions on Signal Processing. 2007 Jul;**55**(7):3172-3184

[5] Chen H, Varshney PK. Theory of the stochastic resonance effect in signal detection: Part II—Variable detectors. IEEE Transactions on Signal Processing. 2008 Oct;**56**(10):5031-5041

[6] Kay S, Michels JH, Chen H, Varshney PK. Reducing probability of decision error using stochastic resonance. IEEE Signal Processing Letters. 2006 Nov;**13**(11):695-698

[7] Kay SM. Fundamentals of Statistical Signal Processing, Volume II: Detection Theory. Upper Saddle River: Prentice Hall; 1998

[8] He D. Optimization of quartic double-well bistable stochastic resonance system. Proceedings of IEEE International Symposium on Circuits and Systems 2011 (ISCAS 2011); May 2011. pp. 1972-1975

[9] Haykin S. Cognitive radio: Brain-empowered wireless communications. IEEE Journal on Selected Areas in Communications. 2005 Feb;**23**(2):201-220

[10] Akyildiz IF, Lee WY, Vuran MC, Mohanty S. NeXt generation/dynamic spectrum access/ cognitive radio wireless networks: A survey. Computer Networks. 2006;**50**:2127-2159

[11] Cabric D, Mishra SM, Brodersen RW. Implementation issues in spectrum sensing for cognitive radios. Conference Record of the Thirty-Eighth Asilomar Conference on Signals, Systems and Computers; 2004 Nov 1. pp. 772-776

[12] Digham FF, Alouini MS, Simon MK. On the energy detection of unknown signals over fading channels. IEEE Transactions on Communications. 2007 Jan;**55**(1):21-24

[13] Cordeiro C, Challapali K, Birru D. IEEE 802.22: An introduction to the first wireless standard based on cognitive radios. The Journal of Communication. 2006 Apr;**1**(1):38-47

[14] DVB BlueBook A092r3. Implementation guidelines for DVB handheld services. http://www.dvb.org/technology/standards/a092r3.dTR102377.V1.4.1.DVB-H_impl_guide.pdf; April 2009

[15] He D, Lin YP, He C, Jiang LG. A novel spectrum sensing technique in cognitive radio based on stochastic resonance. IEEE Transactions on Vehicular Technology. 2010 May; **59**(4):1680-1688

[16] Peh ECY, Liang YC, Guan YL, Zeng YH. Optimization of cooperative sensing in cognitive radio networks: A sensing-throughput tradeoff view. IEEE Transactions on Vehicular Technology. 2009 Sep;**58**(9):5294-5299

[17] Quan Z, Cui S, Poor HV, Sayed AH. Collaborative wideband sensing for cognitive radios. IEEE Signal Processing Magazine. 2008 Nov;**25**(6):60-73

[18] He D. Chaotic stochastic resonance energy detection fusion used in cooperative spectrum sensing. IEEE Transactions on Vehicular Technology. 2013 Feb;**62**(2):620-627

[19] Zhang YL, Zhang QY, Melodia T. A frequency-domain entropy-based detector for robust spectrum sensing in cognitive radio networks. IEEE Communications Letters. 2010 Jun;**14**(6):533-535

Chapter 6

Resonances in Left-Handed Waves Developed in Nonlinear Electrical Lattices

Koichi Narahara

Additional information is available at the end of the chapter

http://dx.doi.org/10.5772/intechopen.70739

Abstract

We investigate resonant interactions in a specific electrical lattice that supports left-handed (LH) waves. The impact of LH waves on the three-wave mixing process, which is the most fundamental resonant interaction, is illustrated. In contrast to the ordinary right-handed (RH) waves, the phase of the LH wave moves to the different direction from its power. This exotic property together with the lattice's dispersive features results in the resonant phenomena that are effectively utilized for practical electrical engineering, including the significant harmonic wave generation via head-on collisions, harmonic resonance, and short pulse generation driven by soliton decay. These resonances are quantified by the asymptotic expansion and characterized by numerical and/or experimental methods, together with several design criteria for their practical utilization. To cope with dissipation, a field-effect transistor (FET) is introduced in each cell. In particular, we characterize the stationary pulse resulting from the balance between dissipation and FET gain.

Keywords: three-wave mixing, soliton decay, harmonic resonance, left-handedness, electrical lattices, composite right- and left-handed transmission lines, traveling-wave field-effect transistors, coherent structures

1. Introduction

Resonances have been utilized as the powerful tool to achieve harmonic wave generation in electrical engineering. This chapter introduces left-handedness to the interacting waves and discusses its impact in that field. In ordinary, that is, right-handed (RH) media, the wave vector directs to the same direction as the Poynting vector, so that the phase and power move to a common direction. In left-handed (LH) media, the situation is reversed.

To achieve strong resonant interactions, frequencies and wave numbers must be preserved. For example, when a wave of frequency ω_1 and wave number k_1 interacts with one of ω_3 and k_3, a significant third wave of ω_2 and k_2 will be generated if the conditions $\omega_1 + \omega_3 = \omega_2$ and $k_1 + k_3 = k_2$ are satisfied. The sum of the wavenumber decreases for the head-on collision, because the wave vector of the left-moving wave has the opposite sign as that of the right-moving one. In contrast, the sum frequency increases. The phase and power are transferred with the phase and group velocities, respectively. In addition, the group velocity is given by the slope of the dispersion curve, so that the frequency increases at least locally as the wavenumber decreases in LH media, so that they can satisfy the resonant conditions for head-on colliding waves.

To investigate resonances involving LH waves, we introduce nonlinearity to composite right- and left-handed (CRLH) transmission lines. CRLH transmission lines have been investigated in electrical engineering community as the practical and broadband platform to support LH waves [1–4]. The line has noteworthy dispersive property that the propagating wave exhibits LH (RH) properties when its carrier frequency is greater (less) than the line's characteristic frequencies. Furthermore, several activities have clarified the wave dynamics in CRLH lines with nonlinearity introduced by voltage-controlled devices [5–13]. In our case, the shunt capacitor each cell of a CRLH line contains is replaced with the Schottky varactor [10, 14]. The three-wave resonant interaction (3WRI) equations have been derived from the transmission equations of that nonlinear CRLH line via the derivative expansion method and is used to characterize the head-on collision of LH waves.

Even when $\omega_1 = \omega_3$ and $k_1 = k_3$, the significant energy is transferred from the fundamental to the second harmonic when the conditions $\omega_2 = 2\omega_1$ and $k_2 = 2k_1$ are satisfied. This process, termed harmonic resonance, is a special case of the three-wave resonant interaction, resulting from resonance of two identical waves. The dispersion of a nonlinear CRLH line can cause harmonic resonance for the LH fundamental and RH second harmonic waves. The phase of the LH fundamental wave advances toward the input end. Accordingly, that of the second harmonic wave should also move to the input. Because the fundamental wave increases to that direction, the harmonic resonance generates the second harmonic wave more when it travels longer. The generation efficiency of the second harmonic waves becomes enhanced through this behavior via supplemental cavity resonance. It should be noted that the fundamental wave is spontaneously converted into its second harmonic one without the aid of pump waves.

Similar spontaneous resonant interaction is expected in nonlinear CRLH lines. The soliton decay is realized for three waves having different group velocities. It requires the situation where the wave having the middle group velocity is incident to the line. Then, a soliton contained in the incident wave decays into the fast and slow solitons spontaneously. Inevitably, the slow soliton(s) occupies the LH branch for the nonlinear CRLH line; therefore, it starts to travel to the opposite direction to the incident and fast solitons, leading to the shortening of the fast soliton. By solving the eigenvalue problem of the Zakharov-Shabat (ZS) equation relating with the 3WRI equation, it is found that the fast soliton can become shorter for longer incident wave. Through these observations, we can utilize the soliton decay in the nonlinear CRLH line for generating broadband envelope pulses.

The use of nonlinear CRLH lines is sometimes limited because of wave attenuation caused by finite electrode resistance and substrate current leakage. In order to achieve loss compensation, a traveling-wave field-effect transistor (TWFET) is considered [15]. For the voltage waves

http://dx.doi.org/10.5772/intechopen.70739

traveling over FET electrodes, two CRLH lattices are required, which are, respectively, loaded with the gate and drain in each cell. The unit-cell FET can be biased via the LH inductors. In addition, the inter-cell direct current flow is cut off by the LH capacitors. The device introduces LC resonant pairs in each cell, which can operate as nonlinear oscillators with the aid of FET gain; therefore, the device can be considered as a kind of spatially extended oscillator systems. Hereafter, we call the device as the CRLH-TWFET. In the case of supercritical Andronov-Hopf bifurcation, the oscillation amplitude gradually increases when the bifurcation parameter passes a critical value. Then, the relaxation time needed to initiate autonomous oscillation becomes sufficiently large; therefore, it succeeds in effectively suppressing autonomous oscillation to guarantee the loss-compensated propagation of LH pulse waves. On the other hand, the amplitude grows to become discontinuously finite in subcritical cases, where the system affords the coexistence of an oscillatory region with a nonoscillatory region in addition to the homogeneous oscillatory state [16]. The resulting coherent structures function as the building blocks of the spatiotemporal patterns appearing in the system. When both boundaries at the ends of the oscillatory region preserve their relative positions, the oscillatory region preserving this envelope is called a pulse. Possibly, the boundary velocity vanishes, so that the pulse becomes localized and stationary [17, 18]. From the scientific viewpoint, a convenient electronic system to support such solitary waves is valuable for clarifying their interacting dynamics using either numerical or experimental method.

After describing the structure and dispersive properties of the nonlinear CRLH line, the head-on collision of envelope pulses is characterized numerically on that line to illustrate significant generation of harmonic waves through resonances. Next, the process is quantified by the 3WRI equations derived by applying the derivative expansion method to the transmission equations of a nonlinear CRLH line. Subsequently, two spontaneous resonant interactions: harmonic resonance and soliton decay are characterized, where the same 3WRI equations are used to model the wave dynamics. Finally, the development of a stationary pulse in a CRLH-TWFET is discussed.

2. Fundamental properties of nonlinear CRLH TLs

Because the nonlinear electrical lattice we investigate is based on CRLH lines, we first describe their fundamental properties. The unit-cell structure is shown at the top of **Figure 1(a)**, where C_R, L_R, C_L, and L_L represent the shunt capacitor, series inductor, series capacitor, and shunt inductor, respectively. It is shown that two different frequencies are allowed to be supported on the line for a wavenumber k. As shown below, the high frequency mode exhibits a RH property and the low frequency one becomes left-handed; therefore, we denote the dispersion relationships of the two as $\omega = \omega_{\mathrm{RH,LH}}(k)$ (ω_{RH} is for the RH and ω_{LH} for LH). Under the sixth order long wavelength approximation, these two are explicitly given by

$$\omega_{RH}(k) = \sqrt{\omega_x^2(k) + \sqrt{\omega_x^4(k) - \frac{1}{C_L C_0 L_L L_R}}}, \tag{1}$$

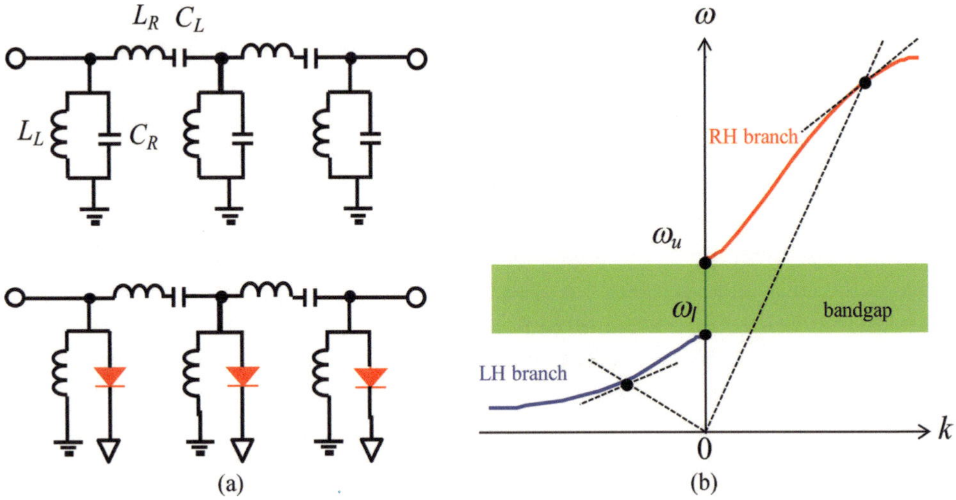

Figure 1. Structure of nonlinear CRLH lines. (a) The cell structures of linear (upper) and nonlinear (lower) CRLH lines and (b) the dispersion curve of CRLH lines.

$$\omega_{LH}(k) = \sqrt{\omega_x^2(k) - \sqrt{\omega_x^4(k) - \frac{1}{C_L C_0 L_L L_R}}}, \tag{2}$$

where $\omega_x(k)$ is defined as

$$\omega_x(k) = \sqrt{\frac{k^6}{720 C_0 L_R} - \frac{k^4}{24 C_0 L_R} + \frac{k^2}{2 C_0 L_R} + \frac{1}{2 C_0 L_L} + \frac{1}{2 C_L L_R}}. \tag{3}$$

Furthermore, $V_g(k)$ represents the group velocity of the line explicitly given by

$$V_g(k) = \frac{k(k^4 - 20k^2 + 120)}{240 C_0 L_R} \frac{\omega(k)}{\omega_x^2(k) - \omega^2(k)}, \tag{4}$$

where $\omega = \omega_{LH}(k)$ for the LH branch and $\omega = \omega_{RH}(k)$ for the RH branch. Typical behavior of $\omega(k)$ is shown in **Figure 1(b)**. There are two essential frequencies that characterize the lines' dispersive nature ω_{se} and ω_{sh} defined by $1/\sqrt{C_L L_R}$ and$1/\sqrt{C_0 L_L}$, respectively. It is found that the line exhibits a LH property at frequencies lower than $\omega_l \equiv \min(\omega_{se}, \omega_{sh})$ and an ordinary RH property at frequencies higher than $\omega_u \equiv \max(\omega_{se}, \omega_{sh})$. When $\omega_{se} = \omega_{sh}$, the LH branch is continuously connected with the RH one, and the line is called *balanced*. On the other hand, when ω_{se} is not coincident with ω_{sh}, a stop band, where all supporting modes become evanescent, appears between ω_l and ω_u, and the line is called *unbalanced*. One of the noteworthy properties of LH waves is that the wavelength becomes longer as the frequency increases at least locally. In addition, the envelop wave (accordingly, the power) moves to the different direction from its carrier wave, because $V_g(k)$ has the opposite sign to the phase velocity.

To introduce nonlinearity, we employ the Schottky varactor in place of C_R as shown at the bottom of **Figure 1(a)**. The Schottky varactor is a special type of a diode, whose capacitance is varied by the terminal voltage that biases reversely. In general, its capacitance voltage relationship is modeled as

$$C(V) = C_0\left(1 + \frac{V_0}{V_J}\right)^m \left(1 + \frac{V}{V_J}\right)^{-m}, \tag{5}$$

where C_0, V_J, and m are the zero-bias junction capacitance, junction potential, and grading coefficient, respectively. In addition, the cathode of the Schottky varactor is biased at V_0. Using this representation, the transmission equations are given by

$$L_R \frac{d^2 I_n}{dt^2} = -\frac{I_n}{C_L} - \frac{d}{dt}(V_n - V_{n-1}), \tag{6}$$

$$C_R \frac{d^2 V_n}{dt^2} = -\frac{V_n}{L_L} + \frac{d}{dt}(I_n - I_{n+1}) - \frac{dC_R}{dV}\left(\frac{dV_n}{dt}\right)^2, \tag{7}$$

where I_n and V_n are the current and voltage at the nth cell, respectively.

3. Head-on collision of LH waves

It is well known that the efficiency of resonant interactions between two waves is maximized, when the phase-matching condition: $k_2 = m_1 k_1 + m_3 k_3$, $\omega_2 = m_1\omega_1 + m_3\omega_3$, where $k_{1,3}$ and $\omega_{1,3}$ represent the wavenumbers and angular frequencies of interacting waves, and k_2 and ω_2 represents those of the wave generated by the interaction. Moreover, $m_{1,3}$ are integers that are specified by the order of the generated harmonics. When the incident pulses have a common carrier frequency and are traveling in opposite directions, it results in the condition $k_1 = -k_3$. Hence, the maximal second harmonic generation can be observed when $k_2 = 0$. Similarly, for the third harmonic generation, k_2 has to be close to k_1. For RH waves, the higher the frequency, the shorter the wavelength; therefore, it is impossible to satisfy this condition. On the other hand, when the carrier frequencies of the interacting waves are both set to $\omega_l/2$, any CRLH lines can generate second harmonic waves effectively via head-on collisions because the second harmonic frequency ω_l corresponds to zero wavenumber. **Figure 2** shows the head-on collision of envelop pulses whose carrier frequencies correspond to $\omega_l/2$ (=1.6 GHz). To obtain **Figure 2**, we set C_0, C_L, L_R, and L_L to 1.0 pF, 1.0 pF, 2.5 nH, and 2.5 nH, respectively, so that the line becomes balanced with $\omega_u = \omega_l$ = 3.2 GHz. In **Figure 2(a)**, the dispersion curve is shown, where P_1, P_2, P_3, and P_4 represent the positions on the dispersive curve the fundamental, second, third, and fourth harmonic waves occupy, respectively. Note that the wavenumber at P_2 is equal to zero, and either P_3 or P_4 exhibits coincident wavenumber with that of P_1. **Figure 2(b)** shows the calculated waveforms, where six spatial waveforms are recorded in 60-ns increments.

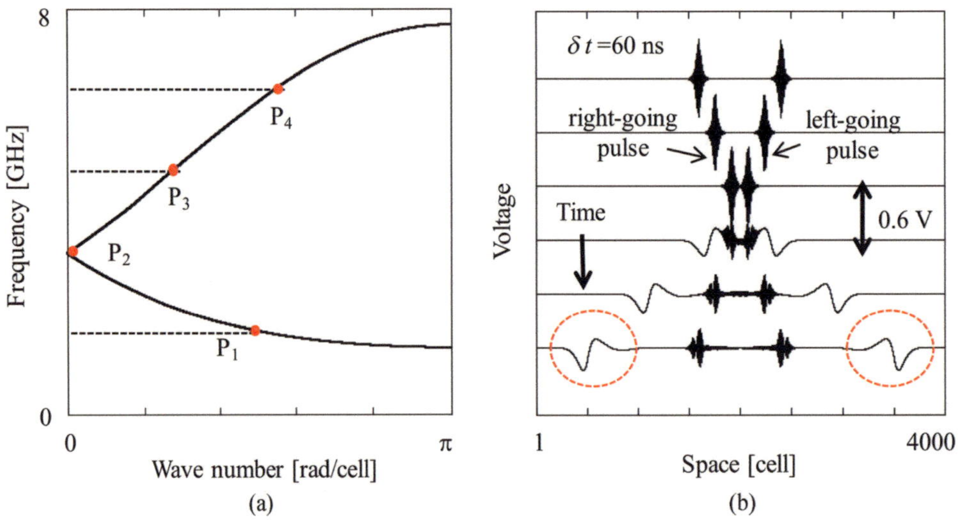

Figure 2. Second harmonic generation via head-on collision of LH waves. (a) The dispersive properties of interacting waves and (b) the numerically obtained time-domain waveforms.

Long wavelength envelope pulses result from the head-on collision as indicated by red circles. Another example is shown in **Figure 3**. The carrier frequency of the colliding pulses is set to 1.9 GHz, such that the wavenumber of the second harmonic becomes nonzero, and the wavenumber of the third harmonic becomes close to that of the fundamental wave; therefore, the resonance conditions can be satisfied for $(m_1, m_3) = (1, 2)$ and/or (2, 1). As expected, we can see that the wavelengths of the collision-induced pulses are comparable to that of the incident ones in **Figure 3(b)**. Actually, the spectral peak of the collision-induced pulses is located at

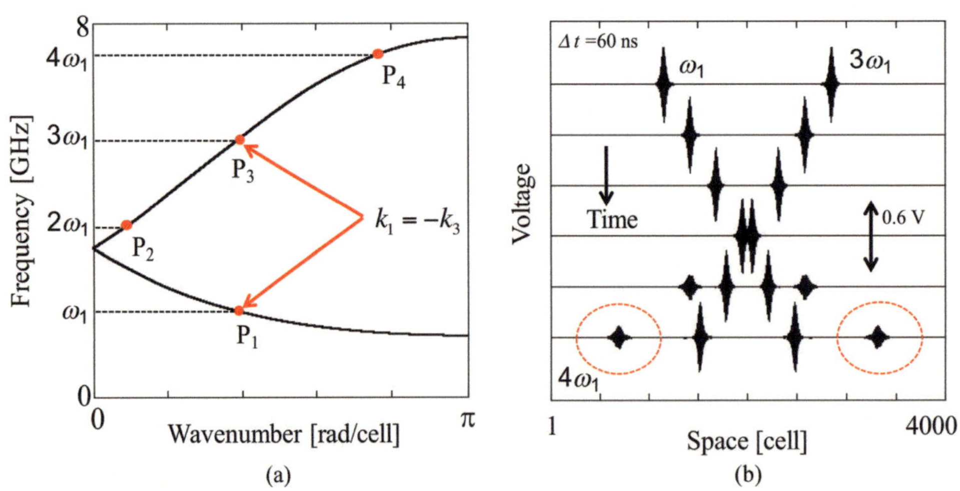

Figure 3. Fourth harmonic generation via head-on collision of LH waves. (a) The dispersive properties of interacting waves and (b) the numerically obtained time-domain waveforms.

5.6 GHz, being close to the third harmonic. Note that P_3 occupies the RH branch, so that the LH waves are converted into the RH ones through resonances.

The resonance is briefly discussed for the two colliding pulses having different carrier frequencies [19]. Let the carrier frequency of the left (right)-moving pulse denote as $\omega_{1(2)}$. Then, we set ω_1 slightly higher than $\omega_l/2$, while ω_2 is fixed at $\omega_l/2$. The resulting amplitude of the wavenumber of the right-moving pulse surpasses that of the left-moving one. Both of incident pulses exhibit left-handedness; therefore, the wave vector directs to the left for the second harmonic wave. Because the second harmonic wave is carried by the RH mode, the collision-induced envelope pulse moves to the left. Similarly, only the right-moving envelope pulse develops, if ω_2 is set slightly higher than $\omega_l/2$, while ω_1 is fixed at $\omega_l/2$. These expectations were validated experimentally using bread-boarded test circuit [20].

In the next section, the evolution equations of the envelope functions of the incident and collision-induced pulses are obtained by the application of the derivative expansion method to the transmission equation of a nonlinear CRLH line [21]. In particular, the generation efficiency of the second-harmonic wave is formulated for the case when the left- and right-moving pulses have a common frequency and wavelength.

4. Three-wave mixing of LH waves

In the present study, we consider the case where the pulse spreads over many cells, and the lattice is regarded as being homogeneous, such that the discrete spatial coordinate n can be replaced by a continuous one x. Then, by series-expanding Eqs. (6) and (7), the evolution equation of the continuous counterpart of the line voltage $\psi = \psi(x, t)$ is given by

$$C_R C_L L_R L_L \frac{\partial^4 \Psi}{\partial t^4} + 4C_L L_R L_L \frac{dC_R}{dV} \frac{\partial \Psi}{\partial t} \frac{\partial^3 \Psi}{\partial t^3} + 3C_L L_R L_L \frac{dC_R}{dV} \left(\frac{\partial^2 \Psi}{\partial t^2}\right)^2 + 6C_L L_R L_L \frac{d^2 C_R}{dV^2} \left(\frac{\partial \Psi}{\partial t}\right)^2 \frac{\partial^2 \Psi}{\partial t^2} + C_L L_R L_L \frac{d^3 C_R}{dV^3} \left(\frac{\partial \Psi}{\partial t}\right)^4 + (C_L L_R + C_R L_L) \frac{\partial^2 \Psi}{\partial t^2} + \Psi - C_L L_L \frac{\partial^4}{\partial t^2 \partial x^2} \left(\Psi + \frac{1}{12} \frac{\partial^2 \Psi}{\partial x^2} + \frac{1}{360} \frac{\partial^4 \Psi}{\partial x^4}\right) = 0, \tag{8}$$

where $C_R = C(\psi - V_0)$. To quantify the resonant nonlinear processes in a nonlinear CRLH line, we apply the derivative expansion method [22] to that evolution equation. It leads to the evolution equations of envelop functions of the involved waves. We first expand the spatial and temporal derivatives as

$$\frac{\partial}{\partial x} = \frac{\partial}{\partial x_0} + \epsilon \frac{\partial}{\partial x_1} + \epsilon^2 \frac{\partial}{\partial x_2} + \cdots, \tag{9}$$

$$\frac{\partial}{\partial t} = \frac{\partial}{\partial t_0} + \epsilon \frac{\partial}{\partial t_1} + \epsilon^2 \frac{\partial}{\partial t_2} + \cdots, \tag{10}$$

for $\varepsilon \ll 1$. For describing the three-wave mixing process of two waves having a wave number of k_1 and k_3, then the wave number of the resulting wave k_2 satisfies the condition $k_2 = k_1 + k_3$.

As mentioned above, for efficient three-wave mixing, the frequencies must satisfy the resonant condition, that is, $\omega(k_2) = \omega(k_1) + \omega(k_3)$. The voltage variable is then assumed to have a form of

$$\Psi(x,t) = \sum_{j=1}^{3} A_j e^{i(\omega_j t_0 - k_j x_0)} + c.c., \tag{11}$$

where $\omega_j \equiv \omega(k_j)$ and A_i denotes the envelope function of variables x_1, x_2,... and t_1, t_2,... Substituting Eq. (11) into Eq. (8), the terms proportional to $e^{i(k_j x_0 - \omega_j t_0)}$ ($j = 1, 2, 3$) of each order of ε are collected to be vanished. From $O(\varepsilon^2)$ terms, the evolution equations of envelope functions are governed by the 3WRI equations given by

$$\frac{\partial A_j}{\partial t} + V_g(k_j)\frac{\partial A_j}{\partial x} = G_j A^*_{j+1} A^*_{j+2}, \tag{12}$$

where $j = 1, 2, 3$, mod 3, and the coupling coefficients are given by

$$G_j = \frac{-i180m\omega_j C_0 L_L\left(-1 + C_L L_R \omega_j^2\right)}{(V_0 + V_J)\left\{-C_L L_L k_j^2\left(360 - 30k_j^2 + k_j^4\right) - 360C_L L_R + 360C_0 L_L\left(-1 + 2C_L L_R \omega_j^2\right)\right\}}, \tag{13}$$

whose denominator becomes zero only at $\omega_2 = (C_0 L_R C_L L_L)^{\frac{1}{4}} \in (\omega_l, \omega_u)$ so that G_2 does not exhibit any diverging behavior for frequencies in either the RH or LH branches. In particular, the head-on collision of two envelope pulses having common wavenumber, there are two cases $\omega_2 = \omega_{LH}(0)$,

$$G_2 = \begin{cases} -i\dfrac{m}{2(V_0 + V_J)}\dfrac{1}{\sqrt{C_0 L_L}}, & \omega_{se} > \omega_{sh}, \\ 0, & \omega_{se} < \omega_{sh}. \end{cases} \tag{14}$$

For $\omega_2 = \omega_{RH}(0)$

$$G_2 = \begin{cases} 0, & \omega_{se} > \omega_{sh}, \\ -i\dfrac{m}{2(V_0 + V_J)}\dfrac{1}{\sqrt{C_0 L_L}}, & \omega_{se} < \omega_{sh}. \end{cases} \tag{15}$$

In summary, the value of G_2 becomes finite only when the second harmonic frequency is matched to ω_{sh}. In contrast, for a balanced CRLH line,

$$G_2 = -i\frac{m}{4(V_0 + V_J)}\frac{1}{\sqrt{C_0 L_L}}. \tag{16}$$

Based on this G_2 property, a scheme can be proposed for converting the carrier frequency of the incident pulsed wave into its second-harmonic wave without deteriorating pulse duration. **Figure 4(a)** shows the circuit configuration of the generator creating the pulsed second harmonic waves. The nonlinear CRLH line is divided into two segments. The first and second

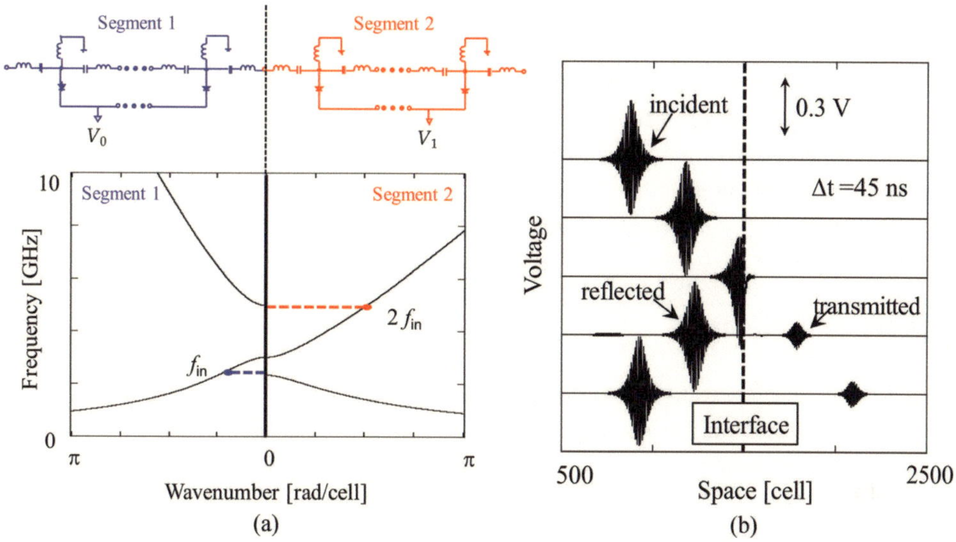

Figure 4. Effective method of second harmonic generation. (a) The device structure (upper), dispersive property of each segment (lower), and (b) the numerically obtained time-domain waveforms.

L_R (nH)	2.8	C_L (pF)	1.0	L_L (nH)	2.5
C_0 (pF)	1.0	V_J (V)	2.0	m	2.0

Table 1. Parameter values used to obtain **Figure 4**.

segments are represented by black and grey elements, respectively. The line parameter values used in the present demonstration are listed in **Table 1**. The biasing voltage to shunt varactors is the unique difference between the segments, which are labeled as V_0 and V_1 for the first and second segments, respectively. Increasing V_0 decreases the capacitance of the Schottky varactors and then increases ω_{sh}. The first segment is then arranged for V_0 to be sufficiently large to satisfy the condition $\omega_{sh} > \omega_{se}$. An envelope pulse, whose carrier frequency fin is half as high as $\omega_{sh}/2\Pi$, is then inputted to the first segment. In contrast, V_1 is set to be small in order to lower ω_{sh} such that the stop band includes f_{in}. The typical dispersion that the segments must have is shown in **Figure 4(b)**. Here, V_0 and V_1 are set to 2.7 and 0.2 V, respectively. The left- and right-side dispersion curves are for the first and second segments, respectively. The incident pulse cannot be transmitted into the second segment because f_1 is designed to be in the stop band. It is then reflected at the interface. The reflected pulse interacts with the incident pulse in the same manner as the oppositely traveling pulse. The condition $\omega_{sh} > \omega_{se}$ guarantees that G_1 becomes finite. Consequently, the second-harmonic wave develops in the first segment at the vicinity of the segments interface. Because the group velocity at $\omega_{RH}(0)$ is zero in the first segment, the second-harmonic wave remains around the interface. This stationary oscillation is partially transmitted into the second segment, resulting in the pulsed second harmonic wave moving to the right on the second segment. The second harmonic pulse is uniquely obtained at the end of the second segment. **Figure 4(c)** shows the numerically obtained evolution of a single soliton having a carrier frequency of f_{in}. Five spatial waveforms recorded at 45 ns intervals are plotted. We can observe that the right-moving incident pulsed wave is reflected at interface *P*, and a small envelope pulse is transmitted into the second segment. The transmitted pulse has only one-fifth the amplitude of the incident pulse; however, it preserves pulse shape and successfully doubles its carrier frequency.

5. Harmonic resonance

In this section, we investigate harmonic resonance in a nonlinear CRLH line [23]. As discussed in Section 1, the harmonic resonance becomes significant when the phase velocities of the fundamental and second harmonic waves are coincident. **Figure 5(a)** shows the typical dispersion of a CRLH line, where L_R, L_L, C_L, and C_0 are set to 2.5 nH, 2.5 nH, 1.0 pF, and 0.6 pF, respectively. For convenience, we also define $\alpha \equiv C_L L_R / C_0 L_L$. Notice that the line is balanced when $\alpha = 1.0$. Two points P_1 and P_2 in **Figure 5(a)** correspond to the fundamental and second harmonic waves, respectively, for significant harmonic resonance. Both points are placed on a common line passing through the origin, so that the second harmonic wave has the same phase velocity as the fundamental. With k_f and ω_f as the wave number and angular frequency of the fundamental wave, harmonic resonance becomes eminent when the second harmonic wave satisfies the two conditions $k_s = 2k_f$ and $\omega_s = 2\omega_f$, where k_s and ω_s represent the wave number and angular frequency of the second harmonic wave, respectively. The second harmonic wave must occupy the RH branch. Thus, the latter condition is more precisely written as $\omega_{RH}(2k_f) = 2\ \omega_{LH}(k_f)$. Note that both P_1 and P_2 exhibit relatively small wave numbers; the

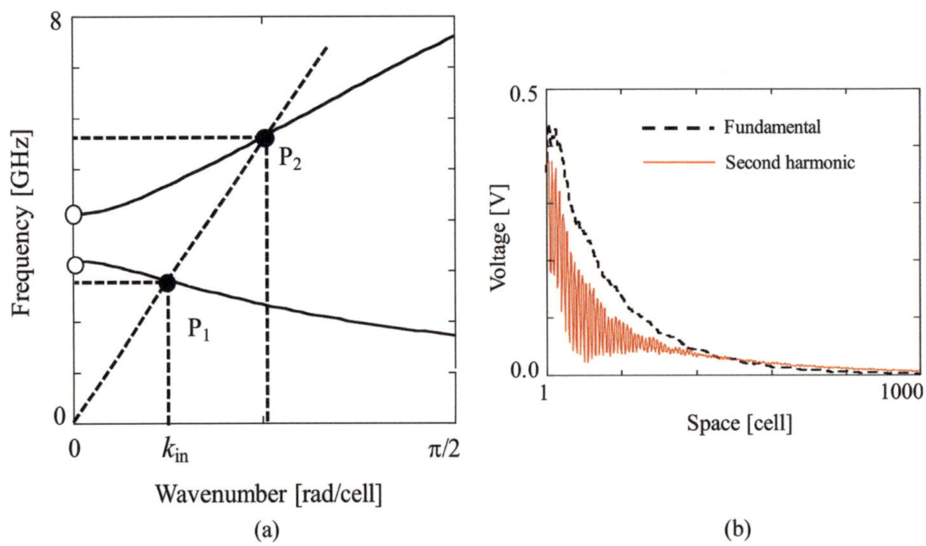

Figure 5. Harmonic resonance in nonlinear CRLH lines. (a) The operating points in dispersion curve and (b) the steady-state voltage profiles of fundamental and second harmonic waves.

second-order long-wavelength approximation suffices to describe the processes involved; therefore, the equation $\omega_{RH}(2k_f) = 2\,\omega_{LH}(k_f)$ is explicitly solved for k_f to give

$$k_f = \frac{1}{2}\sqrt{\frac{C_0}{5C_L}}\sqrt{\frac{-4\alpha^2+17\alpha-4}{\alpha+1}}, \tag{17}$$

$$\omega_f = \sqrt{\frac{5}{4C_0L_L}}\sqrt{\frac{1}{\alpha+1}}. \tag{18}$$

Note that α must be in (1/4, 4) for the real k_f. The fundamental and second harmonic waves are then shown to have the characteristic impedance $Z_f = \sqrt{L_L/C_L}\sqrt{(4-\alpha)/(4\alpha-1)}$ and $Z_s = \sqrt{L_L/C_L}\sqrt{(4\alpha-1)/(4-\alpha)}$, respectively. Note that $Z_f = Z_s$ at $\alpha = 1.0$. According to the derivative expansion method mentioned above, the 3WRI equations that describe the fundamental and second harmonic envelope functions are described as

$$\frac{\partial A_f}{\partial t_1} + v_{gf}\frac{\partial A_f}{\partial x_1} = i\rho_f A_s A_f^* + \gamma_f A_f, \tag{19}$$

$$\frac{\partial A_s}{\partial t_1} + v_{gs}\frac{\partial A_s}{\partial x_1} = i\rho_s A_f^2 + \gamma_s A_s, \tag{20}$$

where v_{gf} and v_{gs} are the group velocities of the fundamental and second harmonic waves, respectively, explicitly given by

$$v_{gf} = -\frac{1}{\sqrt{C_0L_R}}\frac{5\sqrt{\alpha(4\alpha-1)(4-\alpha)}}{16\alpha^2+7\alpha+16}, \tag{21}$$

$$v_{gs} = \frac{1}{\sqrt{C_0L_R}}\frac{5\sqrt{\alpha(4\alpha-1)(4-\alpha)}}{-\alpha^2+23\alpha-1}. \tag{22}$$

Note that v_{gf} becomes negative because the fundamental wave is left-handed. The strength of harmonic resonance is determined by the coupling coefficients $\rho_{f,\,s}$. Because of the term A_f^2, the fundamental wave is spontaneously converted into the second harmonic. Ordinarily, the product $\rho_f\rho_s$ is negative, so that the increase of A_s results in the reduction of A_f. This negative feedback stabilizes both waves. On the other hand, the coupling coefficients are presently given by

$$\rho_f = \frac{5\sqrt{5}m}{4(V_0+V_J)\sqrt{C_0L_L}}\frac{4-\alpha}{16\alpha^2+7\alpha+16}\sqrt{\frac{1}{\alpha+1}}, \tag{23}$$

$$\rho_s = \frac{5\sqrt{5}m}{4(V_0+V_J)\sqrt{C_0L_L}}\frac{4\alpha-1}{-\alpha^2+23\alpha-1}\sqrt{\frac{1}{\alpha+1}}. \tag{24}$$

Both ρ_f and ρ_s are then shown to be positive for $\alpha \in (1/4, 4)$, such that the developing A_s enhances A_f. The second harmonic envelope wave travels backward because the phase of the fundamental wave travels in the opposite direction to its envelope. This means that the amplitude of both the fundamental and second harmonic waves increases as the phase advances. **Figure 5(b)** demonstrates the principle of operation, where the numerically obtained steady-state profile of the voltage envelopes of the fundamental and second harmonic waves. The cell number is set to 2000. In addition, the input and output impedances are set to the characteristic impedances of the second harmonic and fundamental waves, respectively. The second harmonic wave generated by the harmonic resonance should travel to the input end. The reflection of the second harmonic wave at the input end was suppressed via the matched impedance, so the effect of the fundamental's left-handedness on the profile of the second harmonic could be seen. Small line resistors were used to suppress multiple reflections. In addition, α and λ_f were set to 1.5 and 20 cells, respectively. We applied a 0.5-V sinusoidal voltage at the left end (f_f = 1.0 GHz). Through Fourier transformation, filtering, and inverse transformation the calculated spatial voltages are separated into each wave component. The second harmonic wave was superposed in-phase and gained amplitude in the direction to the input end, as clearly shown in **Figure 5(b)**.

By setting f_0 and Z_{in} to f_f and Z_f, respectively, we achieve effective second harmonic generation. By the matched impedances, the fundamental waves can travel along the line without reflections at the ends. On the other hand, the second harmonic wave begins to travel to the input (left) end and is reflected significantly in a line that satisfies the condition $Z_s >> Z_f$. The load impedance also differs from Z_s, such that the second harmonic wave exhibits multiple reflections. Hence, the second harmonic wave becomes resonant in cavity when the cell size of the line is an integer multiple of $\lambda_f/2$, as illustrated in **Figure 6**. This cavity resonance makes the nonlinear CRLH line become an effective platform for second harmonic wave generation together with the above-mentioned positive feedback.

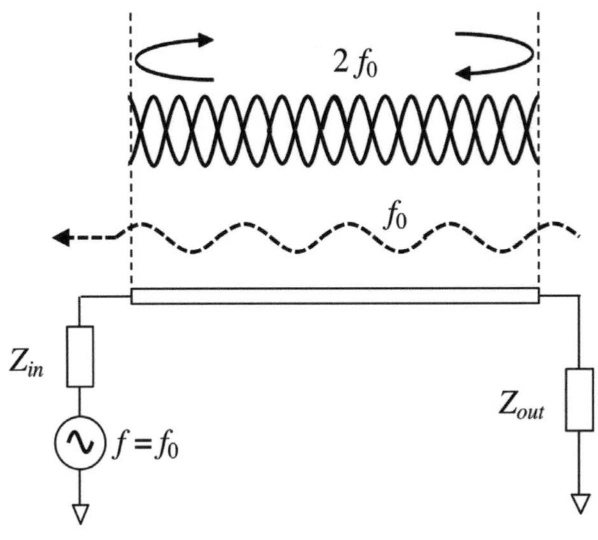

Figure 6. Practical structure for second harmonic generation using harmonic resonance.

6. Soliton decay

To describe the soliton decay in a nonlinear CRLH line, we again consider the 3WRI equations of a nonlinear CRLH line. By introducing $Q_j = i\sqrt{|G_{j+1}||G_{j+2}|}A_j$, Eq. (12) is transformed into the standard 3WRI equation, that is,

$$\frac{\partial Q_j}{\partial t} + V_g(k_j)\frac{\partial Q_j}{\partial x} = \gamma_j Q_{j+1}^* Q_{j+2}^*, \tag{25}$$

where $\gamma_{1,3} = 1$ and $\gamma_2 = -1$. In what follows, an envelope having a carrier frequency of ω_j is called ω_j-envelope for brevity. When a ω_2-envelope is uniquely applied to the line and the group velocities satisfy $V_g(k_1) < V_g(k_2) < V_g(k_3)$, its evolution is predicted by solving the eigenvalue problem of the following ZS equation in the framework of the inverse scattering transform:

$$\frac{\partial u_1}{\partial x} + i\lambda u_1 = q u_2, \tag{26}$$

$$\frac{\partial u_2}{\partial x} + i\lambda u_2 = -q u_1, \tag{27}$$

where λ and $(u_1, u_2)^{\mathrm{T}}$ are the eigenvalue and corresponding eigenvector, respectively [24, 25]. In addition, $q = q(x)$ is defined by

$$\mathrm{q}(x) = -\frac{Q_2^{(0)}(x)}{\sqrt{\left(V_g(k_2) - V_g(k_1)\right)\left(V_g(k_3) - V_g(k_1)\right)}}, \tag{28}$$

for the spatial waveform $Q_2^{(0)}(x)$ of the incident ω_2-envelope. The stability of the ω_1- or ω_3-envelope solitons is shown to be secured, that is, the original envelopes never lose the solitons, while the ω_2-envelope solitons are always unstable, which decay into both the slow and fast envelope ones. The latter phenomenon is called soliton decay. When $Q_2^{(0)}(x)$ evolves into N solitons, the ZS equation must have N pure imaginary eigenvalues in the upper half plane, whose norms are inversely proportional to the spatial width of the corresponding soliton. Let $\lambda_m^{(2)}(m = 1, \cdots, \mathrm{N})$ be such eigenvalues of Eqs. (26) and (27). Then, it is shown that

$$\lambda_m^{(1)} = \frac{V_g(k_3) - V_g(k_2)}{V_g(k_3) - V_g(k_1)}\lambda_m^{(2)}, \tag{29}$$

$$\lambda_m^{(3)} = \left(1 - \frac{V_g(k_3) - V_g(k_2)}{V_g(k_3) - V_g(k_1)}\right)\lambda_m^{(2)}, \tag{30}$$

where $\lambda_m^{(j)}(j = 1, 3)$ defines the eigenvalue corresponding to the soliton in the ω_j-envelope resulting from the decay of the soliton in the ω_2-envelope corresponding to $\lambda_m^{(2)}$. For example,

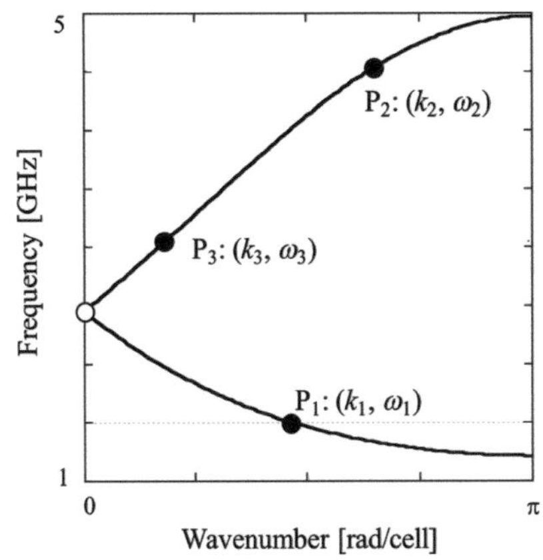

Figure 7. Dispersive properties of waves involved by soliton decay.

the line can be designed to exhibit dispersive property shown in **Figure 7**, where the incident envelope occupies the region in the neighborhood of P_2. Then, due to the resonant conditions, $\omega_{1,3}$-envelope is shown to be around $P_{1,3}$ uniquely. Notice that group velocities satisfy $V_g(k_1) < V_g(k_2) < V_g(k_3)$ and P_1 is on the LH branch. Due to the negative $V_g(k_1)$, $\lambda_m^{(1)}$ takes a small value, while $\lambda_m^{(3)}$ becomes rather large. As a result, the solitons in ω_1-envelope start to travel backward with a relatively wide width. Conversely, the ω_3-solitons become short.

We validate the analysis with the numerical integration of Eqs. (6) and (7). The line is designed to be balanced by setting C_L, L_L, C_0, and L_R to 1.0 pF, 2.5 nH, 1.69 C_L, and 1.69 L_L, respectively. In addition, m, V_J, V_0, and ω_2 are set to 2.0, 2.0 V, 1.0 V, 4.54 GHz, respectively. **Figure 8(a)**

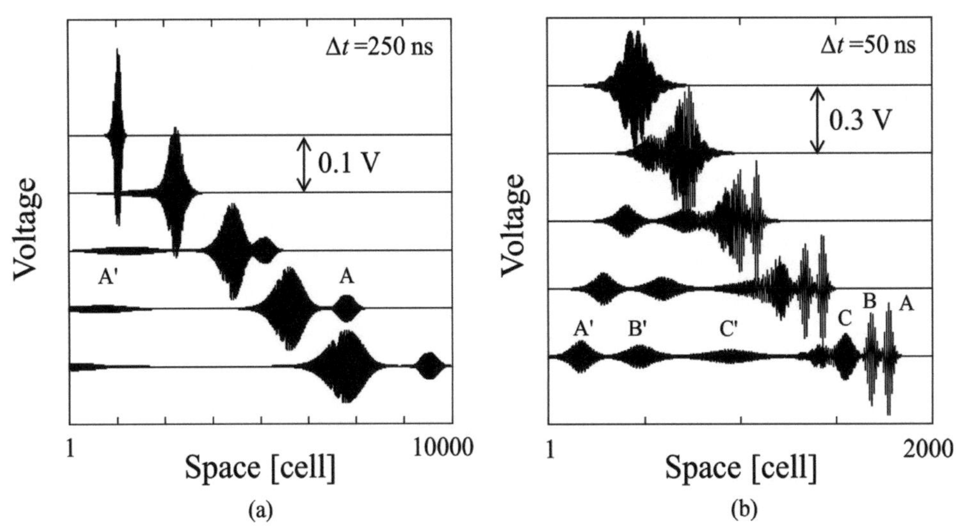

Figure 8. Numerically obtained waveforms exhibiting soliton decay. The dynamics are shown for (a) short and (b) wide envelope pulse incidences.

shows calculated waveforms on the line, where five spatial waveforms recorded at 250-ns increments are plotted. A 0.25 V hyperbolic secant envelope with 3.5-ns duration is applied at the left end. The incident ω_2-envelope decays into a unique pair of the fast and slow solitons, which are labeled at the fourth waveform as A and A', respectively. The duration of the incident ω_2-envelope is varied to be 10.5 ns in **Figure 8(b)**. Three times wider pulse is inputted for **Figure 8(b)** than one for **Figure 8(a)**. The incident ω_2-envelope decays into three pairs of the fast and slow solitons, which are labeled as (A, A′), (B, B′), and (C, C′). As expected, the widths of the emitted solitons become narrower in **Figure 8(b)** than those in **Figure 8(a)**.

As a broadband pulse generator, it suffices for a nonlinear CRLH line to succeed in the emission of the first pair of solitons. To output the short envelope pulse uniquely, we only set up a band-pass filter extracting frequencies around ω_3 in the subsequent stage [26].

7. CRLH-TWFETs

Figure 9(a) shows the structure of a CRLH-TWFET. Two coupled transmission lines are periodically loaded with FETs in such a way that one of the lines is connected to the gate and the other to the drain [15]. The gate line consists of the series inductor, series capacitor, shunt inductor, and shunt varactor, whose values are respectively denoted as L_{Rg}, C_{Lg}, L_{Lg}, and the Schottky varactor modeled by Eq. (5) is assigned to C_{gs}, which is introduced to control bifurcation property of the line via V_{SD}. The biasing voltage V_{GG} is applied to each transistor through the shunt inductance. On the other hand, L_{Rd}, C_{Ld}, L_{Ld}, and C_{ds} configure the unit cell of the drain line. The biasing voltage V_{DD} is applied to the drain of each transistor through L_{Ld}. Each inductor has finite parasitic resistances, which are denoted as R_{Rg}, R_{Rd}, R_{Lg}, and R_{Ld} for L_{Rg}, L_{Rd}, L_{Lg}, and L_{Ld}, respectively. The gate and drain lines are coupled via the gate-drain capacitor denoted as C_{gd}. Because of the couplings, there are at most two different modes for each frequency. Moreover, the lowest and second lowest frequency modes exhibit a LH property, whereas the other two modes exhibit right-handedness.

As in the case of nonlinear CRLH lines, the device can generate long wavelength harmonic wave via head-on collision of LH waves. Interestingly, such collision-induced wave evolves to a stationary pulse. **Figure 9(b)** demonstrates that, for the varactor, m and V_J are set to 1.5 and 5.0 V, respectively. We then set C_0 to the value, for which C_{gs} becomes 140 pF at $V = V_0 = V_{GG}$. The other reactance values are listed in **Table 2**. In general, the resistances tend to be proportional to the corresponding inductances. V_{SD} is set to 18.0 V to guarantee subcritical bifurcation. The cell size is 500. Both ends are excited by a sech-shaped envelope pulse whose carrier frequency is 7.7 MHz. The inset of **Figure 9(b)** shows the steady-state profile of the stationary solitary wave, which has a flattop waveform with a width of 30 cells.

In practice, the line parameter values fluctuate, such that finite disorder is introduced to the lattice dynamics, which effectively serves the Pieres-Nabarro potential to the wave dynamics. When the pulse cannot overcome the potential, it is partially reflected to become a stationary pulse via resonance. Thus, the stationary pulse is expected to develop more frequently on the line when the fluctuation increases. To examine the property of the practical line, we fabricated

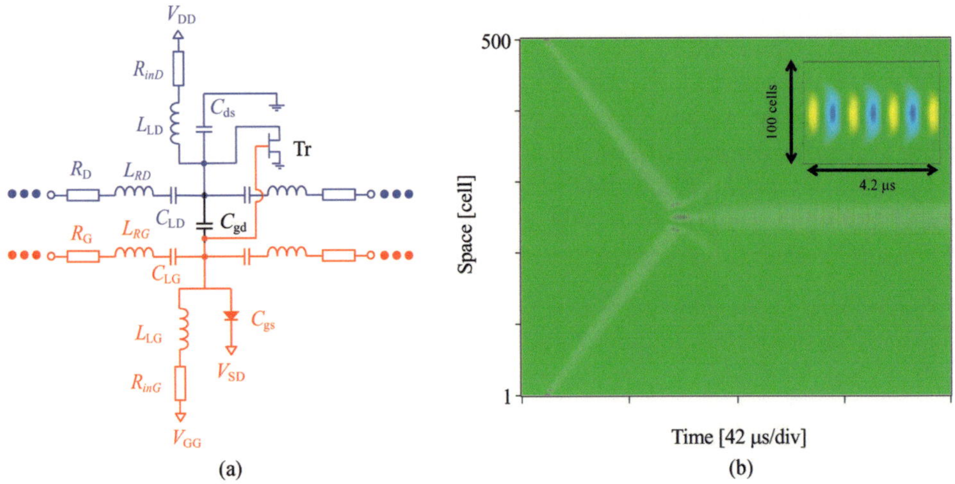

Figure 9. Head-on collision of envelop pulses in a TWFET. (a) The unit-cell structure and calculated spatiotemporal profile is shown in (b). No fluctuation of device parameter values is assumed.

C_{Lg} (pF)	22.0	C_{Ld} (pF)	22.0	L_{Lg} (μH)	10.0	L_{Ld} (μH)	4.7	L_{Rg} (μH)	4.7
L_{Rd} (μH)	10.0	R_{Lg} (Ω)	9.7	R_{Ld} (Ω)	4.5	R_{Rg} (Ω)	4.5	R_{Rd} (Ω)	9.7
C_{ds} (pF)	47.0	C_{gd} (pF)	13.0	C_{gs0} (pF)	137.0	V_J (V)	4.96	m	1.5

Table 2. Parameter values used to obtain **Figure 9(b)**.

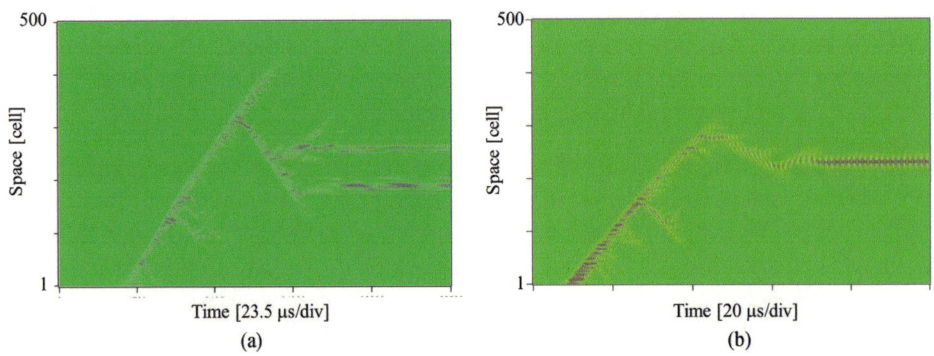

Figure 10. Envelope pulses in disordered lattice. The spatiotemporal voltage profile obtained by (a) the measurement and (b) calculation.

a test line on print circuit board. Actually, the parameter values used to obtain **Figure 9(b)** simulate those of the test line. **Figure 10(a)** shows the measured spatiotemporal voltage profile. A sech-shaped envelope pulse was inputted only at the near end. The pulse moving to the far end was significantly reflected near the 300th cell and two different stationary pulses developed after reflection. **Figure 10(b)** shows the calculated voltage profile to simulate the measured result, where the fluctuation has 7% standard deviation. The device fluctuation cannot be modeled exactly. However, it successfully demonstrates both the reflection and the development of a stationary pulse. With the balance between the dissipation and FET gain in a disordered lattice, resonant interactions lead to this interesting wave dynamics.

8. Conclusions

We first describe the three-wave mixing process in nonlinear CRLH lines. The head-on collision of LH waves results in a significant amount of harmonic waves, whose efficiency is accurately predicted by the asymptotic method.

The CRLH dispersion allows us two spontaneous resonant processes to generate harmonic waves: the harmonic resonance and soliton decay. The harmonic resonance in a nonlinear CRLH line succeeds in generating second-harmonic waves even under the presence of finite line resistance, when the line is designed for the second-harmonic waves to cause cavity resonance. The left-handedness of the fundamental wave guarantees that both the fundamental and second harmonic waves can gain amplitude as phase advances. The soliton decay in a nonlinear CRLH line gives the effective way for generating broadband envelope pulses. The incident envelope spontaneously emits several pairs of the fast and slow solitons. In general, slow solitons exhibit left-handedness to travel backward and their fast counterparts become shorter than the incident pulse. In addition, the wider the incident pulse, the narrower the fast solitons.

A CRLH-TWFET is shown to support stationary nonlinear oscillatory pulse waves, which is generated by the collision of two counter-moving waves through resonance. The presence of disorder helps the development of stationary pulses. The bias voltage of varactor in each cell can be set independently and control the position and number of such stationary pulses.

Author details

Koichi Narahara

Address all correspondence to: narahara@ele.kanagawa-it.ac.jp

Kanagawa Institute of Technology, Atsugi, Japan

References

[1] Caloz C, Itoh T. Electromagnetic Metamaterials: Transmission Line Theory and Microwave Applications. New York: Wiley; 2006

[2] Gomez-Diaz JS, Gupta S, Alvarez-Melcon A, Caloz C. Investigation on the phenomenology of impulse-regime metamaterial transmission lines. IEEE Transactions on Antennas and Propagation. 2009;**57**:4010-4014

[3] Chi P, Itoh T. Dispersion engineering with CRLH metamaterials. In: Proceedings of IEEE International Symposium on Radio-Frequency Integration Technology. 2009. pp. 128-131

[4] Zong B-F, Wang G, Liang J, Zhang X, Wang D. Wideband frequency-scanning phased-array feed network using novel composite right/left-handed unit cell. Electronics Letters. 2016;**52**:55-57

[5] Kozyrev AB, van der Weide DW. Nonlinear wave propagation phenomena in left-handed transmission-line media. IEEE Transactions on Microwave Theory and Techniques. 2005;**53**:238–245. DOI: 10.1109/TMTT.2004.839305

[6] Simion S, Marcelli R, Bartolucci G, Sajin G, Craciunoiu F. Nonlinear composite right/left-handed transmission line for frequency doubler and short pulse generation. In: Proc. Metamaterials 2008. 2008. pp. 492-494

[7] Gupta S, Caloz C. Dark and bright solitons in left-handed nonlinear transmission line metamaterials. In: Proceedings IEEE MTT-S International Microwave Symposium. 2007. pp. 979–982

[8] Kafaratzis A, Hu Z. Envelope solitons in nonlinear left handed transmission lines. In: Proc. Metamaterials 2007. 2007. pp. 771–773.

[9] Kozyrev AB, van der Werde DW. Trains of envelope solitons in nonlinear left-handed transmission line media. Applied Physics Letters. 2007;**91**:254111–254113.

[10] Ogasawara J, Narahara K. Short envelope pulse propagation in composite right- and left-handed transmission lines with regularly spaced Schottky varactors. IEICE Electronics Express. 2009;**6**:1576-1581. DOI: 10.1587/elex.6.1576

[11] Gharakhili FG, Shahabadi M, Hakkak M. Bright and dark soliton generation in a left-handed nonlinear transmission line with series nonlinear capacitors. Progress in Electromagnetics Research. 2009;**96**:237-249. DOI: 10.2528/PIER09080106

[12] Veldes GP, Cuevas J, Kevrekidis PG, Frantzeskakis DJ. Coupled backward- and forward-propagating solitons in a composite right- and left-handed transmission line. Physical Review E. 2013;**88**:013203-013216

[13] Powell DA, Shadrivov IV, Kivshar YS. Asymmetric parametric amplification in nonlinear left-handed transmission lines. Applied Physics Letters. 2009;**94**:084105-084107

[14] Ogasawara J, Narahara K. Experimental characterization of left-handed transmission lines with regularly spaced Schottky varactors. IEICE Electronics Express. 2010;**7**:608-614

[15] Narahara K. Numerical characterization of nonlinear oscillatory waves in a composite right- and left-handed traveling-wave field-effect transistor. International Journal of Circuit Theory and Applications. 2016. DOI: 10.1002/cta.2245

[16] van Saarloos W. Fronts, pulses, sources and sinks in generalized complex Ginzburg-Landau equations. Physica D: Nonlinear Phenomena 1992;**56**:303–367.

[17] Akhmediev NN, Afanasjev VV, Soto-Crespo JM. Singularities and special soliton solutions of the cubic-quintic complex Ginzburg-Landau equation. Physical Review E. 1996; **53**:1190-1201

[18] Descalzi O, Argentina M, Tirapegui E. Stationary localized solutions in the subcritical complex Ginzburg-Landau equation. International Journal of Bifurcation and Chaos. 2002;**12**:2459-2465

[19] Narahara K. Collision of nonlinear envelope pulses developed in composite right- and left-handed transmission lines periodically loaded with Schottky varactors. Progress in Electromagnetics Research C. 2011;**21**:1-12

[20] Narahara K, Yamane Y. Experimental observation of collision of nonlinear envelope pulses in left-handed transmission lines periodically loaded with Schottky varactors. Progress in Electromagnetics Research C. 2012;**26**:59-70

[21] Narahara K. Efficiency of three-wave mixing in nonlinear composite right- and left-handed transmission lines. IEICE Electronics Express. 2014;**11**:1-8

[22] Jeffrey A, Kawahara T, editors. Asymptotic Methods in Nonlinear Wave Theory. London: Pitman; 1982

[23] Narahara K. Harmonic resonance in a composite right-handed and left-handed transmission line periodically loaded with Schottky varactors. International Journal of Circuit Theory and Applications. 2016;**44**:492-503

[24] Zakharov VE, Manakov SV. The theory of resonance interaction of wave packets in nonlinear media. Soviet Physics, JETP. 1975;**42**:842-850

[25] Kaup DJ, Reiman A, Bers A. Space-time evolution of nonlinear three-wave interactions. I. Interaction in a homogeneous medium. Reviews of Modern Physics. 1979;**51**:275-309

[26] Narahara K. Soliton decay in composite right- and left-handed transmission lines periodically loaded with Schottky varactors. IEICE Electronics Express. 2014;**11**:1-10

Chapter 7

Introduction to Parametric and Autoparametric Resonance

Lukas C. Kurmann

Additional information is available at the end of the chapter

http://dx.doi.org/10.5772/intechopen.70521

Abstract

This chapter will give an introduction to linear and nonlinear oscillators and will propose literature to this topic. Most importantly, hands on examples with numerical simulations are illustrating oscillations and resonance phenomena and where useful, also analytical methods to treat nonlinear behavior are given.

Keywords: parametric resonance, autoparametric resonance, nonlinear vibration, Mathieu equation, Hopf bifurcation, Strutt diagram, nonlinear natural frequency, instability domain, basepoint excited primary and secondary system

1. Introduction

When a mechanical system has at least two vibrating components, the vibration of one of the components may influence the other component. This influence effect which might stabilize or destabilize the system is called autoparametric resonance. This chapter will introduce autoparametric resonance by examining hands on examples for such systems. In particular, basepoint excited systems are analyzed. Beside purely mechanical systems, also examples of an electrical system with two coupled resonators are investigated.

There are three main types of oscillation: (1) free oscillation, (2) forced excited oscillation and (3) self-excited oscillation.

Free oscillation is defined as temporal fluctuations of the state variables of a system. Such temporal fluctuations can be defined as deviations from a mean value. Vibrations are present in many mechanical systems and occur always in feedback systems. The concept of free oscillation is misleading since nearly all physical systems are subject to attenuation. However, it depends on the size (and thus the time). Exceptions could be, for example, orbit oscillations of planets (macroscopic) or oscillations of electrons (microscopic). The two systems mentioned are also subjected to a type of damping, since both systems cannot remain stable indefinitely, but for an extremely long time.

A forced excited spring mass system might be a mechanically forced oscillator. Such systems of translational motions are discussed in Sections 2 and 3. Beside translatory oscillations, rotatory oscillations and resonance is of vital interest to design engineers of aircraft turbines, etc. Unbalanced rotating machine parts are sources of unwanted vibrations and might resonate when excited accordingly.

Self-excited oscillation, also called as self-oscillation, self-induced, maintained or autonomous oscillation is known in electronics as parasitic oscillation and in mechanical engineering literature as hunting. Such systems are discussed in Section 3.

Table 1 depicts relevant parameters for characterization motion in translational and rotational structures. The parameters for displacement, velocity and acceleration have been written as absolute values – knowing that depending on the application, they might be vectors, depending on the chosen frame of reference. In the most general case, they form a four vector. The force is written as mass times acceleration (Newton's second law) and therefore force is also a vector. That brings us to Newton's first law, which states that an object that is at rest will stay at rest unless a force acts upon it or inversely an object will not change its velocity unless a force acts upon it. For completeness, also Newton's third law shall be given: Actio et Reactio – all forces between two objects exist in equal magnitude and opposite direction. A treaty to Newton's laws of dynamics can be found, for example in chapter 9 of volume I [1].

D'Alembert's principle is a statement of the fundamental classical laws of motion. It is the dynamic analogue to the principle of virtual work for applied forces in a static system and in fact is more general than Hamilton's principle, avoiding restriction to holonomic systems[1].

Translational			**Rotational**		
Symbol	Description	SI Unit	Symbol	Description	SI Unit
s	Displacement	m	φ	Angle	rad
$v=\frac{ds}{dt}$	Velocity	$\frac{\text{m}}{\text{s}}$	$\omega=\varphi\frac{d}{dt}$	Angular velocity	$\frac{\text{rad}}{\text{s}}$
$a=\frac{dv}{dt}$		$\frac{\text{m}}{\text{s}^2}$	$\alpha=\omega\frac{d}{dt}$	Angular acceleration	$\frac{\text{rad}}{\text{s}^2}$
m	Mass	kg	J	Inertia	kg m^2
$F=m\,a$	Force	N	$T=J\,\alpha$	Torque	Nm
$I=m\,v$	Momentum	Ns	$L=J\,\omega$	Angular momentum	Nms
$T=\frac{1}{2}m\,v^2$	Kinetic energy	Nm	$T=\frac{1}{2}J\,\omega^2$	Kinetic Energy	Nm
$U=\frac{1}{2}k\,y^2$	Potential energy	Nm	$U=\frac{1}{2}c\,\varphi^2$	Potential energy	Nm
$W=\int Fds$	Work	J	$W=\int Td\varphi$	Work	J
$P=F\,v$	Power	W	$P=J\,\omega$	Angular power	W

Table 1. Comparison of translational and rotational motion parameter characteristics.

[1]A holonomic constraint depends only on the coordinates and time and does not depend on velocities.

If the negative terms in accelerations are recognized as inertial forces, the statement of d'Alembert's principle becomes "the total virtual work of the impressed forces plus the inertial forces vanishes for reversible displacements". The principle does not apply for irreversible displacements, such as sliding friction, and more general specification of the irreversibility is required. A derivation of the Lagrangian equation of motion is well explained in Chapter 1 of [2] or [3]. In (1), the non-conservative energy term defined as the Lagrangian (L) is, composed of the kinetic energy T and the potential energy U.

$$L = T - U \tag{1}$$

In (2) the Lagrangian equation is given with generalized coordinates q_i of a dynamic system and dissipative generalized forces Q_i.

$$\frac{\mathrm{d}}{\mathrm{d}t}\left(\frac{\partial L}{\partial \dot{q}_i}\right) - \frac{\partial L}{\partial q_i} = Q_i \tag{2}$$

The sum of all kinetic energies T in the system, whether translational or rotational character (see also **Table 1**) needs to be included. The sum of all potential energies U in the system, whether it stems from the gravitation or energy from linear or nonlinear springs or whatever scalar field. Elastic potential energy from any linear or nonlinear spring can be obtained calculating its potential energy. The Lagrangian formalism can also be used for mechanical systems with mass explicitly dependent on position, see for example [4]. In the following chapters, all analyzed dynamical systems are derived using this elegant and powerful method.

Looking at translational (classic) mechanical springs, the displacement dependent force $F(q_t)$ can be written as shown in (3) using spring stiffness k_1 $\left(\frac{N}{m}\right)$, the generalized translational coordinates q_t and having also introduced a nonlinear spring term k_n and an exponent n for setting nonlinearity of spring. For linear springs, where the force is proportional to the displacement (Hooke's law), this nonlinear spring term will be zero.

$$F(q_t) = k_1\, q_t + k_n\, {q_t}^n \tag{3}$$

The elastic energy $E_{elastic}$ of the spring is obtained by integrating the exerted force over its covered path s.

$$E_{elastic}(s) = \int_0^s F(q_t)\mathrm{d}q_t = \int_0^s k_1\, q_t + k_n\, {q_t}^n = \frac{1}{2}k_1 + \frac{k_n}{1+n}\, {q_t}^{1+n}\Big|_0^s$$

$$E_{elastic}(s) = \frac{1}{2}k_1\, s^2 + \frac{k_n}{1+n}s^{1+n} \; with \; n > -1 \tag{4}$$

The elastic energy $E_{elastic}$ (a potential U) for a linear translational spring is given also in **Table 1**. For (rotational) torsion springs, the procedure is the same and a given spring torque can be expressed as shown in (5) with torsion coefficient D_1 $\left(\frac{Nm}{rad}\right)$, the generalized translational coordinates q_r and having also introduced a nonlinear torsion spring term D_n.

$$T(q_r) = D_1\, q_r + D_n\, q_r{}^n \tag{5}$$

The elastic rotational energy can be expressed as follows (6):

$$E_{elastic}(\varphi) = \int_0^{\varphi} T(q_r)\mathrm{d}q_r = \int_0^{\varphi} D_1\, q_r + D_n\, q_r{}^n\ \mathrm{d}q_r = \frac{1}{2} D_1\, q_r{}^2 + \frac{k_n}{1+n} D_n\, q_r{}^{1+n} \Big|_0^{\varphi}$$

$$E_{elastic}(\varphi) = \frac{1}{2} D_1\, \varphi^2 + \frac{1}{1+n} D_n \varphi^{1+n}\ \ with\ n > -1 \tag{6}$$

In **Figure 1**, sketches on the top show translational springs (from left to right linear, nonlinear and nonlinear unsymmetrical) and sketches on the bottom depict rotational springs (from left to right linear, nonlinear and nonlinear unsymmetrical). The origin is depicted with an O and the spring displacement is depicted as y and φ, respectively. For the translational magnetic spring systems, the lower and upper magnets are fixed to the reference frame with origin O. The nonlinear symmetric magnetic spring uses three identical block or disk magnets. Such magnetic springs have a nonlinear term $k_n > 0$ – forming so called hardening springs if $k_n < 0$ in literature referred to softening springs, see for example [5]). The nonlinear unsymmetrical translational magnetic spring is also shown in the neutral position and the displacement around the origin is unsymmetrical.

The rotational magnetic spring systems have a ferromagnetic stator fixed to the reference frame and a rotating hollow shaft carrying two permanent magnets, in this scenario also made of ferromagnetic material. The spring is drawn in the unstable equilibrium position. Exemplarily spring characteristics for such translational and rotational spring systems are depicted in **Figure 2**.

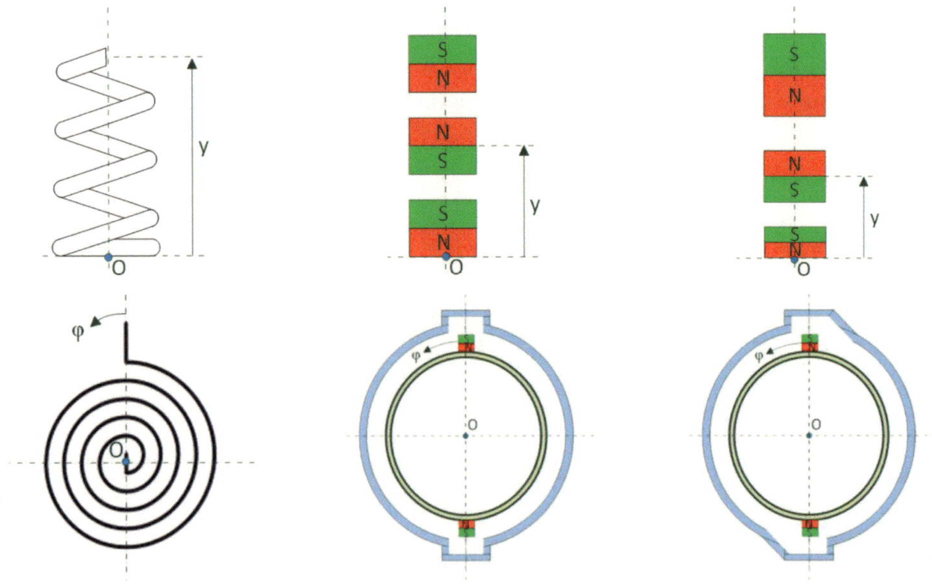

Figure 1. Sketches of translational and rotational spring systems.

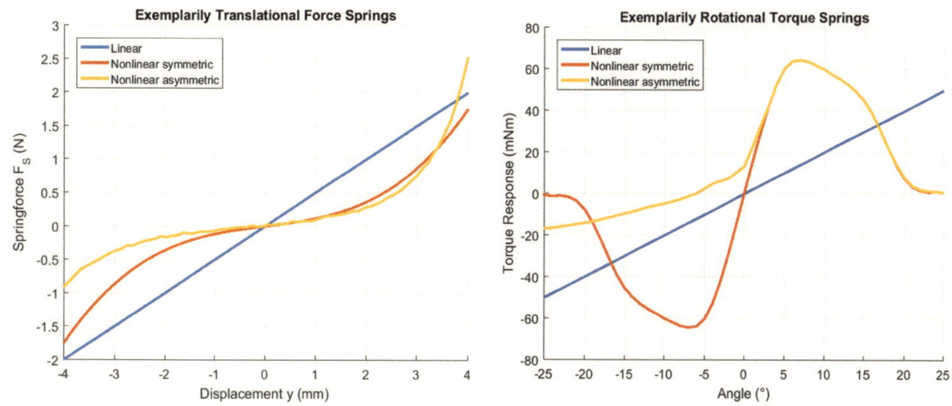

Figure 2. Exemplarily displacement-force signals (l) and angular displacement-torque signals (r) of the shown spring systems of **Figure 1**.

Note that integrating the force or torque response will add up to zero for rotational and translational spring systems. The given exemplarily spring characteristics of drawn spring systems in **Figure 1** are shown in **Figure 2**. On the left-hand side, exemplarily translational spring characteristics are shown and on the right-hand side rotational spring characteristics are depicted. The linear case where the force and torque are proportional to the displacement is shown in blue. In red, symmetric nonlinear – here for translational and rotational systems a permanent magnet system is shown, but also mechanical spring systems could be envisaged. The curves of the nonlinear asymmetric cases are depicted in orange. In the appendix A.1, more simulations have been depicted for translational symmetric spring systems using ring magnets.

Equivalence of electrical and mechanical systems are shown in **Table 2**. On the left-hand side, a mechanical system with only one degree of freedom in y direction is shown and its equivalent electrical structure with charge q and current i on the right-hand side. Kinetic energies are denoted with T_T (translational kinetic energy) and T_L (inductive kinetic energy), potentials are written as U_S (spring potential) and U_C (capacitive potential) and the gravitational potential denoted as U_G and the DC battery voltage is U_B. Non-conservative components are in the mechanical system the viscous damping force and in the electrical system the electrical resistor.

The Lagrange energy function is shown for the mechanical system in (7) and its equivalent electrical system in (8).

$$L_{mech} = T_T - (U_S + U_G) = \frac{1}{2}\, m\, y'^2 + \frac{1}{2}\, k\, y^2 - m\, g\, y \tag{7}$$

$$L_{el} = T_L - (U_C + U_B) = \frac{1}{2}\, L\, q'^2 + \frac{1}{2}\frac{1}{C}\, q^2 - U_0\, q \tag{8}$$

Applying the Lagrangian formalism (1) and (2) to these SDoF systems, will lead to the resulting DE's as shown in (9) and (10). Note that the sign of the (viscous) damping must be introduced always with a negative sign using the generalized coordinates – as its velocity is

Mechanical System

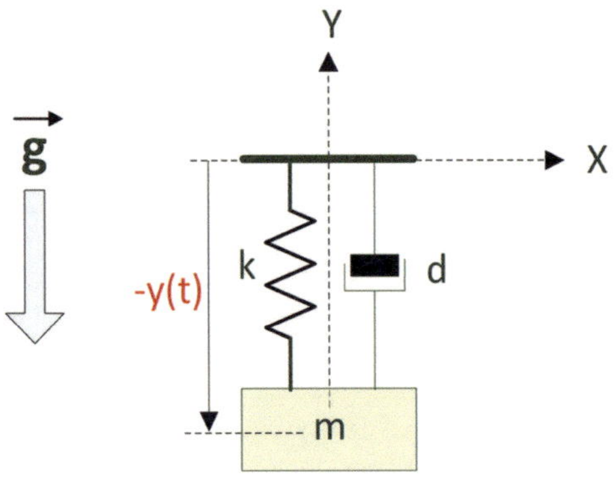

Electrical System

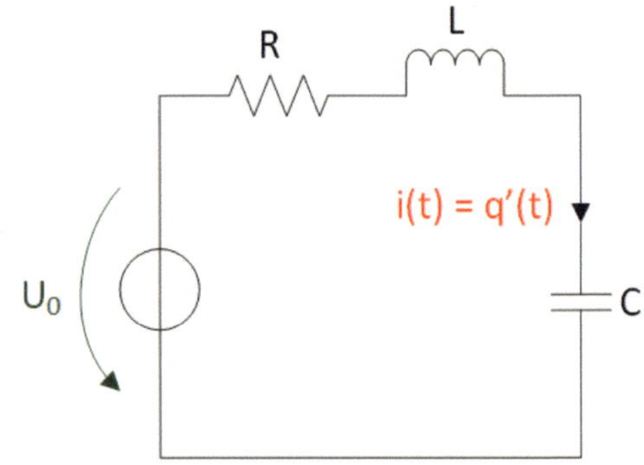

Mechanical System	Electrical System
Mechanical kinetic energy:	Electrical kinetic energy:
$T_T = \frac{1}{2} m\, y'^2$	$T_L = \frac{1}{2} L\, q'^2$
Mechanical potential energies:	Electrical potential energies:
$U_S = \frac{1}{2} k\, y^2$	$U_C = \frac{1}{2}\frac{1}{C}\, q^2$
$U_G = m\, g\, y$	$U_B = U_0\, q$
Mechanical damping force:	Electrical damping voltage:
$F_d = d\, y'$	$V_d = R\, q'$

Table 2. Equivalence of electrical and mechanical systems.

always opposing the system velocity. In the electrical circuit, having the flowing charge velocity q' for example, current i defined in clockwise direction, the battery voltage, as it is a source, must act in the opposite direction and therefore this potential energy must be introduced with a negative sign.

$$m\, y'' + d\, y' + k\, y = g\, m \tag{9}$$

$$L\, q'' + R\, q' + \frac{1}{C}\, q = U_0 \tag{10}$$

Both systems (9) a force DE, (10) a voltage DE, belong to the same class of ordinary linear second-order DE. Resonance frequency for the mechanical system (9) is $\omega_{mech}{}^2 = \frac{k}{m}$ and for the electrical system (10) $\omega_{el}{}^2 = \frac{1}{L\,C}$.

2. Linear resonance systems

2.1. Linear single degree of freedom systems

In this section, a linear basepoint excited single degree of freedom systems is discussed. The lumped parameter model for the examined system (**Figure 3**) consists of a linear oscillator with

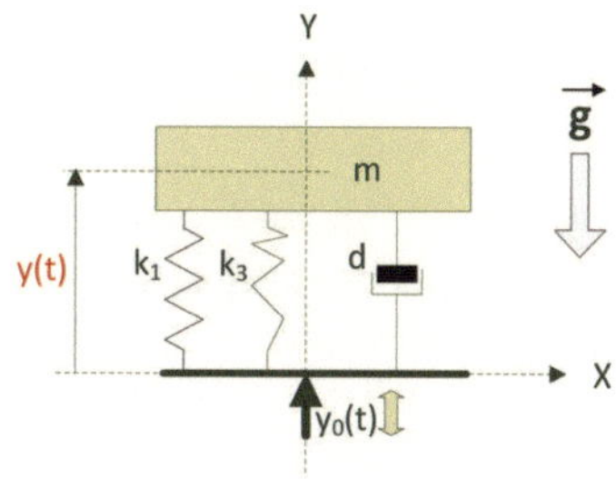

Figure 3. Linear single degree of freedom (SDoF) spring mass damper model of a resonant harmonic basepoint excited oscillator.

mass m damping factor d, a linear spring with a spring rate k_1 and an external basepoint excited harmonic force with amplitude $y_0(t)$. System coordinate origin is placed at the basepoint excitation.

The kinetic energy (11) of this system and the potential energy (12) form the non-dissipative energy of this system. The dissipative force F_d of this system – we consider only viscous friction – is shown in (13) and the driving force F_0 (14) assuming a harmonic basepoint excitation with amplitude A and driving frequency ω.

$$T = \frac{1}{2}\, m\, y'^2 \tag{11}$$

$$U = m\, g\, y + \frac{1}{2}\, k_1\, y^2 \tag{12}$$

$$F_d = \mathrm{d}\ \ y' \tag{13}$$

$$F_0 = m\frac{\mathrm{d}^2}{\mathrm{d}t^2}(A\ \cos\,\omega t) = -mA\omega^2 \cos\,\omega t \tag{14}$$

Applying the Lagrangian formalism (1) and (2), we deal with SDoF system, will lead to the resulting DE shown in (15). Note that the sign of the viscous damping must be introduced with a negative sign using the generalized coordinates. The driving force F_0, as it is a harmonic signal, can be introduced with a positive or a negative sign, resulting in a phase shift of 180°.

$$m\, y'' - (-mg - k_1 y) = -F_d - F_0 \tag{15}$$

$$m\, y'' + d\, y' + k_1 y + mg = mA\omega^2 \cos\,\omega t \tag{16}$$

Introducing dimensionless notation, by using a dimensionless time τ, a dimensionless system resonance frequency Ω, the damping factor ξ_1 and the gravity offset term ϱ (17) and setting the dimensionless displacement u (18).

$$\tau = \mathrm{t}\,\omega_1;\, \Omega = \frac{\omega}{\omega_1};\, {\omega_1}^2 = \frac{\mathrm{k}_1}{\mathrm{m}};\, \xi_1 = \frac{d}{2\, m\, \omega_1};\, \varrho = \frac{g}{A\, {\omega_1}^2} \tag{17}$$

$$\text{path } u(\tau) = \frac{y(t)}{A} \tag{18}$$

By replacing parameters of (16) with (17, 18), we obtain (19). We can drop the gravity offset term ϱ, as it will only add a non-time dependent offset to the solution u (20).

$$u'' + 2\xi_1 u' + u + \varrho = \Omega^2 \cos(\Omega\tau) \tag{19}$$

$$u'' + 2\xi_1 u' + u = \Omega^2 \cos(\Omega\tau) \tag{20}$$

Frequency domain behavior is obtained by applying the Laplace Transformation (21–24) and by replacing $s = j\Omega$ we obtain the frequency response (25).

$$U(s) = \mathcal{L}\{u(\tau)\} = \int_0^\infty u(\tau)e^{-s\tau}\mathrm{d}\tau \tag{21}$$

$$\mathcal{L}\left\{u'' + 2\xi_1 u' + u\right\} = -\frac{\mathrm{d}^2}{\mathrm{d}t^2}\cos(\Omega\tau) \tag{22}$$

$$U(s)s^2 + 2\xi_1 U(s)s + U(s) = s^2 Y_0(s) \tag{23}$$

$$G(\mathrm{s}) = \frac{U(s)}{Y_0(s)} = \frac{\mathrm{s}^2}{1 + 2\xi_1\mathrm{s} + \mathrm{s}^2} \tag{24}$$

$$G(\xi_1, \Omega) = \frac{U(j\Omega)}{Y_0(j\Omega)} = \left|\frac{\Omega^2}{1 + \Omega^2 + j2\xi_1\Omega}\right| \tag{25}$$

G represents the relative motion of the oscillation. As long as the excitation frequency can be represented by a Fourier series of harmonic functions, this obtained solution is valid and a very powerful result. (24) and (25) are represented in **Figure 4**. The advantage of the Bode Plot

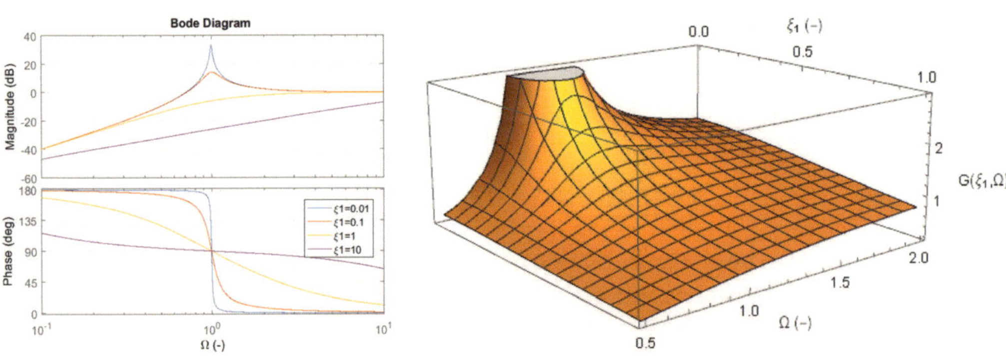

Figure 4. Representation of frequency response of a linear SDoF system using (24) Bode diagram (left) and absolute value representation of (25) (right).

http://dx.doi.org/10.5772/intechopen.70521

representation is to have also the phase shown. As smaller the dimensionless damping ξ_1 become, as larger becomes the scaled resonance at the dimensionless frequency ration Ω.

2.2. Linear two degree of freedom systems (2DoF systems)

In this section, a linear basepoint excited two degree of freedom systems is discussed. The lumped parameter model for the examined system (**Figure 5**) consists of two linear oscillators with mass m_1 and m_2, damping factors d_1 and d_2, linear springs with spring rates k_1 and k_2 and an external basepoint excited harmonic force with amplitude $y_0(t)$. System coordinate origin is placed at the basepoint excitation.

$$T = \frac{1}{2} m_1 {y_1'}^2 + \frac{1}{2} m_2 {y_2'}^2 \tag{26}$$

$$U = m_1 g y_1 + m_2 g y_2 + \frac{1}{2} k_1 {y_1}^2 + \frac{1}{2} k_2 (y_2 - y_1)^2 \tag{27}$$

$$F_d = d_1 y_1' + d_2 y_2' \tag{28}$$

$$F_0 = (m_1 + m_2) \frac{d^2}{dt^2} (A \cos \omega t) = -(m_1 + m_2) A\omega^2 \cos \omega t \tag{29}$$

Applying the Lagrangian formalism (1) and (2) to this 2DoF problem, we obtain the coupled DE system shown in (30) and (31).

$$m_1 {y_1}'' + d_1 y_1' + k_1 y_1 - k_2 (y_2 - y_1) + m_1 g = (m_1 + m_2) A\omega^2 \cos \omega t \tag{30}$$

$$m_2 {y_2}'' + d_2 y_2' + k_2 (y_2 - y_1) + m_2 g = 0 \tag{31}$$

DE system shown in (30), (31) is represented dimensionless in equation DE system (32), (33) using the dimensionless parameters of (34), (35).

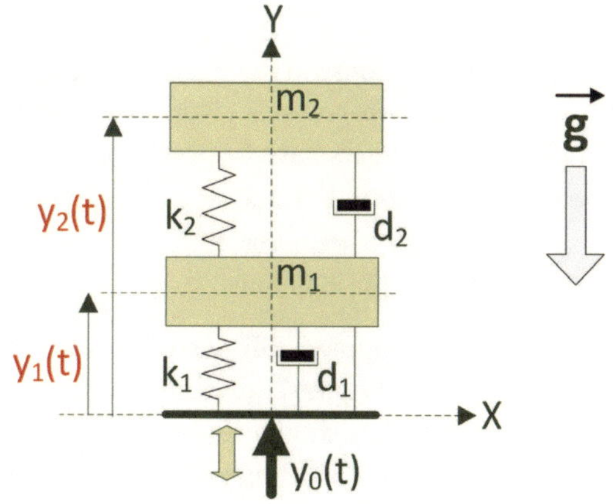

Figure 5. Linear 2DoF spring mass damper model of a resonant harmonic basepoint excited oscillator.

$$u''(\tau) + 2\xi_1 u'(\tau) + u(\tau) + \lambda_m {\Omega_0}^2 u(\tau) - \lambda_m {\Omega_0}^2 v(\tau) + \varrho = \Omega^2 \cos(\tau\Omega) \tag{32}$$

$$v''(\tau) + 2\xi_2 \Omega_0 v'(\tau) + {\Omega_0}^2 v(\tau) - {\Omega_0}^2 u(\tau) + \varrho = 0 \tag{33}$$

Similar to dimensionless parameters of (17), (18), the dimensionless time τ, a dimensionless system resonance frequency Ω, damping factors ξ_1 and ξ_2, a gravity offset term ϱ, system resonance frequencies of each oscillator ω_1 and ω_2 plus a mass ratio λ_m and an oscillator frequency ratio Ω_0 plus dimensionless displacements u and v.

$$\tau = t\,\omega_1; \Omega = \frac{\omega}{\omega_1}; {\omega_1}^2 = \frac{k_1}{m_1}; {\omega_2}^2 = \frac{k_2}{m_2}; \xi_1 = \frac{d_1}{2\,m_1\,\omega_1}; \xi_2 = \frac{d_2}{2\,m_2\,\omega_2}; \Omega_0 = \frac{\omega_2}{\omega_1}; \lambda_m = \frac{m_2}{m_1}; \varrho = \frac{g}{A\,{\omega_1}^2} \tag{34}$$

$$\text{path } u(\tau) = \frac{y_1(t)}{A} \text{ and path } v(\tau) = \frac{y_2(t)}{A} \tag{35}$$

The frequency response of this coupled oscillator system can again be obtained using the Laplace transformation introduced in (21). The system in the frequency domain is shown in (36) and (37) using the same steps as shown in the SDoF system (22)–(25).

$$U(s)s^2 + 2\xi_1 U(s)s + U(s)\left(1 + \lambda_m {\Omega_0}^2\right) - \lambda_m {\Omega_0}^2 V(s) + \varrho = s^2 Y_0(s) \tag{36}$$

$$V(s)s^2 + 2\xi_2\Omega_0\, V(s)s + V(s){\Omega_0}^2 - U(s){\Omega_0}^2 + \varrho = 0 \tag{37}$$

As (36) and (37) represent two algebraic equations, $U(s)$ and $V(s)$ can be separated, resulting in (38) and (39). Note that the gravity term ϱ in the numerator will introduce an additional damping of the transfer function.

$$U(s) = -\frac{-\rho\lambda_m {\Omega_0}^2 + \left(s^2 + 2s\xi_2\Omega_0 + {\Omega_0}^2\right)\left(-\rho + s^2 Y_0(s)\right)}{\lambda_m {\Omega_0}^4 - \left(s^2 + 2s\xi_2\Omega_0 + {\Omega_0}^2\right)\left(1 + s^2 + 2s\xi_1 + \lambda_m {\Omega_0}^2\right)} \tag{38}$$

$$V(s) = \frac{-\rho\left(1 + s^2 + 2s\xi_1 + (1 + \lambda_m){\Omega_0}^2\right) + s^2{\Omega_0}^2 Y_0(s)}{s^4 + {\Omega_0}^2 + 2s^3(\xi_1 + \xi_2\Omega_0) + 2s\Omega_0\left(\xi_2 + \xi_1\Omega_0 + \lambda_m\xi_2{\Omega_0}^2\right) + s^2(1 + \Omega_0(4\xi_1\xi_2 + \Omega_0 + \lambda_m\Omega_0))} \tag{39}$$

Figure 6 depicts the relative oscillation response in the frequency (left) and time domain (right) of the derived 2DoF system. The frequency response is given as a dimensionless ratio Ω, see also (34). The dimensionless simulation parameters have been set exemplarily to $\xi_1 = \xi_2 = 0.021$, $\lambda_m = 0.42$ and $\Omega_0 = 1$. As we have two resonators with same system frequencies $\omega_1 = \omega_2 = 169\,\frac{rad}{s}$, two resonances will occur.This system reaches resonances at 0.71 Ω (19Hz) and at 1.41 Ω (38Hz) for $\Omega_0 = 1$ and 0.82 Ω and at 6.1 Ω for $\Omega_0 = 5$ (dashed lines).

The time-domain response from this coupled DE system with lumped parameter model **Figure 5** and (32) and (33) is shown in **Figure 7**. On the left-hand side, the ^dimensionless basepoint acceleration signal is given and its dimensionless response signals of first (blue) and second (red) DoF, simulating 50 periods and starting with settled initial conditions (amplitude

http://dx.doi.org/10.5772/intechopen.70521

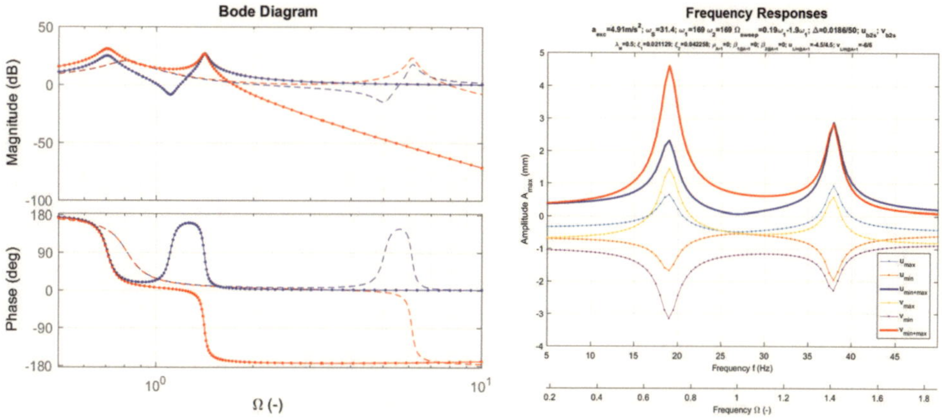

Figure 6. Bode diagram of a linear 2DoF system represented by (38) and (39) (left); first oscillator with mass m_1 in blue and second with m_2 in red with $\Omega_0 = 1$ and $\Omega_0 = 5$ (dashed lines) and constructed frequency response using time domain signals (right).

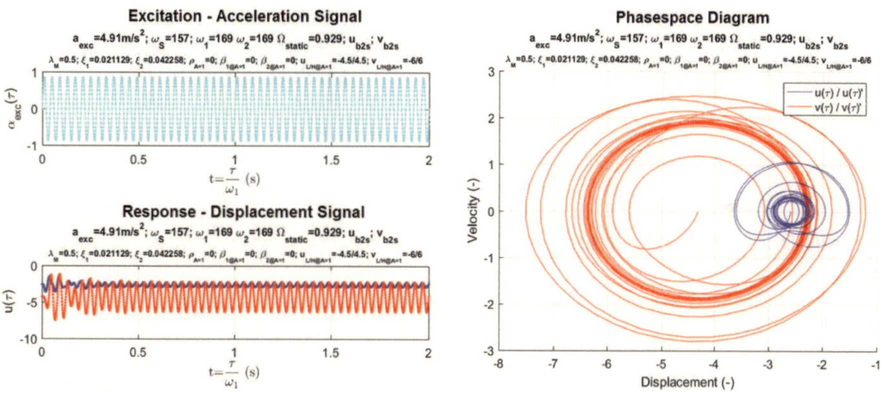

Figure 7. Numerical simulation results of the linear basepoint excited 2DoF system shown in (32) and (33) using a constant acceleration of 0.5g and a basepoint excitation of $\omega = 25$ ***Hz***.

of $\hat{y}_{1start} < 1\mu m$ and $\hat{y}_{2start} < 1\mu m$). The main simulation parameters are shown in the heading, a variant of this setup using nonlinear springs is given in the appendix A.3.

3. Nonlinear resonance systems

In the introduction, Section 1, we distinguished three cases of vibration. The class forced excitation will be further investigated in this section. In **Figure 8** five systems are depicted that can potentially exhibit parametric resonance effects. The term parametric means that of cases where the external excitation appears as a time varying modification of a system parameter. A "normal" forced excitation system whether linear or nonlinear, will respond to the excitation

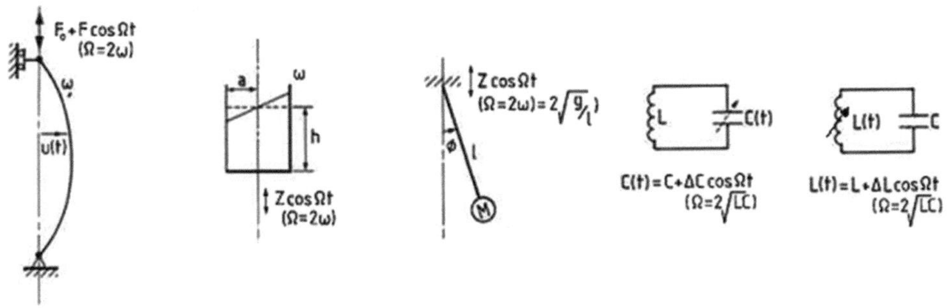

Figure 8. Examples of physical systems exhibiting potential parametric resonance effects, adapted after [2].

with or without resonance using the energy fed into it and no time varying modification of a system parameter might excite additionally the system.

The five depicted systems in **Figure 8** might show an exponential amplitude growth when excited externally in presence of a system damping factor. In the two electrical systems on the right-hand side, any of the three components R (here not drawn), L or C that is parametrically excited will respond with an exponential amplitude growth, if the mathematical physical system model has at least one degree of freedom of the Mathieu DE (40) or the Hill DE (41).

$$q'' + q(a + b\cos \Omega t) = 0 \tag{40}$$

The Hill differential equation is a generalized form of (40), in which the harmonic function is replaced with any periodic function, shown in (41).

$$q'' + q\left(a + f_p(t)\right) = 0 \tag{41}$$

It is most interestingly that any system parameter including also damping factors with time varying influence of a system parameter will result in an exponential growth of the response amplitude. To give a concrete example of this behavior, we consider here the example from Section 3.2 and inspect the resulting (dimensioned) DE system with (62) as primary system and (63) as secondary system of such a behavior.

The primary system has no such configuration, but the secondary system (63) is of Mathieu type. To simplify the treated system, we use instead of the basepoint excitation y_0 the primary system y, compare also the lumped parameter model in **Figure 12** (in an experiment we would simply make the stiffness k of the system very large, for example, replacing the spring with a fixed stiff rod). Now the new induced basepoint excitation y will excite the secondary system directly. As y is appearing in (63) as acceleration, we adjust this basepoint excitation simply in form of an acceleration (42).

$$y = A\cos(\omega t) \rightarrow y'' = A\,\omega^2 \cos(\omega t) \tag{42}$$

Writing (63) as an acceleration DE (dividing by $m_2\, l$) and inserting it the acceleration of (42), it is read (43).

$$l\,\varphi'' + \frac{D}{m_2\,l}\,\varphi' + g\sin\,\varphi + A\,\omega^2\cos\,(\omega t)\,\sin\,\varphi = 0 \tag{43}$$

Rearranging the terms and setting $\sin\varphi \cong \varphi$, we get (44), which is a Mathieu type DE with parameters $a = \frac{g}{l}$ and $b = -\frac{A\ \omega^2}{l}$.

$$\varphi'' + \frac{D}{m_2}\,\varphi' + \sin\,\varphi\left(\frac{g}{l} + \frac{A\,\omega^2}{l}\cos\,(\omega t)\right) = 0 \tag{44}$$

For generating parametric resonances, the (natural) system frequency needs to be coupled with the excitation frequency ω. Using the same nomenclature as in (64), we define the pendulum system frequency $\omega_2{}^2 = \frac{g}{l}$. For generating parametric resonances for which the angle $\varphi(t)$ is growing exponentially, a frequency ratio $\omega : \omega_2 = 1:1$ is sufficient (as well as the ratio $\omega : \omega_2 = 2:1$), see also left-hand side of **Figure 9**. In case of letting the displacement term be

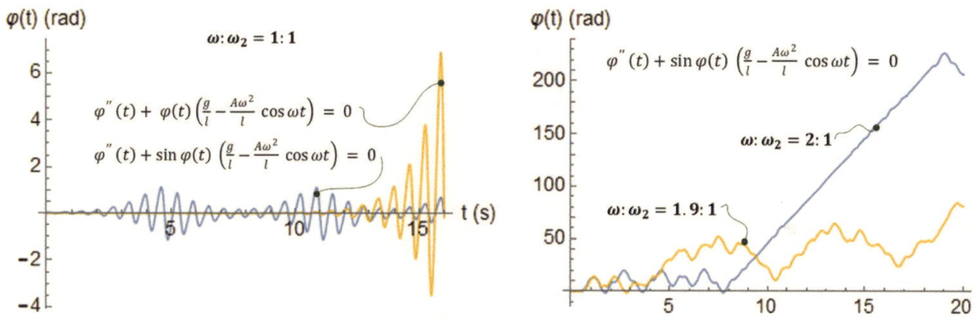

Figure 9. Nonlinear single degree of freedom (SDoF) spring mass damper model of a resonant harmonic basepoint excited oscillator.

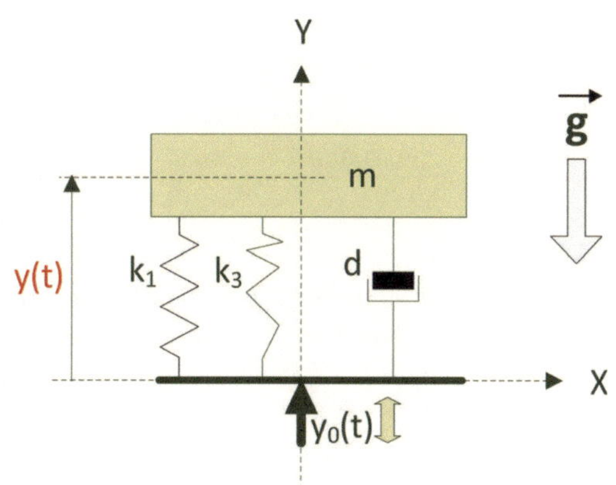

Figure 10. Response signals of a parametrically excited pendulum examining DE (47) with keeping damping term $D = 0$, $l = 108.1$ mm, $A = 100$ mm and $\omega_1 = 10\ rad$.

harmonic (left-and right-hand side of **Figures 9**), the frequency ratio must be very close to a 2:1 ratio to have a large amplitude response. The ratio tolerance for having a large growth has a band width of ca. 1 rad to keep a large amplitude growth going.

Note that the response signal in orange on the left-hand side of **Figure 10** is using an approximated linear displacement function $\sin(\varphi(t)) \rightarrow \varphi(t)$ and is scaled down by factor 10^{-4}.The generated beat frequency signal is obtained using the exact harmonic displacement function.

3.1. Nonlinear single degree of freedom systems

Similar to the case in Section 2.1, also a SDoF system will be discussed, but this time a linear and a nonlinear spring will be present. The nonlinearity of this spring shall have the form shown in (3) having a nonlinear exponent $n = 2$ and $k_3 > 0$, a parameterization like that is generally used for a magnetic spring (see also top middle sketch in **Figure 1** and appendix A.1). The lumped parameter model for the examined system (**Figure 10**) consists beside this spring system with linear spring rate k_1 and nonlinear spring rate k_3 of an oscillator mass m, a viscous damping factor d and an external basepoint excited harmonic force with amplitude $y_0(t)$. System coordinate origin is placed at the basepoint excitation.

The elastic energy of this nonlinear spring system with $n = 2$ will lead to the following spring energy, see also derivation in (4).

$$E_{elastic}(y) = \frac{1}{2}k_1 y^2 + \frac{1}{4}k_3\, y^4 \tag{45}$$

Adding up all kinetic energies and all potential energies, disturbances in form of a viscous damping and a basepoint excited force is given in (46)–(49).

$$T = \frac{1}{2}\, m\, y'^2 \tag{46}$$

$$U = m\, g\, y + \frac{1}{2}\, k_1\, y^2 + \frac{1}{4} k_3\, y^4 \tag{47}$$

$$F_d = d\ y' \tag{48}$$

$$F_0 = m\frac{d^2}{dt^2}(A\ \cos\,\omega t) = -mA\omega^2 \cos\,\omega t \tag{49}$$

Applying the Lagrangian formalism (1) and (2), we deal again with a SDoF system, will lead to the resulting DE shown in (50), similar to the result derived in Section 2.1 – but here we have now introduced a nonlinear spring system.

$$m\, y'' + d\, y' + k_1 y + k_3 y^3 + mg = mA\omega^2 \cos\,\omega t \tag{50}$$

Introducing dimensionless notation, by using a dimensionless time τ , a dimensionless system resonance frequency Ω, the damping factor ξ_1 and the gravity offset term ϱ (51) and setting the dimensionless displacement u (52).

$$\tau = t\,\omega_1; \Omega = \frac{\omega}{\omega_1}; \omega_1{}^2 = \frac{k_1}{m}; \beta = \frac{k_3}{k_1}A^2; \xi_1 = \frac{d}{2\,m\,\omega_1}; \varrho = \frac{g}{A\,\omega_1{}^2} \tag{51}$$

$$\text{path } u(\tau) = \frac{y(t)}{A} \tag{52}$$

By replacing parameters of (50) with (51), (52) we obtain (53), including the dimensionless gravity term ϱ as well.

$$u'' + 2\xi_1 u' + u + \beta u^3 + \varrho = \Omega^2 \cos\,(\Omega\tau) \tag{53}$$

DE (53) is nonlinear, as we have also a path dependent function to the power of 3 and a dimensionless factor β which is generally small. If this factor β is positive, we deal with a nonlinear spring hardening system, if β is negative, it is a spring softening system. Unfortunately, such a system cannot be examined using the Laplace or Laplace-like transformation, such as [6], as this transformation can deal only with linear functions, respectively, nonlinear quadratic functions. For solving this nonlinear so-called Duffing DE, there are several methods available, such as averaging method or the harmonic balancing method. The averaging method assumes that a solution of the DE can be obtained using harmonic functions In Ref. [7], chapter 9.3, a general solution for nonlinearity terms with a positive integer exponent βu^n is obtained using the averaging method. Another method to get analytic solutions is, as said, the harmonic balance method, which is well explained in the textbook [8], chapter 2.3.4; the DE case of (53) is discussed in the same book, chapter 4.1 and there are many research papers to discuss this nonlinear DE, see for example [9, 10]. Note that this case is of nonlinear nature, as it includes the nonlinear term βu^3, but cannot exert parametric resonance. However, there are also many research papers where such nonlinearities coupled with a Mathieu DE are discussed, see for example [11].

The depicted **Figure 11** shows simulation results of DE (53). The time domain behavior (top left) and its dimensionless phase space behavior (top right) is shown with simulating 50 periods with settled initial conditions (amplitude of $\hat{y} < 1\mu m$). The top row depicts one simulation point in the bottom row, where a frequency sweep has been done, sweeping the basepoint excitation from ω = 5…40Hz and keeping the acceleration signal constant at 0.5 g. To make sure that only non-transient amplitudes are selected to create the frequency response, only in the last 5 periods (out of 50) the maximal and minimal value is selected). The top row is using a constant angular excitation of 25 Hz and depicts only one simulation point of the generated frequency response. The main simulation parameters are shown in the heading, a variant of this simulation is given in the appendix A.3. Note the shown simulated data are taken from a validated electromagnetic SDoF vibration energy harvester system by the author.

3.2. Nonlinear two degree of freedom systems

Let us consider the lumped parameter model in **Figure 12**. A pendulum with a stiff rod of length l and mass m_2 suspended on a spring damper system with mass m_1 and stiffness k and

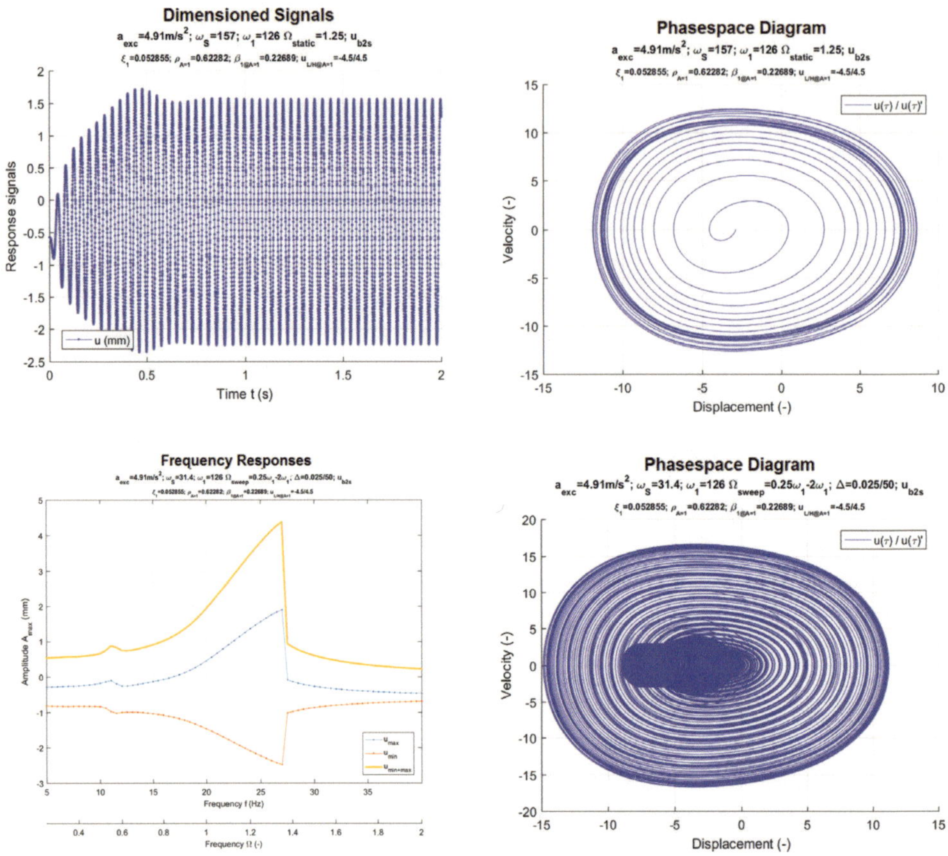

Figure 11. Numerical simulation results of the nonlinear basepoint excited SDoF system shown in (53) using a constant acceleration of 0.5 *g* and a basepoint excitation of ω = 25 **Hz** (top row) and its sweep behavior ω = 5…40 **Hz** (bottom row).

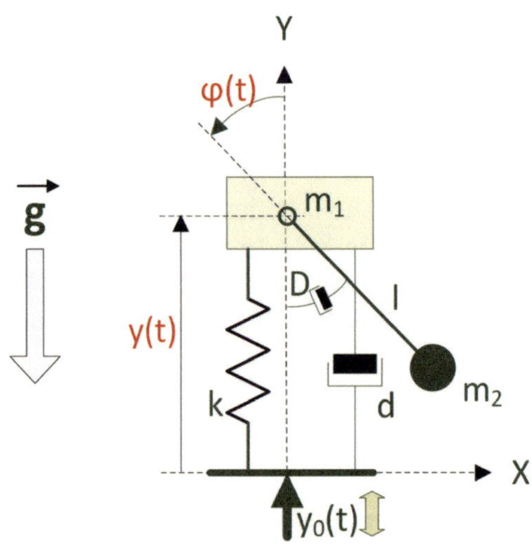

Figure 12. Nonlinear two degree of freedom (2DoF) spring mass damper model.

http://dx.doi.org/10.5772/intechopen.70521

damping factor d. Mass m_1 can move only in the depicted y direction and pendulum only in the X-Y plane.

Governing equations are derived using again the Lagrange formalism. Considering the frame of reference at the origin shown in **Figure 12** and defining in Cartesian coordinates first the two degrees of freedom vector $\boldsymbol{r}_y$ and $\boldsymbol{r}_\varphi$ (54).

$$\boldsymbol{r}_y = \begin{pmatrix} 0 \\ y(t) \end{pmatrix} \text{ and } \boldsymbol{r}_\varphi = \begin{pmatrix} l \sin \varphi(t) \\ y(t) - l \cos \varphi(t) \end{pmatrix} \tag{54}$$

The kinetic energy for both degrees of freedom are shown in (55, 56).

$$T_y = \frac{1}{2} m_1 \left(\left(\frac{d}{dt} r_{y.x} \right)^2 + \left(\frac{\boldsymbol{d}}{\boldsymbol{dt}} r_{y.y} \right)^2 \right) = \frac{1}{2} m_1 y'^2 \tag{55}$$

$$T_\varphi = \frac{1}{2} m_2 \left(\left(\frac{d}{dt} r_{\varphi.x} \right)^2 + \left(\frac{\boldsymbol{d}}{\boldsymbol{dt}} r_{\varphi.y} \right)^2 \right) = \frac{1}{2} m_2 l^2 \cos(\varphi)^2 \varphi'^2 + \frac{1}{2} m_2 (y' + l \sin \varphi \, \varphi')^2 \tag{56}$$

The potential energies derived from the same vectors lead to (57) and (58).

$$U_y = m_1 g \, y + \frac{1}{2} k \, y^2 \tag{57}$$

$$U_\varphi = m_2 g \, (y - l \cos \varphi) \tag{58}$$

The Lagrange energy function L becomes:

$$L = T_y + T_\varphi - (U_y + U_\varphi) \tag{59}$$

The viscous friction for both degree of freedoms is given in (60) and the basepoint excited driving force is given in (61).

$$F_{dy} = d \, y' \text{ and } T_{d\varphi} = D \, \varphi' \tag{60}$$

$$F_0 = -(m_1 + m_2) \frac{d^2}{dt^2} (A \cos \omega t) = (m_1 + m_2) A \omega^2 \cos \omega t \tag{61}$$

Applying the Lagrange formalism (2) for both degrees $q_1 = y$ and $q_2 = \varphi$ lead to the coupled DE system of (62), (63), representing a force DE respectively a torque DE. On the right-hand side of those DE's, the defined viscous frictions of (60) and the basepoint excitation (61) is present.

$$(m_1 + m_2) y'' + g(m_1 + m_2) + ky + l m_2 \left(\varphi'^2 \cos \varphi + \varphi'' \sin \varphi \right) = -d \, y' + F_0 \tag{62}$$

$$m_2 l^2 \varphi'' + g l m_2 \sin \varphi + l m_2 y'' \sin \varphi = -D \, \varphi' \tag{63}$$

The parameters for non dimensionalization are given in (64), (65). Note that the reference system frequency ω_1 is set to the 1. DoF (also called primary, the mass spring system) and the second system frequency ω_2 to the 2. DoF (the pendulum – also called secondary system). The excitation frequency is associated to ω.

$$\tau = t\,\omega_1;\,\lambda_m = \frac{m_2}{m_1 + m_2} = \frac{m_2}{m};\,\Omega = \frac{\omega}{\omega_1};\,{\omega_1}^2 = \frac{k}{m};\,{\omega_2}^2 = \frac{g}{l};\,{\Omega_0}^2 = \frac{{\omega_2}^2}{{\omega_1}^2};\,\lambda_l = \frac{l}{A};$$

$$\xi_1 = \frac{d}{2\,m_1\,\omega_1};\,\xi_2 = \frac{D}{2\,m_2\,l^2\,\omega_2};\,\varrho = \frac{g}{A\,{\omega_1}^2} \tag{64}$$

$$\text{path } u(\tau) = \frac{y(t)}{A} \text{ and angle } \theta(\tau) = \frac{\varphi(t)}{\varphi_0} \tag{65}$$

$$u'' + 2\xi_1 u' + u + \lambda_m \lambda_l \left(\theta'^2 \cos\,\theta + \theta'' \sin\,\theta\right) + \varrho = \Omega^2 \cos\,(\Omega\tau) \tag{66}$$

$$\theta'' + 2\xi_2 \Omega_0\,\theta' + \sin\,\theta\left({\Omega_0}^2 + {\lambda_l}^{-1}\,u''\right) = 0 \tag{67}$$

Figure 13 depicts left the excitation (magenta) and the time response signals of y (blue) and φ (red) and its phase space behavior on the right-hand side.The parameters are chosen in such a way,that the pendulum starts rotate.The excitation primary system has a resonance ratio of $\omega : \omega_1 = 1 : 1$ and secondary primary frequency ratio is $\omega_2 : \omega_1 = 2 : 1$. The stability of such a pendulum is described in chapter 4.4 of [7], where so called semi trivial and nontrivial solutions for this system are discussed.

A treaty of such a system, a kinetic energy harvesting device, with additionally a nonlinear spring system on the primary and an electromagnetic harvester on the secondary system is given in Ref. [12]. Note that the derivation of the system equations there have been made without the Lagrangian formalism and the found system equations are equivalent.

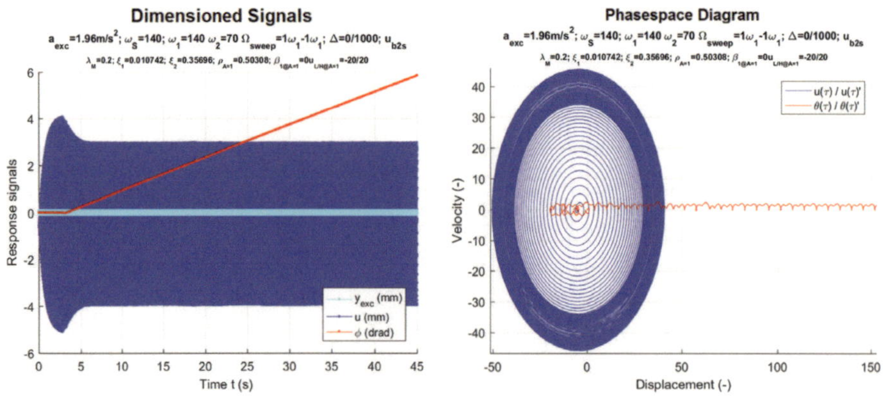

Figure 13. Numerical simulation results of the nonlinear basepoint excited 2DoF system shown in (66, 67) using a constant acceleration of 0.2 g and a constant basepoint excitation of f_{exc} = 22.22 Hz.

4. Conclusions

In the introduction, we showed the equivalence of rotary and translatory mechanical systems as well as the equivalence of mechanical and electrical resonance systems. Also, a brief introduction to the Lagrangian formalism is given. In preparation to nonlinear resonance systems, also rotational and translational springs are discussed. Three classes of spring systems have been identified: linear springs nonlinear symmetric springs and nonlinear asymmetric springs. Throughout the chapter further readings are proposed.

In Section 2, linear resonance systems with one and two degrees of freedom have been investigated using basepoint excited systems. Using the Laplace Transformation is most useful to analyze any linear resonance system with a periodic excitation.

Section 3 deals with nonlinear resonance systems. When in such a dynamical system one of the resulting DE's is of Mathieu or Hill type, the response amplitude of such a system might grow exponentially. This is exemplary demonstrated in Section 3.2 identifying the system differential equations of a basepoint excited two degree of freedom system. Some dynamic properties of such a system is demonstrated.

A. Appendix

A.1. Nonlinear symmetric spring systems

Using instead of disk magnets ring magnets, strong nonlinearities can be generated. The following series in **Figure A1** depicts a few simulation cases. Some of shown simulation cases have been validated and proven experimentally.

A.2. Variant of linear 2DoF system

Instead of using linear springs, magnetic nonlinear springs can be used (see also a selection of such spring characteristics in A.1). Using nonlinear springs and making the system nonlinear (instead of having only a linear spring term, we have for each spring also a term of the form $\beta_1 u^3$ and $\beta_2 v^3$). The relative response signals of such a system is depicted in **Figure A2**.

It is interesting, that the relative motion of the 2. DoF is responding with resonance between 19…25.5 Hz.The first degree of freedom has a nonlinear spring hardening behavior, reaching A_{pp} = 7.9 mm at 25.5 Hz.

A.3. Variant of nonlinear SDoF system

A created tool by the author in Matlab/Simulink has been used to simulate many basepoint excited SDoF or nDoF systems with rotational or translational or mixed structures.

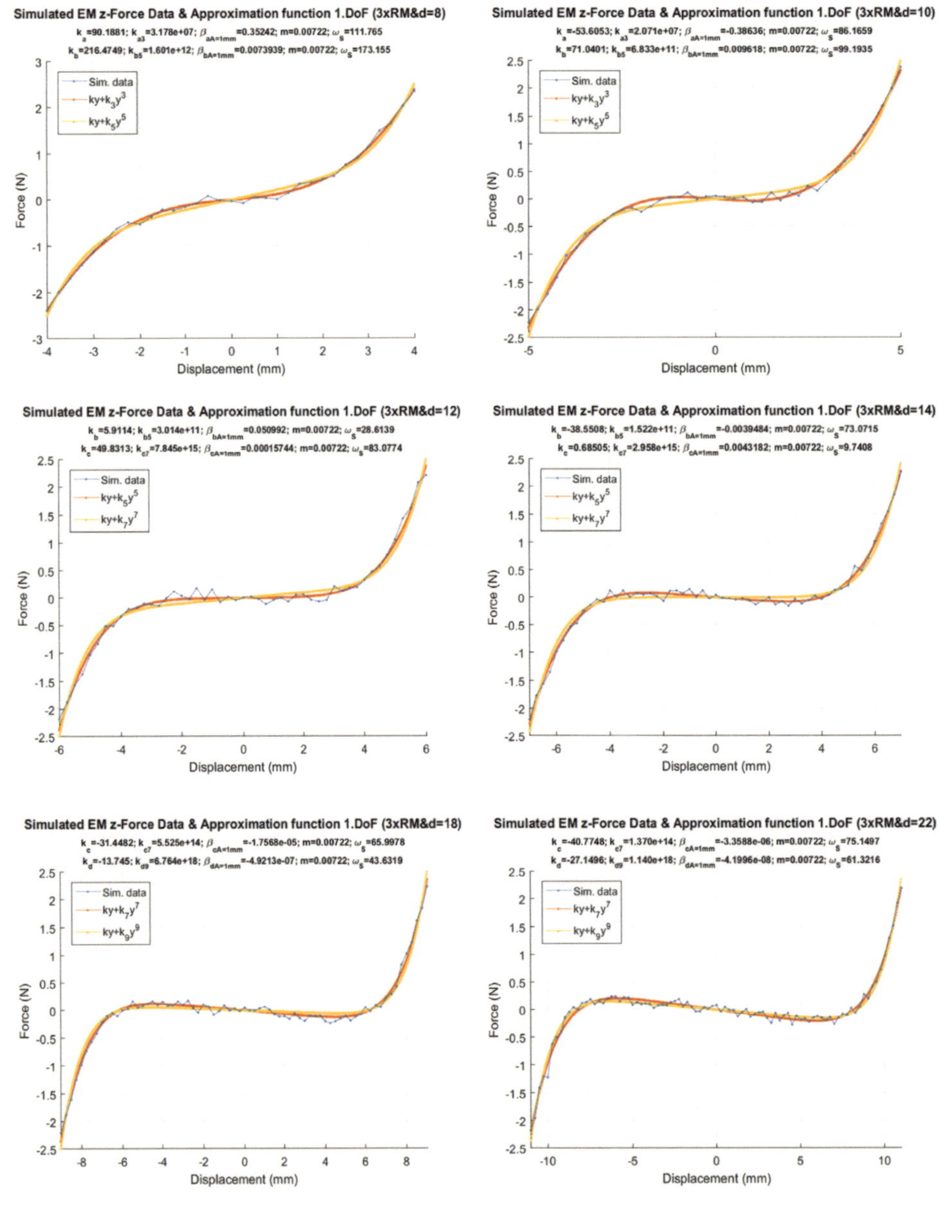

Figure A1. Spring force behavior of ring magnets using different distances of non-movable magnets.

It allows to simulate such systems with constant amplitude or constant acceleration, can handle hard or soft-impact of the oscillating proof mass(es). In addition, one sided spring characteristics can be simulated, see also **Figure A3** – a feature that is especially interesting in relation with magnetic springs. Main disadvantage of such one-sided bound springs is the fact, that they need to be installed upright. The behavior of such a one-sided magnetic spring is depicted in **Figure A3**. It has a frequency response similar to a softening spring. The maximal

http://dx.doi.org/10.5772/intechopen.70521

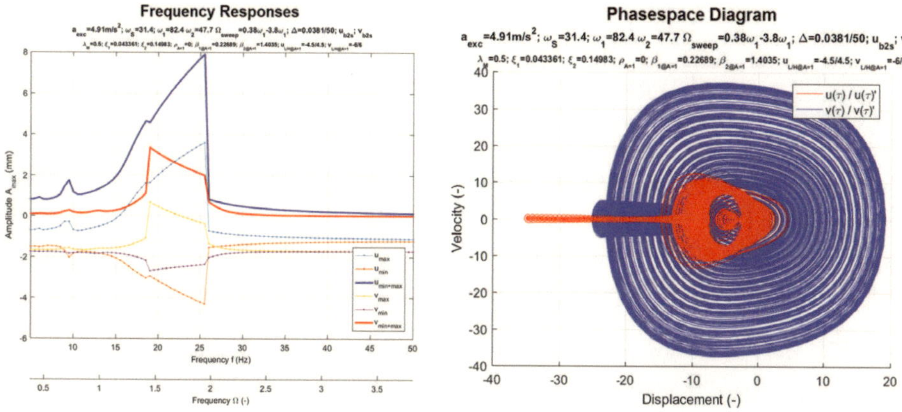

Figure A2. Response signals of a 2DoF system using lumped parameter model in **Figure 5**: Instead of having linear springs, also nonlinear springs are present.

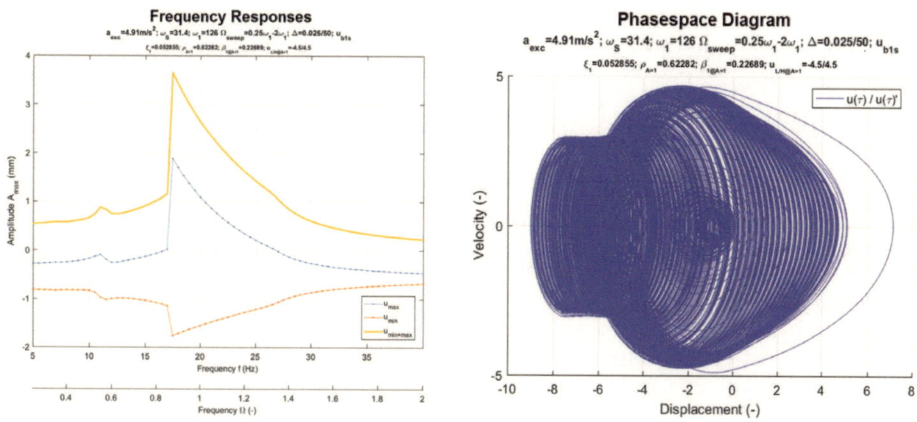

Figure A3. Response signals of a one-sided bound magnetic spring.

amplitude of 3.65 mm occurs at 17.5 Hz (the two-sided classical hardening magnetic spring reaches an amplitude maximum of 4.3 mm at 27 Hz). Such a spring system could also be analytically described, by introducing for example continuous piecewise functions.

Author details

Lukas C. Kurmann

Address all correspondence to: lukas.kurmann@gmail.com

University of Applied Sciences and Arts Northwestern Switzerland, Switzerland

References

[1] Feynman RP, Leighton RB, Sands ML. The Feynman Lectures on Physics. New York: Basic Books, 2011; 2010

[2] M. Cartmell, Introduction to linear, parametric and nonlinear vibrations. London, UK: Chapman and Hall, 1990

[3] Calkin MG. Lagrangian and Hamiltonian Mechanics: Hamiltonian and Lagrangian Formalism. Singapore, River Edge NJ: World Scientific; 1996

[4] Pesce CP. The application of Lagrange equations to mechanical systems with mass explicitly dependent on position. Journal of Applied Mechanics. 2003;**70**(5):751

[5] Spreemann D, Manoli Y. Electromagnetic Vibration Energy Harvesting Devices: Architectures, Design, Modeling and Optimization. Dordrecht: Springer Netherlands; 2012

[6] Elzaki TM. On the New Integral Transform "ELzaki Transform" Fundamental Properties Investigations and Applications. Global Journal of Mathematical Sciences. 2012;**4**(1):1-13

[7] Tondl A. Autoparametric Resonance in Mechanical Systems. Cambridge, UK: Cambridge University Press, 2000

[8] Nayfeh AH, Mook DT. Nonlinear Oscillations. Weinheim: Wiley; 2007

[9] Kurmann L et al. Autoparametric resonance Systems for vibration-based energy harvesters. Journal of Physics: Conference Series. 2015;**660**:12070

[10] Mann BP, Sims ND. Energy harvesting from the nonlinear oscillations of magnetic levitation. Journal of Sound and Vibration. 2009;**319**(1–2):515-530

[11] A. C. Luo and D. O'Connor, Stable and unstable periodic solutions to the mathieu-duffing oscillator: 6 - 11 Aug. 2012, Budapest, Hungary ; [proceedings]. Piscataway, NJ: IEEE, 2012.

[12] Kurmann L, Jia Y, Manoli Y, Woias P. Magnetically levitated autoparametric broadband vibration energy harvesting. Journal of Physics: Conference Series. 2016;**773**:12006

Chapter 8

Resonance Effect of Nanofibrous Membrane for Sound Absorption Applications

Klara Kalinova

Additional information is available at the end of the chapter

http://dx.doi.org/10.5772/intechopen.70361

Abstract

Nanofibrous layers have unique acoustic properties due to the large specific surface area of the nanofibers, where viscous losses may occur and also the ability of the nanofiber layer to resonate at its own frequency. The resonance membrane is then, upon impact of sound waves of low frequency, brought into forced vibrations, whereby the kinetic energy of the membrane is converted into thermal energy by friction of individual nanofibers, by the friction of the membrane with ambient air, and possibly with other layers of material arranged in its proximity, and part of the energy is also transmitted to the frame, by which means the vibrations of the resonance membrane are damped. When sound waves hit the nanofiber membrane, they introduce forced vibrations in the case of resonance which have maximal amplitude. The principle of the technology is achieved by the synergy of perforated plate in the form of a cavity resonator with nanofibrous layer in the form of resonant membrane. The parameters of the resonant nanofibrous membrane together with the shape and volume of the perforations then determine which sound frequencies will be damped and to what extent.

Keywords: membrane, nanofibers, sound absorption, foil

1. Introduction

The confusion between sound insulation and sound absorption is often phenomenon. Sound-absorbing materials play an indispensable part in controlling noise generated within a room or in reverberant areas. Although such materials are highly effective as sound absorbers, they are relatively poor sound insulators because of their soft, porous, and lightweight construction. Sound insulation prevents sound traveling from one place to another such as between apartments in a building. A part of sound energy is absorbed, the next part is reflected, and the rest is transmitted to the second room. The sound attenuation is due to the air viscosity, nonreversible deformation of material, and the thermal conduction between the fibers and the air. The sound absorption also depends on structure characteristics, that is, fiber arrangement, fiber fineness, density of fiber, porosity, and so on. The sound absorption of lower frequencies becomes the main problem of this acoustic section.

Nanofibrous layers have unique acoustic properties due to the large specific surface area of the nanofibers, where viscous losses may occur and also the ability of the nanofiber layer to resonate at its own frequency. The resonance membrane is then, upon impact of sound waves of low frequency, brought into forced vibrations, whereby the kinetic energy of the membrane is converted into thermal energy by friction of individual nanofibers, by the friction of the membrane with ambient air, and possibly with other layers of material arranged in its proximity, and part of the energy is also transmitted to the frame, by which means the vibrations of the resonance membrane are damped. When sound waves hit the nanofiber membrane, they introduce forced vibrations in the case of resonance which have maximal amplitude.

1.1. Membrane resonators

Materials based on resonance principle can be divided into three groups: arrangements behaving as vibrating membranes, arrangements behaving as vibrating plates, and arrangements consisting in the principle of Helmholtz resonators.

The work [1] uses a mechanic analogy of an acoustic resonance system consisting of an acoustic mass m_a connected to an acoustic plasticity c_a, the movement of which is dampened by an acoustic resistance R_a. The behavior of the membrane (plate) can be compared to the behavior of a corpus with a certain mass flexibly connected to a spring (represented by an air cushion, of by the air in material pores). Assuming that the elements representing the mass are perfectly stiff and the elements representing the flexibility have no mass, this problem can be compared to the theory of linear circuits in the field of electrical engineering, where the coils are considered as having no capacity, condensers having no inductivity, and resistors being purely ohmic [2]. As in the field of electrical engineering, where the notion of electrical impedance is introduced, which is defined as the ratio between the voltage and current, a similar variable can be introduced for acoustic systems—the acoustic impedance Z. It is defined as the ratio of the pressure affecting the system and the volumetric rate at which the system vibrates thanks to the effect of the abovementioned force. For individual elements, apply the following:

$$Z_m = j\omega m_a,\ Z_R = R_a,\ Z_c = \frac{1}{j\omega c_a} \tag{1}$$

where j is an imaginary unit, ω the angular frequency, in s^{-1}.

Assuming that the system is not damped ($R_a = 0$), it meets the equation

$$j\omega m_a + \frac{1}{j\omega c_a} = 0, \tag{2}$$

the resonance of the system according to Ref. [1] then occurs at the frequency

http://dx.doi.org/10.5772/intechopen.70361

$$f_r = \frac{1}{2\pi}\sqrt{\frac{1}{m_a c_a}} = \frac{1}{2\pi}\sqrt{\frac{\rho}{m_{sq} d}}, \tag{3}$$

where c is the speed of sound propagation through the environment, in m s^{-1}, ρ is the air density, in kg m^{-3}, m_{sq} is the surface density of the membrane, in kg m^{-2}, and d is the thickness of the air cushion in m.

The influence of the surface density of the membrane on the resonance frequency of the system, or the sound absorption coefficient that increases with this characteristics, is also apparent. We can also notice the shift in the maximums of the sound absorption coefficient toward lower frequencies.

A thin circular membrane is defined in Ref. [3] as a structure that arises by stretching, for example, a thin homogeneous elastic film with a constant surface density to a rigid circular frame. The membrane gains its rigidity by means of this stretching induced by radially acting force. The resonance frequency of a thin circular membrane is defined using the relation

$$f_i = \frac{a_{0,i}}{2\pi R}\sqrt{\frac{\nu}{m_{sq}}}, \tag{4}$$

where $a_{0,i}$ are constants of symmetric vibrations of the circular membrane for i-modes, R is the membrane diameter, in m, and ν radially acting stretching force related to the unit of the frame diameter (membrane tension), in N m^{-1}.

From the above, it shows that the resonance frequency of the membrane decreases with its increasing surface density. Membrane resonators based on a resonant principle of a nanofibrous layer function effectively as slim lightweight absorbing solutions. Contrary to conventionally used microscale sound absorbers, sound-absorbing membranes based on submicron fibers show a higher absorption abilities—due to the possibility of resonating on its own resonant frequency, the nanofibrous membrane is able to absorb critical lower sound frequencies. These unique properties come from the nature of nanofibrous layers, that is, small fibrous diameter (respectively, high specific surface area) and high porosity. This makes it possible to reach higher viscous loss inside the material and consequently to dissipate the acoustic energy. Nanofibrous elements and optimal rigidity of the membrane itself then allow an acoustic system to vibrate more efficiently [4, 5]. Resonant nanofibrous membranes of insignificant thickness are prepared from different polymer solutions in the form of electrospun nanofibers captured on a substrate layer via electrospinning method.

The theoretical bases of sound absorption characteristics that the paper deals with are studies performed by Sakagami et al. The study [6] focuses on a membrane-type sound absorber. To analyze the absorption mechanism, the solution is rearranged in a form which points out the contribution from each element of the membrane. The effects of the parameters of the sound absorption system are discussed in the light of the calculated results. Also, the method used for predicting the peak frequency and the peak value of the oblique-incident absorption coefficient of the membrane-type sound absorber is presented and satisfactorily explains the relationship between the absorption characteristics and the parameters.

Resonant behavior of a microperforated panel for various perforation ratios in comparison with a panel-/membrane-type absorber is presented in Ref. [7], considering back-wall surface effect. The effectiveness of a fiber-based sound absorbance material involves several parameters such as porosity, tortuosity, fiber diameter, surface density, and thickness [8]. The optimal material types and structural characteristics of such membranes are in the deep interest of researchers, and although some have been proposed, it still remains as a subject of research. Kalinová has demonstrated that the resonance frequency of polyvinyl alcohol (PVA) nanofibrous acoustic membranes decreases with an increasing surface density and the average diameter of the nanofibers [5]. Rabbi et al. sandwiched polyacrylonitrile (PAN) and polyurethane (PUR) nanofibrous membrane between two nonwoven layers of polyester (PET) and wool. All materials with electrospun membrane(s) were found to significantly increase its absorbance. Moreover, the effect of nanofiber layer's number and its surface density was investigated [9]. Asmatulu et al. tested the sound absorbance property of electrospun polyvinyl chloride (PVC) mat of different thickness and with fiber diameters ranging from a few hundred nanometers to a few microns. When the fiber diameter goes beyond 500 nm, the sound absorbance shift toward the lower frequency with a thicker mesh but absorption coefficients remain the same [4].

1.2. Helmholtz's resonators

Helmholtz's resonators are acoustic systems that consist of a swinging air plug and a connected air volume. It can have a variety of forms: an empty wine bottle, corpus of a string instrument, bass reflex enclosures of loudspeakers, and wall coverings made of perforated panel. These acoustic systems can be arranged either separately or jointly to the perforated board, which is mounted to a certain distance from the wall [10].

In the study [11], the variable system of sound absorption power by the chairs in the low-frequency range was examined. As the results of scale model experiments (1/10 scale) in the reverberation room, the absorption power was controlled in the low-frequency range by the opening and closing of holes of the resonator. The diameter of holes, a neck's length, and a cavity volume of the seat were evaluated. The result was obtained for 125 or 250 Hz by changing the cavity volume of the seat in the experiment.

An acoustical structure consisting of a large-scale isolated resonator with a large-diameter cylindrical cavity has been studied in the work [12]. This resonator differs from the classical Helmholtz's resonator where the cavity is only several millimeters in diameter and lined with a sound-absorbing material. The impedance of the cavity and the impedance of the volume of the resonator are calculated. Calculations show that the sound energy is absorbed by resonators made of sound-reflecting materials. Absorption is of a resonant character with the resonant frequency at 60 Hz. A resonator measuring 200 × 200 cm, with the cavity diameter of 50 cm and the distance to the rigid surface being 30 cm, absorbs 3.5 m^2 of sound energy at the resonant frequency. At very low frequencies, changes in the imaginary parts of both cavity and radiation impedances occur along with the increase in the cavity diameter and frequency.

The study [13] deals with the effect of orifice geometry on the resonance frequency of Helmholtz's resonators. Helmholtz's theoretical formula for calculating resonant frequency f_H is as follows:

$$f_H = \frac{c}{2\pi}\sqrt{\frac{a}{V}}, \tag{5}$$

where c is the sound velocity, a is the diameter of orifice, and V is the volume of cavity. Further, Sondhauss's calculation of resonant frequency f_S with the correction $\delta = 4a/3\pi$ is then

$$f_H = \frac{c}{2\pi}\sqrt{\frac{A}{V(l+2\delta)}}, \tag{6}$$

where A is the orifice area and l is the orifice thickness.

1.3. Measuring methods

Vibration phenomena can be investigated by the noninvasive optical methods. One of the most widely used methods for vibration measurement and analysis is laser vibrometry that can be combined with the high-speed camera. This approach can be seen in different application fields, for example, the development and monitoring of high-speed milling devices [14–21]. Nabavi describes the utilization of the particle image velocimetry technique to measure the velocity of the standing waves within an air-filled rigid-walled square channel subjected to acoustic standing waves. The data were compared with the analytical results obtained from the time-harmonic solution of the wave equation [22].

The resonant effect of nanofibrous membrane has been studied by means of high-speed digital camera in the author's paper [23]. The study attempted to predict the sound absorption behavior of the PVA nanofibrous membrane in comparison with the homogeneous membrane structure using an experimental setup involving a high-speed camera. The membrane has been exposed to plane sinusoidal sound wave and its deflection was picked by the high-speed digital camera. The resonant peaks of oscillating nanofibrous membrane as well as homogeneous membrane occur (see **Figure 1**). The recent study [24] shows how except for the lowest frequencies (first resonance peak), the resonant behavior of the membrane is affected by the resonance of the tube when the effect of mass per unit area on resonance frequencies of the membrane placed in an open and closed tubes is investigated.

Two-microphone impedance measurement tube type 4206 is used to measure the absorption coefficient in the frequency ranges from 100 Hz to 6.4 kHz. This is achieved by measuring the incident and reflected components of random noise, which is generated inside the tube. From the incident and reflected components of the sound pressure at two microphone positions, the frequency response functions are calculated due to the cross-spectrum of the two microphone signals. Using these values, the sound absorption coefficient can be determined. An apparatus is used to determine the sound absorption coefficient of laboratory circular samples with a

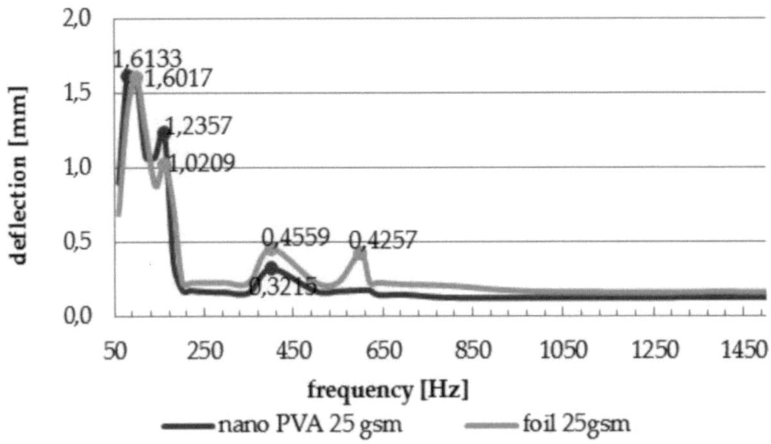

Figure 1. The resonance peaks of nanofibers compared with foil (took over author's paper 23).

diameter of 100 mm for a frequency range of 100–1600 Hz and 29 mm for a frequency range of 500–6400 Hz, according to the standard ASTM E1050-08.

Due to the fact that the quad perforations in the plate were almost the same size as the diameter of the small tube of the measuring apparatus, the samples were measured only in a large tube with a diameter of 100 mm for a limited frequency range of 100–1600 Hz. These frequencies, however, cover the area particularly focusing on middle and lower frequencies.

2. Acoustic element design

2.1. Production of nanofibrous layer for membrane resonator design

For the production of nanofibrous membranes, roller electrospinning method (nanospider machine) was used. In this method, there is a roller that is connected to a high voltage supplier, and at the top of the roller there is a counterelectrode that was grounded. Taylor cones are created on the roller surface toward counterelectrode (**Figure 2**). Individual nanofibrous layer of very low basis weight of about 0.1–2 g/m^2 is not self-supporting. That is why the nanofibers are deposited on a thin supporting textile. This carrier has to be sound permeable with a low basis weight of about 20–50 g/m^2. Process parameters such as roller speed, distance between the electrodes, voltage, and so on are set for an optimal nanofiber diameter and the basis weight of nanofibrous membrane.

For the production of PA6 nanofibrous membranes, the cord electrospinning method was used [25]. In this method, the cord was connected to a high voltage supply, and at the top of the cord there was a counterelectrode, which was grounded. The liquid polymeric material is applied onto the cord around its whole circumference, and then the application means moving reversibly along the active spinning zone of the cord and the process of electrostatic spinning of the liquid polymeric material is started. Taylor cones were created on the cord surface toward the counterelectrode.

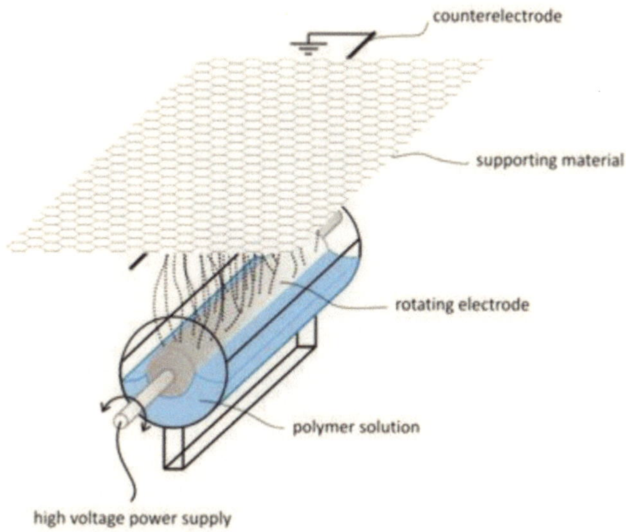

Figure 2. Schematic diagram of roller electrospinning method used for PVA nanofibrous membrane production.

2.2. Cavity resonator together with nanofibrous resonant membrane

The principle of the technology is achieved by the synergy of the perforated plate in the form of a cavity resonator with nanofibrous layer in the form of a resonant membrane. The resonant nanofibrous membrane is arranged on the surface of the cavity resonator, to which it is fixedly attached, for example, glued or laminated, and so on. Its parts, which overlap the orifices leading into the cavities of the cavity resonator, constitute separate resonant surfaces, whereby the resonant frequency of each of them is determined, apart from the overall properties of the resonant membrane, also by their size and shape. Upon impact of sound waves, these resonant surfaces are brought into forced vibrations, which are subsequently damped by friction in the inner structure of the resonant membrane, by the friction of the resonant membrane against ambient air, and possibly against other layers of the material arranged in its proximity, wherein part of the kinetic energy of the resonating membrane is transmitted to the cavity resonator. Moreover, friction in the inner structure of the resonant membrane is further increased by the fact that the neighboring resonant surfaces can vibrate with mutually different periods or deviation.

At the same time, it is possible—while maintaining the thickness of the acoustic element—to damp sound frequencies which could be normally damped by the cavity resonator with extremely large air gap. In order to obtain the required sound-absorbing properties, the resonant membrane can be arranged on both opposing surfaces of the cavity resonator.

The acoustic element is based on a quad hollow plate (see **Figure 3**) whose reverse side is covered by a thin carrier layer with a nanofibrous membrane which to a certain extent protects the frame against mechanical damage. For the final application in the room acoustic, the space between the nanofibrous membrane covering the thin perforated plate and the wall or ceiling

Figure 3. Components used to design the acoustic elements—aluminum quad hollow plate 9/11 (size of perforation is 9 × 9 mm, span of perforation is 11 × 11 mm) with a thickness of 1 mm (left) covered with a nanofiber layer (middle) lighted (right).

(20–50-mm air gap in the mentioned experimental) is of huge benefit to the new technology. It can be used for the installation of lighting, audio speakers or heating, and so on. The sound-absorbing means can be used, for example, for the production of acoustic bodies, interior blinds, tiling, ceilings, screens, and separating walls for interiors, or, as the case may be, segment or profile elements for the transportation industry (paneling of cabin).

Figure 3 (on the right) shows the final lighted prototype of acoustic system based on nanofibrous membrane covering the thin perforated plate. The resonance frequency of the acoustic system is then determined especially by dimensions of plate perforations, by the size and shape of the inlet orifices, and by its material and thickness of the plate.

Table 1 shows the calculation of resonant frequency for each of quad perforated plates that have been studied at the experimental section of this work.

Firstly, due to the fact that the quad perforations in the plate were almost the same size as the diameter of the small tube of the measuring apparatus, the samples were measured only in a

Quad hollow plate (quad size in mm/quad span in mm)	f_H (Hz)	f_S (Hz)
3/5	2682	2467
4/6	2581	2462
5/7	2473	2415
8/10	2190	2219
9/11	2112	2155
10/12	2040	2095
25/30	1290	1369

Table 1. Calculated resonant frequency of separate perforated plates based on Helmoltz's (f_H) formula (5) and Sondhauss's (f_S) formula (6).

large tube with a diameter of 100 mm for a limited frequency range of 100–1600 Hz. These frequencies, however, do not cover the resonant frequencies of separate perforated plates calculated in **Table 1**. Secondly, in the case of nanofibrous layer in a form of resonant membrane, the measurement of membrane tension v for resonant frequency calculation according to formula 4 is impossible because of the low tension together with non-homogeneous nanofibrous layer. It is why the resonant frequency of nanofibrous membrane has been determined by the optical method [24] where the first resonant peak was detected around 100 Hz. Then, the results of nanofibers-covering perforated plate and the separate perforated plate are compared only by way of sound absorption curves.

3. Sound absorption results

In this section, the sound absorption measurements of acoustic means with nanofibrous membrane are shown. Two-microphone impedance measurement tube type 4206 was used to measure the sound absorption coefficient in a limited frequency range of 100–1600 Hz.

The following figures show a graphs of sound absorption coefficients α in dependence on the frequency of sound for separate aluminum plate having different size of orifices and spacing between quad orifices, which is deposited in different distances from the wall (i.e., separate Helmholtz resonator), as well as for sound-absorbing means comprising this perforated plate, whose surface is overlapped by the resonant membrane formed by the layer of nanofibers from polyamide 6 (PA6) having a basis weight of 0.2 g m^{-2} deposited on a thin carrier having a basis weight of 25 g m^{-2}. One of the configurations is filled by a foam or a fleece having a thickness of 20 mm.

The individual perforated plate and the same perforated plate covered by a thin carrier with nanofibers have been compared and are shown in **Figure 4**. The huge growth of sound absorption of middle frequencies can be seen. Starting with 500 Hz, the sound absorption curve of nanofibers improved element is constant contrary of the unstable curve of individual

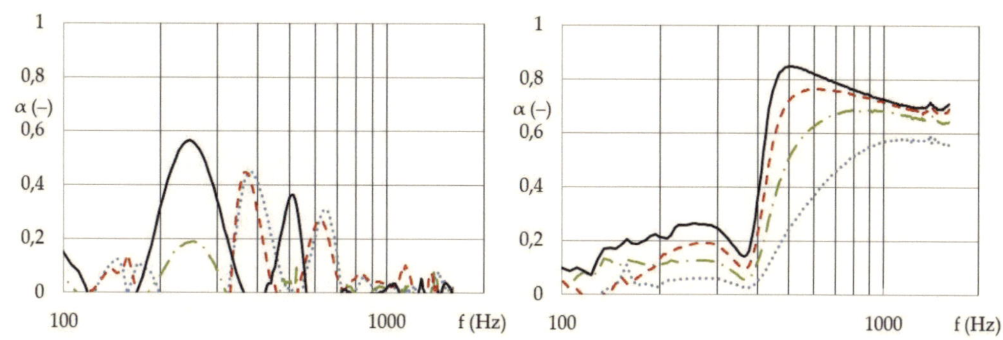

Figure 4. Frequency dependence of the sound absorption coefficient; quad hollow plate 9/11 (side of quad perforation is 9 mm, span of quad perforation is 11 mm) with a thickness of 1 mm with an air gap of 20 (blue - dotted), 30 (green - dash-dotted), 40 (red – dashed), and 50 mm (black) on the left. Nanofibrous membrane of 0.2 g/m^2 on a carrier of 25 g/m^2 covering the same perforated plate (quad 9/11) on the right.

perforated plate. For the verification of nanofibrous membrane efficient, the individual nanofiber carrier without nanofibers has been evaluated and is shown in **Figure 5**. Then, it is evident that the carrier-covering perforated plate improves the sound absorption of high frequencies but it does not provide wide-frequency efficiency as well as nanofibers improving plate.

Quad perforated plate of different sizes and spans has been evaluated and is shown in **Figure 6**. When the size of the perforation is 9 × 9 mm and the span of the perforation is 11 × 11 mm, then it is marked (9/11).

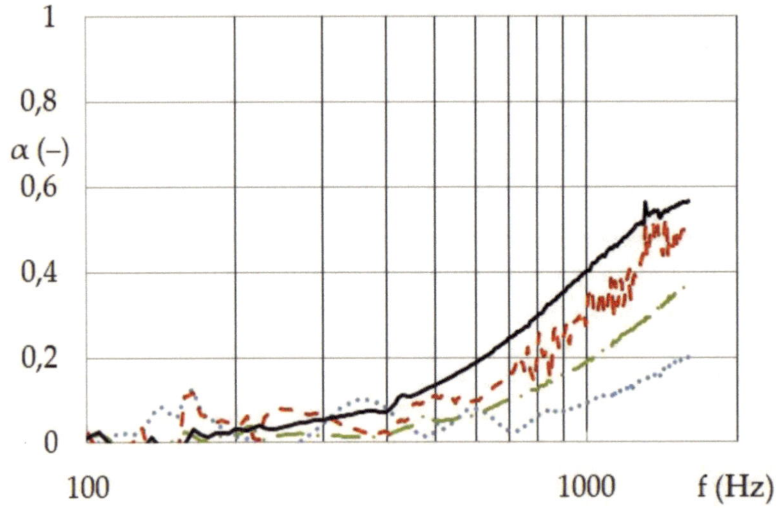

Figure 5. Frequency dependence of the sound absorption coefficient; individual carrier of 25 g/m^2 covering the perforated plate (quad 9/11) with an air gap of 20 (blue - dotted), 30 (green - dash-dotted), 40 (red – dashed), and 50 mm (black) between the acoustic element and the wall.

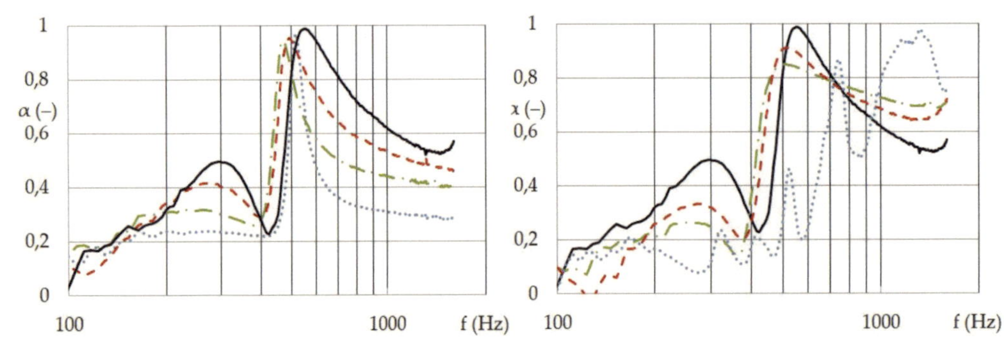

Figure 6. Frequency dependence of the sound absorption coefficient; nanofibrous membrane of 0.2 g/m^2 on a carrier of 25 g/m^2 covering the quad perforated plate of different size with a thickness of 1 mm with an air gap of 50 mm. Quad perforated plate of 3/5 (blue - dotted), 4/6 (green - dash-dotted), 5/7 (red – dashed), and 8/10 (black) of side/span (left). Quad perforated plate of 8/10 (black), 9/11 (green - dash-dotted), 10/12 (red – dashed), and 25/30 (blue - dotted) of side/span (right).

With an increasing size of quad hole, the sound absorption achieves the wide-frequency efficiency generally as can be seen in **Figure 6**. The best arrangement of quad hole seems to be 9-mm side of quad and the span of 11 mm (9/11), where the nanofibrous resonant membrane interacts with the perforated panel to achieve optimal parameters of the acoustic system.

Due to two effects, the large specific surface area of the nanofibers and also the ability of the nanofibrous layer to resonate at its own frequency, the nanofibrous membrane achieves broadband sound absorption compared to the narrowband effect of homogeneous foil on the same perforated plate (see **Figure 7**). Starting with 500 Hz, the sound absorption curve of nanofibers improved element is constant contrary of the unstable curve of foil improved perforated plate.

When the perforated plate is improved by the nanofibrous membrane on each of both sides, then the sound absorption of higher frequencies falls slightly (see **Figure 8**). Then, the

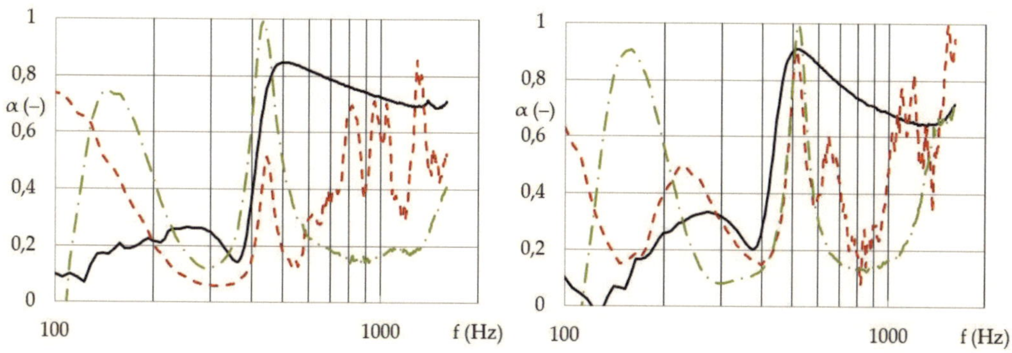

Figure 7. Frequency dependence of the sound absorption coefficient; quad perforated plate of 9/11 (left) and 10/12 (right) is covered by the nanofibrous membrane of 0.2 g/m^2 on a carrier of 25 g/m^2 (black) or foil of 7 g/m^2 (red – dashed) or foil of 40 g/m^2 (green - dash-dotted). The air gap between the 1-mm thick panel and the wall is 50 mm.

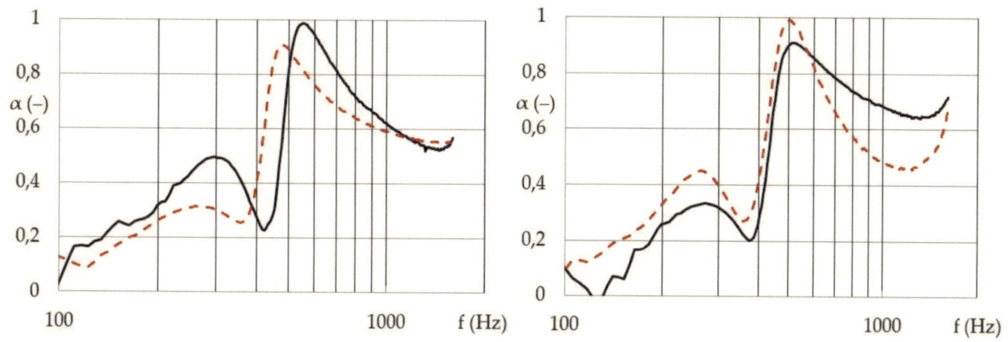

Figure 8. Frequency dependence of the sound absorption coefficient; quad perforated plate of 8/10 (left) and 10/12 (right) is covered at the top by the single nanofibrous membrane of 0.2 g/m^2 on a carrier of 25 g/m^2 (black) or it is covered by the nanofibrous membrane of 0.2 g/m^2 on a carrier of 25 g/m^2 (red – dashed) from both sides. The air gap between the 1-mm thick panel and the wall is 50 mm.

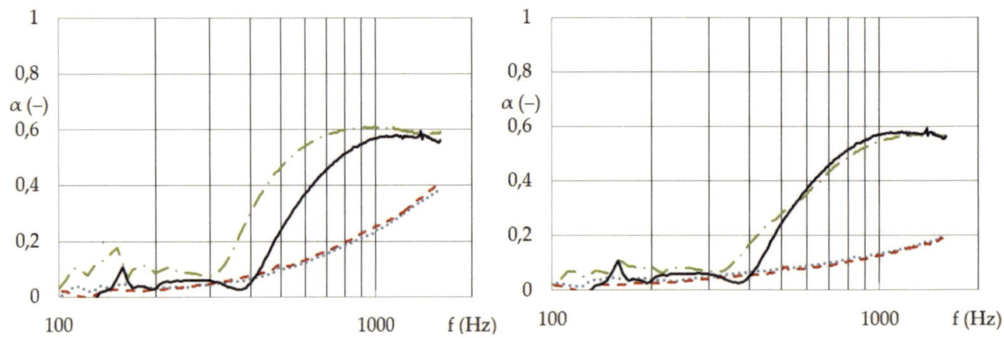

Figure 9. Frequency dependence of the sound absorption coefficient; quad perforated plate (9/11) covered by the nanofibrous membrane of 0.2 g/m^2 on a carrier of 25 g/m^2. The air gap between the 1-mm thick panel and the wall is 20 mm (black): the same nanofibers-covering perforated plate filled (green - dash-dotted) by the foam (left) or fleece (right); the same separated plate filled (red – dashed) by the foam (left) or fleece (right); the separated fillings (blue - dotted) of foam (left) or fleece (right). The thickness of the whole acoustic system is 20 mm in all configurations.

membrane resonators covering the mass of Helmholtz's resonator obstruct the sound absorption inside the cavity.

The individual perforated plates in a form of cavity resonators should be filled for sound absorption of higher frequencies. **Figure 9** shows the comparison of acoustic system consisting of nanofibers-covering quad perforated plate and the same perforated plate without covering but filled. The filling has been chosen from the standard sound absorbers line. The first is melamine foam of 9.5 ± 1.5 kg/m^3 and 20-mm thickness (**Figure 9** on the left) and the second polyester fleece of 24 kg/m^3 $\pm$ 10% and 20-mm thickness (**Figure 9** on the right). From the comparison, it can be seen that the inferior sound absorption results if the perforated plate is filled (red – dashed curve) in comparison with nanofibers covering the same perforated plate without filling (black curve). Then, the resonance capability of nanofibrous membrane more than compensates a mass of filling. Regarding the applicability of nanofibers-covering perforated plate, the gap between the panel and the wall can be used for light or audio installation. If the nanofiber-covering plate is filled (green - dash-dotted curve), then the sound absorption is slightly better than that of non-filled. However, the benefit of air gap outweighs the nominal sound absorption growth.

4. Conclusions

The resonance ability of nanofibrous layer has been verified in the last author's paper. The membrane has been exposed to plane sinusoidal sound wave and its deflection was picked by the high-speed digital camera. The resonant peaks of oscillating nanofibrous membrane as well as homogeneous membrane occur around 70–100, 300–400, and 550–600 Hz depending on their parameters. The calculated resonant frequency of the perforated plate is around 2–2.5 kHz. The sound absorption peaks of nanofibers-covering perforated plate are around 500 Hz. From the comparison of resonant frequencies perforated plate in a form of Helmoltz's resonator, separate nanofibrous layer in a form of membrane resonator, and final nanofibers-covering perforated plate, the major effect of the resonant frequency of the nanofibrous layer together with a distance of the final plate from the wall can be seen.

The diameter of nanofibers, the basis weight, and the polymer of the nanofibrous membrane as well as the shape, size, and span of perforations of Helmholtz's resonator affect the sound absorption behavior of acoustic element.

The two applied nanofibrous membranes have not almost any effect on sound absorption. The improvement would be redundant.

Acknowledgements

The results of this project LO1201 were obtained through the financial support of the Ministry of Education, Youth and Sports in the framework of the targeted support of the "National Programme for Sustainability I" and the OPR&DI project Centre for Nanomaterials, Advanced Technologies and Innovation CZ.1.05/2.1.00/01.0005. Particularly the work was supported by the Technology Agency of the Czech Republic within the project "Broadband components with resonant nanofibrous membrane for room acoustics" no. TH02020524. I also thank my doctoral student Tomas Ulrich for measurement support.

Author details

Klara Kalinova

Address all correspondence to: klara.kalinova@tul.cz

Technical University of Liberec, Czech Republic

References

[1] Kolmer F, Kyncl J. Prostorová Akustika (in Czech). Prague: SNTL Praha; 1980. p. 244. ISBN: 04-514-80

[2] Merhaut J. Teoretické Základy Elektroakustiky (in Czech). Vol. 3. Prague: Academia; 1976. p. 325

[3] Škvor Z. Akustika a Elektroakustika (in Czech). Vol. 1. Prague: Academia; 2001. p. 527. ISBN: 80-200-0461-0

[4] Asmatulu R, Khan W, Yildirim MB. Acoustical Properties of Electrospun Nanofibers for Aircraft Interior Noise Reduction [Online]. B.m.: ASME. 2009;2009:223-227. ISBN: 978-0-7918-4388-8. DOI: 10.1115/IMECE2009-12339

[5] Kalinová K. Nanofibrous resonant membrane for acoustic applications. Journal of Nanomaterials. 2011;**2011**:1687-4129. ISSN: 1687-4110. DOI: 10.1155/2011/265720

[6] Kimihiro S, Kiyama M, Morimoto M, Takahashi D. Sound absorption of a cavity-backed membrane: A step towards design method for membrane-type absorbers. Applied Acoustics. 1996;**49**(3):237-247. ISSN: 0003682X-4. DOI: 10.1016/S0003-682X(96)00025-4

[7] Kimihiro S, Morimoto M, Yairi M. A note on the relationship between the sound absorption by microperforated panels and panel/membrane-type absorbers. Applied Acoustics. 2009;**70**(8):1131-1136. ISSN 0003-682X. DOI: http://dx.doi.org/10.1016/j.apacoust.2009.03.003

[8] Hoda SS. Factors influencing acoustic performance of sound absorptive materials. Australian Journal of Basic and Applied Sciences. 2009;**3**(4):4610-4617. ISSN 1991-8178

[9] Rabbi A, Nasouri K, Shoushtari AM, Haji A. Fabrication of electrospun polyacrylonitrile and polyurethane nanofibers for sound absorption. In: 6th TEXTEH International Conference; October 17–18, 2013; Bucharest, Romania. 2013. pp. 233-241. DOI: https://www.academia.edu/download/45476648/FABRICATION_OF_ELECTROSPUN_POLYACRYLONIT20160509-22247-3x9zsf.pdf

[10] Randeberg RT. Perforated panel absorbers with viscous energy dissipation enhanced by orifice design [PhD thesis, dissertation]. NTNU Trondheim, 2000. Available from: https://brage.bibsys.no/xmlui/bitstream/handle/11250/249798/125365_FULLTEXT01.pdf?sequence=1&isAllowed=y

[11] Kobayashi Y, Hase S. Examination of a resonator system for sound power absorption by hall chairs in the low-frequency range. The Journal of the Acoustical Society of America. 1996;**99**:2459

[12] Stauskis VJ. Sound absorption by a large-size isolated acoustic resonator with a cylindrical cavity. The Journal of the Acoustical Society of America. 1998;**103**(5):306

[13] Chanaud RC. Effects of geometry on the resonance frequency of Helmholtz resonators. Journal of Sound and Vibration. 1994;**178**(3):337-348. DOI: 10.1006/jsvi.1994.1490

[14] Kourosh T, Per G. Measurement of milling tool vibrations during cutting using laser vibrometry. International Journal of Machine Tools and Manufacture. 2008;**48**(3–4):380-387. ISSN: 08906955. DOI: 10.1016/j.ijmachtools.2007.09.009

[15] Morteza K, Lu W, Dobrev I, Cheng JT, Furlong C, Rosowski JJ. Digital holographic measurements of shape and three-dimensional sound-induced displacements of tympanic membrane. Optical Engineering. 2013;**52**(10):101916. ISSN: 0091-3286. DOI: 10.1117/1.OE.52.10.101916

[16] Elisabeth IC, Döllinger M, Schuster M, Eysholdt U, Bohr C. Multiparametric analysis of vocal fold vibrations in healthy and disordered voices in high-speed imaging. Journal of Voice. 2011;**25**(5):576-590. ISSN: 08921997. DOI: 10.1016/j.jvoice.2010.04.004

[17] Weizhuo W, Mottershead JE, Ihle A, Siebert T, Schubach HR. Finite element model updating from full-field vibration measurement using digital image correlation. Journal of Sound and Vibration. 2011;**330**(8):1599-1620. ISSN: 0022460X. DOI: 10.1016/j.jsv.2010.10.036

[18] Ishizu K, Yasui T, Ohashi M, Matsumoto N, Komune S. High-speed video analysis of acoustically oscillated guinea pig stapes. The Journal of Laryngology & Otology. 2015;**129**(2):S33-S37. ISSN: 0022-2151. DOI: 10.1017/S0022215114002357

[19] Qican Z, Su X. High-speed optical measurement for the drumhead vibration. Optics Express. 2005;**13**(8):3110. ISSN: 1094-4087. DOI: 10.1364/OPEX.13.003110

[20] Siebert T, Wood R, Splitthof K. High speed image correlation for vibration analysis. Journal of Physics: Conference Series. 2009;**181**:12064. ISSN: 1742-6596. DOI: 10.1088/1742-6596/181/1/012064

[21] David Asael G-H, Perez Lopez C, Mendoza Santoyo F, Tellez-Quinones A, Aguayo DD. Optical characterization of a semisolid membrane by high speed interferometry. Optoelectronics and Advanced Materials, Rapid Communications. 2014;**8**(5-6):382-386. ISSN: 1842-6573

[22] Majid N, Siddiqui K, Dargahi J. Measurement of the acoustic velocity field of nonlinear standing waves using the synchronized PIV technique. Experimental Thermal and Fluid Science. 2008;**33**(1):123-131. ISSN: 08941777. DOI: 10.1016/j.expthermflusci.2008.07.013

[23] Ozturk MK, Kalinova K, Nergis B, Candan C. Comparison of resonant frequency of nanofibrous membrane and homogenous membrane. Textile Research Journal. 2013;**83**(20):2204-2210. DOI: 10.1177/0040517513490064

[24] Klara K, Ozturk MK, Komarek M. Open and closed tube method for determination of resonance frequencies of nanofibrous membrane. The Journal of the Textile Institute. 2016;**107**(8):1068-1078. DOI: 10.1080/00405000.2015.1083353

[25] Maly M, Sejak P, et al. Method for Application of Liquid Polymeric Material onto Spinning Cords and a Device for Production of Nanofibers Through Electrostatic Spinning. Patent 2012, WO2012139533; 2012

Chapter 9

Modal Analysis of Surface Plasmon Resonance Sensor Coupled to Periodic Array of Core-Shell Metallic Nanoparticles

Nadson Welkson Pereira de Souza,
Jefferson Souza Costa, Rafael Correa dos Santos,
André Felipe Souza da Cruz,
Tommaso Del Rosso and Karlo Queiroz da Costa

Additional information is available at the end of the chapter

http://dx.doi.org/10.5772/intechopen.70522

Abstract

The influence of a dielectric shell on metallic spherical nanoparticles [core-shell nanoparticles (CSNps)] in the resonant modal response of a surface plasmon resonance (SPR)-type sensor is presented. The planar multilayer sensor structure, based on the Kretschmann and surface plasmon coupled emission (SPCE) configurations, is coupled to a periodic array of these nanoparticles. In the first configuration, the CSNps are considered as a homogeneous layer with effective permittivity given by the Clausius-Mossotti mixing formula and polarizability of a core shell for a quasi-static scattering regime. In the second configuration, it performed an evaluation via the discrete complex image method (DCIM). Electromagnetic wave propagation is evaluated by the generalized reflection coefficient for multilayer structures. The analytical results are validated by numerical simulations performed via finite element method and also by experimental data. We observed that the dielectric shell thickness affects considerably the sensibility of the sensor when analyzing the change in other parameters of the CSNps array.

Keywords: SPR sensor, wave propagation, modal analysis, core-shell metallic nanoparticle, Kretschmann sensor, SPCE configuration

1. Introduction

Surface plasmon resonance (SPR) sensor is a photonic device capable to detect sensitive variations in the effective electromagnetic refraction index near its multi-layered structure, which can be related to intermolecular interactions or the detection of immobilized analytes, from the interaction between the analyzed samples and the evanescent field generated by surface plasmon polaritons (SPPs) wave, which propagate in the metal-dielectric interface [1].

Despite the first observations of the SPP that have been referenced at the beginning of the last century [2, 3], only at the beginning of the 1980s, SPP-based devices began to be applied to optical sensors with applications in gas detection and biosensors [4, 5], characterizing and quantifying biomolecular interactions [6], medical diagnostics, and viral monitoring [7], among others. The researches in SPR sensors have been increased mainly due to the development of modern nanofabrication techniques, such as the colloidal lithography, focused ion beam (FIB), and electron beam lithography (EBL) [8].

We evaluate an SPR sensor based on Kretschmann configuration (KR) [9] and surface plasmon coupled emission (SPCE) [10] coupled to the periodic array of (CSNps), which can represent the surface immobilization of metal nanopollutants generated, for example, from the nanocomposites manufacturing process [11]. The former has a structure (**Figure 1**) comprising a multilayer formed by a prism (dielectric), a thin metal film (gold), a dielectric spacer (silicon dioxide), the periodic array of CSNps and air. The second one has a similar structure (**Figure 10**) and differs from the first one by the direct incidence of the optical excitation over the immobilized nanoparticles and by the suppression of a layer.

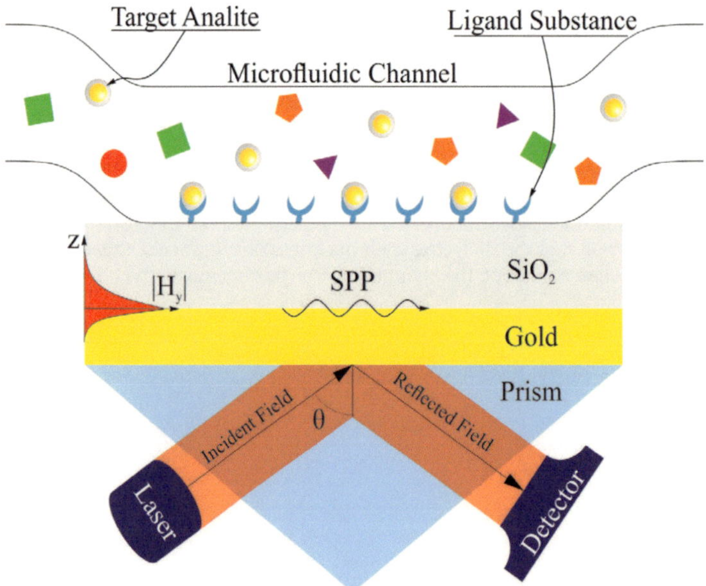

Figure 1. A functional illustration of the SPR sensor based on Kretschmann configuration, coupled to a microfluidic channel with a sample to analyze.

2. Kretschmann configuration

2.1. Functioning description

A functional illustration of the SPR sensor based on the Kretschmann configuration is shown in **Figure 1**, where the structure is coupled to a microfluidic channel with a sample flowing at a controlled rate, while a ligand substance immobilizes only the target nanoparticles (analytes) in the functionalized sensor surface. The optical excitation, coupled through the prism, is linearly polarized on transversal magnetic (TM) or transversal electric (TE) and configured in angular modulation, this is with fixed wavelength $\lambda = 632.8$ nm and variable incidence angle θ [1]. The intensities of the incident and reflected beams are used to determine the angular reflectivity $\Gamma(\theta)$ curve, which is the base information to determine the sensor response.

For TM polarization, the SPP is excited in the gold-SiO_2 interface (**Figure 1**) when the phase condition $\mathrm{Re}\{\beta\} = k_0\sqrt{\varepsilon_p}\sin(\theta)$ matches only for θ greater than the attenuated total reflection (ATR) angle, which implies in a trough point of $\Gamma(\theta)$ [12]. The parameter β is the SPP complex propagation constant, ε_p is the prism electric permittivity, and k_0 is the propagation constant in free space [13].

The alterations in $\Gamma(\theta)$ can be related to the analytes because the coupling conditions of the SPP wave change when the sample material interacts with the sensor field. In this case, we use the angular shifting ($\Delta\theta$) of the minimum points in $\Gamma(\theta)$ as the sensor output, and thus, the sensor sensibility is proportional to $\Delta\theta$ [1].

The extra dielectric layer allows the excitation of multiple resonant wave modes, like guides modes, even in TE polarization [12]. Using both TE and TM $\Gamma(\theta)$ curves, the amount of information about the CSNps increases and improves the estimation of parameters such as surface density, size, and distance between immobilized nanopollutants [1, 14]. This parametric estimation can be performed using the approximated model of Clausius-Mossotti, or even using tools such as Winspal free software [15].

2.2. Theoretical modeling

The SPR sensor in **Figure 1** is modeled by the multilayer planar structure depicted in **Figure 2**. The incident beam, reflected beam, and incidence angle θ refer internally to the prism. The CSNps have a dielectric shell of thickness b, composed of fused silica for this study, to provide stability to a nanoparticle, preventing agglomeration, and decreasing their surface interaction [16]. The periodic planar array of CSNps is described by the geometric period of $d + 2(a + b)$, where a is the nanoparticle core radius and d is the distance between them.

The applied relative permittivity was prism (SF4) $\varepsilon_p = 3.0615$, gold film $\varepsilon_{Au} = -11.66 + 1.35i$[1], and SiO_2 $\varepsilon_d = 2.132$ [15, 17]. The CSNps array is treated as a homogeneous layer with thickness of $h_{CS} = 2(a + b)$ and effective permittivity ε_{eff} in Eq. (1), given by the Clausius-Mossotti

[1]Obtained from the Lorentz-Drude model with one term of interband and time dependence with $\exp(-i\omega t)$.

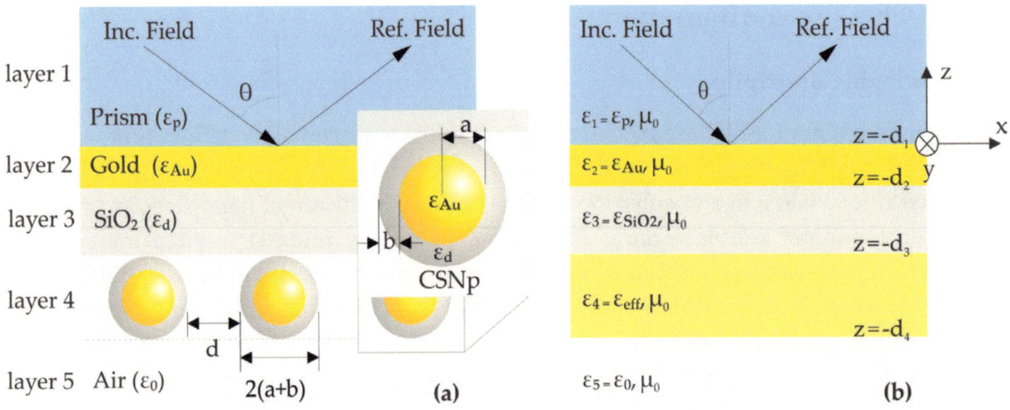

Figure 2. (a) The multilayer structure model of the SPR sensor coupled to the periodic array of CSNps. The inset highlights the CSNps with gold-core (ε_{Au}) of radius *a* and dielectric shell (ε_d) of thickness *b*; (b) Resulting planar structure using the effective layer to approximate the CSNps array.

mixing formula and the core-shell polarizability of [18], set to the quasi-static scattering regime [13, 19]. The resulting planar structure of the sensor is shown in **Figure 2(b)**.

$$\varepsilon_{eff} = \varepsilon_0 \left(\frac{1 - 2f_s \Lambda}{1 - f_s \Lambda} \right) \tag{1}$$

In Eq. (1): $f_s = 2\pi/3[(a+b)/(d+2(a+b))]^2$is the CSNps volume fraction in the planar array; and the parameter Λ is defined in Eq. (2), where $f = a^3/(a+b)^3$ is the core volume fraction in the CSNps [19]. The parameter f_s is zero for no immobilized CSNps ($\varepsilon_{eff} = \varepsilon_0$) and $f_s = \pi/6 \simeq 52.36\%$ when the distance *d* is zero. To eliminate the shell of the CSNps, we set $\varepsilon_d = \varepsilon_0$ and $b = 0$ nm in Eq. (2), obtaining the Maxwell-Garnett mixing formula [20, 21]

$$\Lambda = \frac{f(\varepsilon_{Au} - \varepsilon_d)(\varepsilon_0 + 2\varepsilon_d) + (\varepsilon_{Au} + 2\varepsilon_d)(\varepsilon_d - \varepsilon_0)}{f(\varepsilon_{Au} - \varepsilon_d)(2\varepsilon_d - \varepsilon_0) + (\varepsilon_{Au} + 2\varepsilon_d)(2\varepsilon_d + \varepsilon_0)} \tag{2}$$

The propagation of the electromagnetic wave in the sensor planar structure (**Figure 1(b)**) is performed, in the frequency domain with time dependence of $\exp(-i\omega t)$, by the generalized reflection coefficient in Eq. (3), which considers the multiple reflections and transmissions in all layers [22]. In Eqs. (3) and (5), $R_{n,n+1}$ and $T_{n,n+1}$ are the Fresnel's reflection and transmission coefficients, respectively, set to TM or TE polarization in accordance with the excitation. For TM case, the transverse magnetic field $H_{n,y}$ in the *n*-th layer is given in Eq. (4) and for TE case, Eq. (4) is set to the electric field $E_{n,y}$. In Eq. (4), A_n is the field amplitude in the *n*-th layer, given by Eq. (5).

$$\tilde{R}_{n,n+1} = \frac{R_{n,n+1} + \tilde{R}_{n+1,n+2} \exp\left[i2k_{n,z}(d_{n+1} - d_n)\right]}{1 + R_{n,n+1}\tilde{R}_{n+1,n+2} \exp\left[i2k_{n,z}(d_{n+1} - d_n)\right]} \tag{3}$$

$$H_{n,y} = A_n \left[\exp\left(-ik_{n,z} z\right) + \tilde{R}_{n,n+1} \exp\left(ik_{n,z}\left(z + 2d_n\right)\right) \right] \exp\left(ik_x x\right) \tag{4}$$

http://dx.doi.org/10.5772/intechopen.70522

$$A_n = \frac{T_{n-1,n} A_{n-1} \exp\left[i(k_{n-1,z} - k_{n,z}) d_{n-1}\right]}{1 - R_{n,n-1} \tilde{R}_{n,n+1} \exp\left[i2k_{n+1,z}(d_{n+1} - d_n)\right]} \tag{5}$$

For the recursive expressions, Eqs. (3)–(5): $A_1 = 1$ is the incident field amplitude in prism layer; $k_n^2 = k_0^2 \varepsilon_n$ is the propagation constant in the n-th layer, considering no magnetics in all the materials; $k_x = k_1 \sin(\theta)$ is the wave vector in x-axis direction, which is the same for all layers; and $\tilde{R}_{N,N+1} = 0$ is the reflection coefficient in the last layer, where $N = 5$ is the number of layers in the sensor structure (**Figure 2(b)**) [22]. The angular reflectivity curve is obtained from Eq. (3) for the gold-SiO_2 interface by $\Gamma(\theta) = \left|\tilde{R}_{1,2}\right|^2$.

2.3. Model validation and modal analysis

Herein, we compare the approximate analytical model with the results obtained by numerical simulations and experimental data to achieve the theoretical consistency between the models and study the parametric interval for validation. The numerical results were obtained through the 3D simulation environment COMSOL Multiphysics, based on the finite element method [23]. We obtained the experimental data from the SPR spectrometer described in [24], which uses a He-Ne laser as the excitation source and a rotary base to control the incident angle. The sensor's structure is fabricated by e-beam vacuum deposition process.

In **Figure 3**, we compare the analytical (An.), numerical (Num.), and experimental (Exp.) $\Gamma(\theta)$ curves for no CSNps case. The experimental curves are restricted in θ to the interval of TM2 and TM1 modes in **Figure 3(a)** due to limitations in the SPR spectrometer [15, 24]. The thickness of the sensor structure used in **Figure 3** was estimated by curve fitting using the free software Winspal [15]. The minimum of $\Gamma(\theta)$ highlights in **Figure 3** represents the resonant guide modes of order 1 (TM1 and TE1), order 2 (TM2 and TE2), and the SPP[2] mode [12].

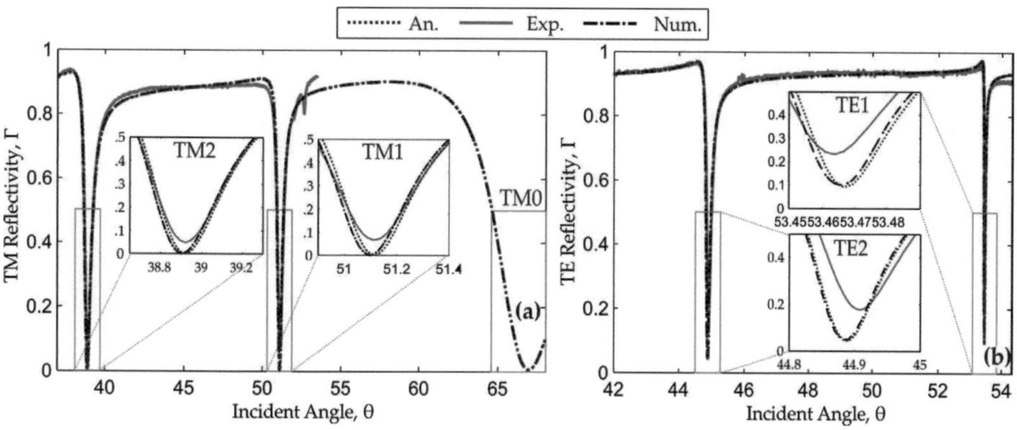

Figure 3. Comparison of the angular reflectivity curves Exp., An., and Num. for no CSNps on the sensor for (a) TM polarization and (b) TE polarization. The layer's thickness in the structure is t_{Au} = 48 nm and t_{SiO_2} = 677 nm.

---

[2]The SPP wave is named TM0, this is a zero-order guide mode in TM polarization.

The deviations An.-Exp. and Num.-Exp. in **Figure 3** are 2.39 and 2.12% for the TM curves, and 1.97 and 2.05% for the TE curves, showing high accuracy for the numerical simulation and the analytical model. The differences may be due to measurement errors and roughness in the fabricated multilayer structure [15, 24].

Figure 4 shows the magnitudes of the transversal fields, in the *z*-axis of **Figure 2(b)**, for the resonant mode highlights in **Figure 3**. One can note in **Figure 4(a)** the high field amplitude in the gold-SiO_2 interface for the TM0 mode and its characteristic evanescent field in both gold and SiO_2 layers [13]. This is also observed for the modes TM1 (**Figure 4(b)**) and TM2 (**Figure 4(c)**), but with a predominant intensity in the SiO_2 layer, like guide modes [12]. For the TE modes TE1 (**Figure 4(d)**) and TE2 (**Figure 4(e)**), there is no surface wave in the gold-SiO_2 interface because the plasmonic wave only exists for TM polarization [13].

To validate the analytical model, in **Figure 5**, we compare it with Num. simulations in three cases for the sensor: (i) No CSNps; (ii) CSNps with $b = 0$ nm; and (iii) CSNps with $b = 10$ nm. The relative deviation An.-Num. in **Figure 5** are 8.88% ($b = 0$ nm) and 9.18% ($b = 10$ nm) for the TM curves, and 0.74% ($b = 0$ nm) and 1.22% ($b = 10$ nm) for the TE curves. Therefore, the deviation An.-Num. tends to increase for tested values of *b*, and a possible cause is the increase of the CSNps size.

The hypothesis that generally increases the relative deviation characterizes the An. model limitation, such as (A) scattering losses [13, 21]; (B) dipole field interaction between CSNps in the array [18]; and (C) the restriction as thin of the effective layer thickness [12, 19]. As (A) grows with the CSNp size (parameters *a* and *b*), the relative deviation tends to increase with *a* and *b*, so, these parameters need to be restricted [19]. The phenomenon (B) and the relative deviation decrease with the distance *d* in the array, because this here is used the minimum

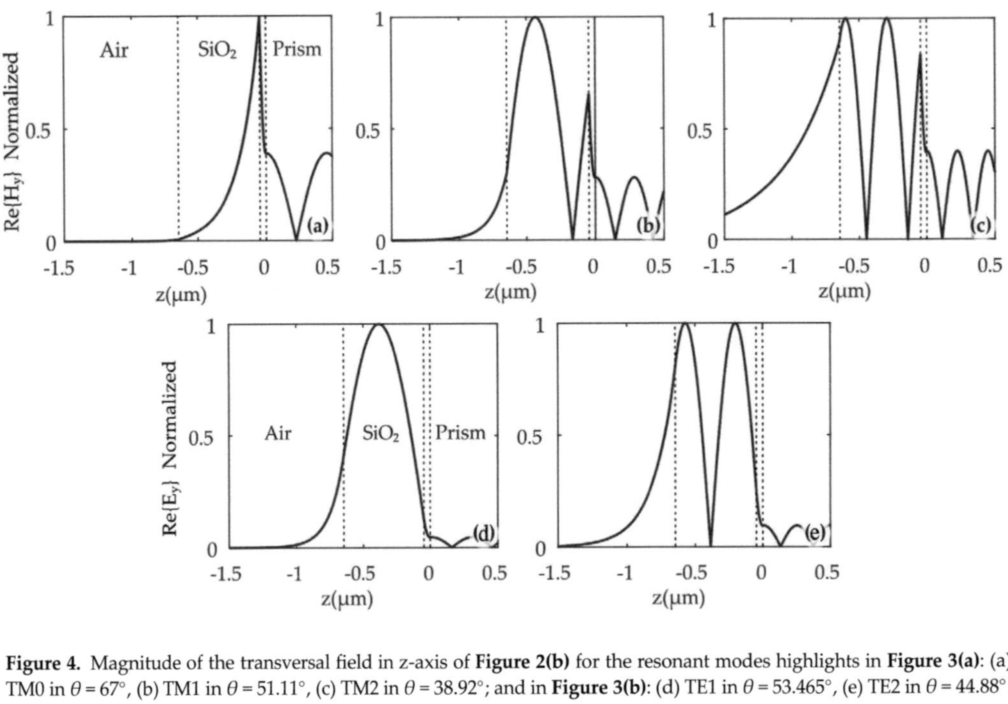

Figure 4. Magnitude of the transversal field in z-axis of **Figure 2(b)** for the resonant modes highlights in **Figure 3(a)**: (a) TM0 in $\theta = 67°$, (b) TM1 in $\theta = 51.11°$, (c) TM2 in $\theta = 38.92°$; and in **Figure 3(b)**: (d) TE1 in $\theta = 53.465°$, (e) TE2 in $\theta = 44.88°$.

http://dx.doi.org/10.5772/intechopen.70522

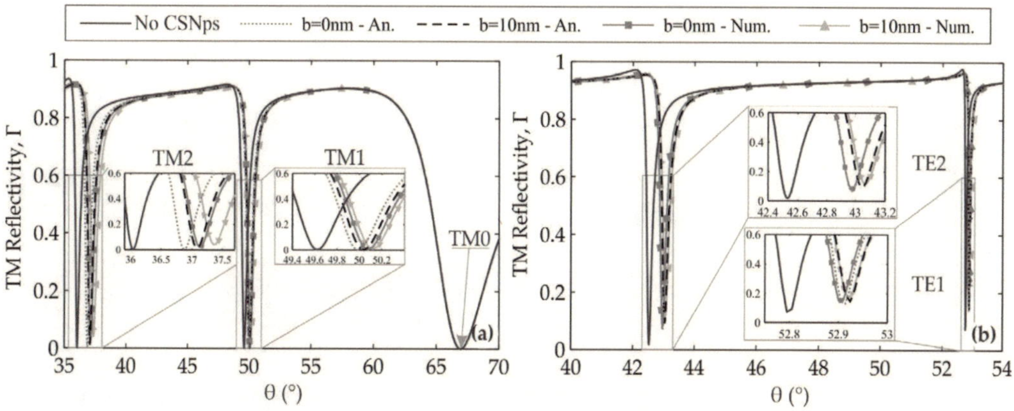

Figure 5. Comparison of the An. and Num. curves for the cases (i) No CSNps; (ii) CSNps with fixe b = 0 nm; and (iii) CSNps with fixe b = 10 nm. (a) TM polarization and (b) TE polarization. For all cases are set the fixed parameters a = 30 nm, d = 100 nm, t_{Au} = 46 nm, and t_{SiO_2} = 600 nm.

value of d = 50 nm. The limitation (C) can modify the direct and inverse behavior in the relative deviation with the limitations (A) and (B), respectively, for other values of a and d.

In **Figure 6**, we compare the An. and Num. real magnetic fields, in the zx-plane of **Figure 2(b)**, for the mode minimum points in the TM curve b = 10 nm of **Figure 5(a)**. Note that, in general, the An. and Num. results are very similar, differing basically due to the high field amplitude in the metal-core surface of the nanoparticles (**Figure 6(b), (d)**).

Based on **Figure 6**, in **Figure 7**, we compare the An. and Num. transversal fields for the TE curve b = 10 nm of **Figure 5(b)**. Different from the TM fields in **Figure 6**, the TE mode fields present a low field amplitude in the shell and a constant electric field in the CSNps metal core. The visual analysis of **Figures 4–7** indicates a greater interaction with the sensing layer for the wave modes in which the minimum region is closest to the ATR angle, even for TE modes,

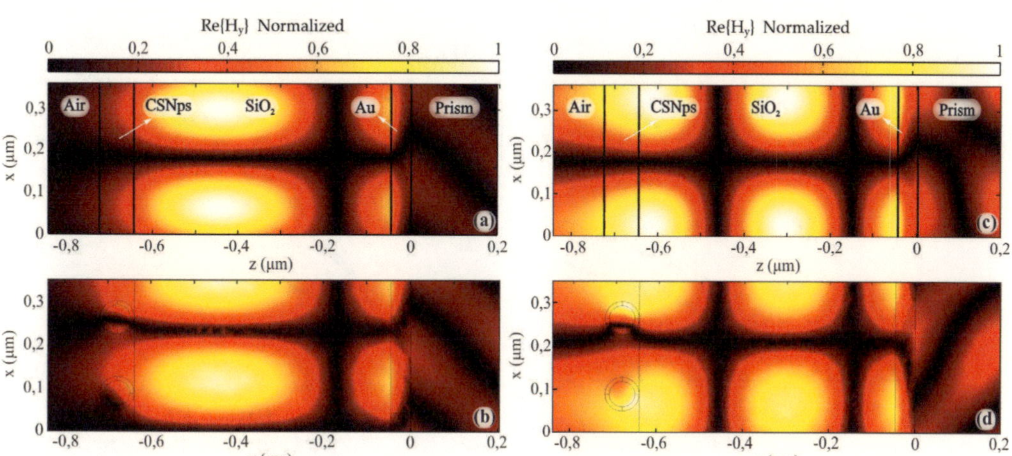

Figure 6. Comparison of the An. and Num. real magnetic fields for the TM modes in **Figure 5(a)**, for the curve b = 10 nm. An.: (a) TM1 in θ = 50.04° and (c) TM2 in θ = 37.09°; and Num.: (b) TM1 in θ = 50.12°; and (d) TM2 in θ = 37.36°.

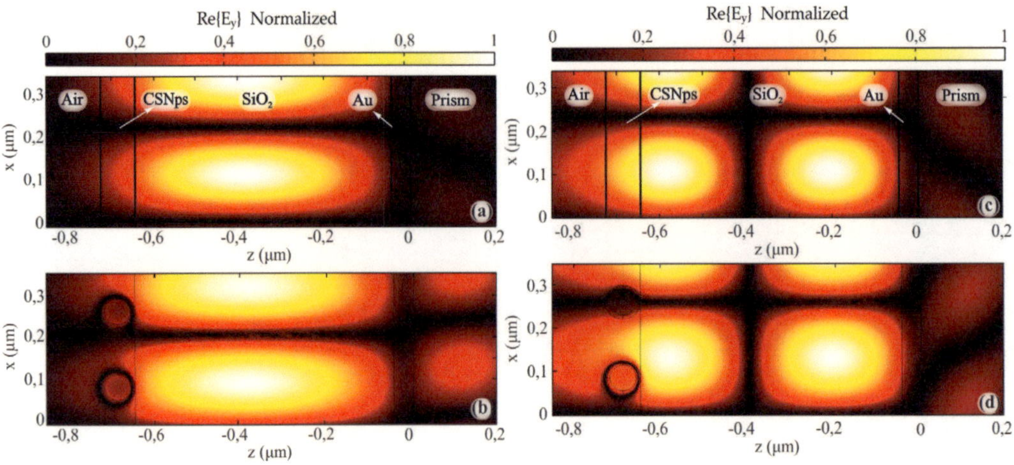

Figure 7. Comparison of the An. and Num. real electric fields for the TE modes in **Figure 5(b)**, curve b = 10 nm. An.: (a) TE1 in θ = 52.93° and (c) TE2 in θ = 43.04°; Num.: (b) TE1 in θ = 52.92° and (d) TE2 in θ = 43.07°.

because of the greater field intensity and the consequent greater interaction with the CSNps [25]. In [17], one can note that the thickness of the SiO_2 layer, in the sensor structure, can regulate the reflectivity minimum point of the resonant wave modes, and the wave order in the same curve, so this parameter can improve the sensor sensibility.

2.4. Sensitivity analysis

To evaluate the sensibility, we vary the CSNps array parameters a, b, and d, and calculate the angular shift $\Delta\theta$ only for the more sensitive TM2 mode in the minimum points of $\Gamma(\theta)$. First,

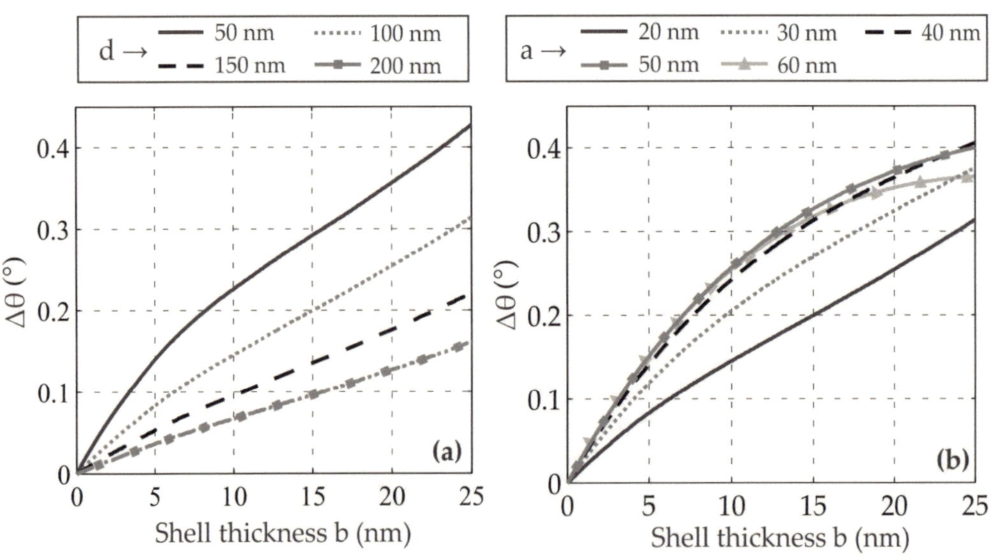

Figure 8. Curves of $\Delta\theta \times b$ for the values of: (a) distance d = 50, 100, 150, and 200 nm, and the fixed parameter a = 20 nm; and (b) radius a = 20, 30, 40, 50, and 60 nm, and the fixed d = 100 nm.

we analyze the sensibility to one of the parameters, and then we verify how other parameters influence in the sensibility, setting some different values. For this analysis, we use only the analytical model, that presents low costs in computing, and the fixed parameters were $t_{Au} = 46$ nm and $t_{SiO_2} = 600$ nm.

2.4.1. *Sensitivity to the shell thickness b*

In **Figure 8**, we present the curves of $\Delta\theta$ in function of b ($\Delta\theta \times b$ curve), relative to the initial curve $b = 0$ nm. In **Figure 8(a)**, each curve is defined for different values of the distance d and in **Figure 8(b)**, they are for values of a. The shifting of the minimum region to right is the positive direction of the $\Delta\theta$.

Note in **Figure 8(a)**, $\Delta\theta \times b$ decreases as d increases, which declines the sensor sensibility. However, higher values of d decrease, in general, the sensor sensitivity and the contrary are true for lower values of d. For the parameter a, the sensitivity by the curves $\Delta\theta \times b$ in **Figure 8(b)** increases for the most cases, but for the curves $a = 50$ nm and $a = 60$ nm, one can visualize the tendency of an inverse behavior. This can turn the CSNps characterizing process more complex because we can obtain, for the same value of $\Delta\theta$, more than one value of b. To avoid this, we can restrict $a < 50$ nm and $b < 20$ nm.

2.4.2. *Sensitivity to the distance d*

Figure 9 presents the $\Delta\theta$ of TM2 mode in function of d ($\Delta\theta \times d$ curve), relative to the initial curve $d = 50$ nm. In **Figure 9(a)**, the curves differ in the values of b and in **Figure 9(b)** in the values of a. The increasing of d, corresponding to the concentration decrease of the immobilized CSNps on the sensor, basically, its shifts the minimum points to left, explaining the negative

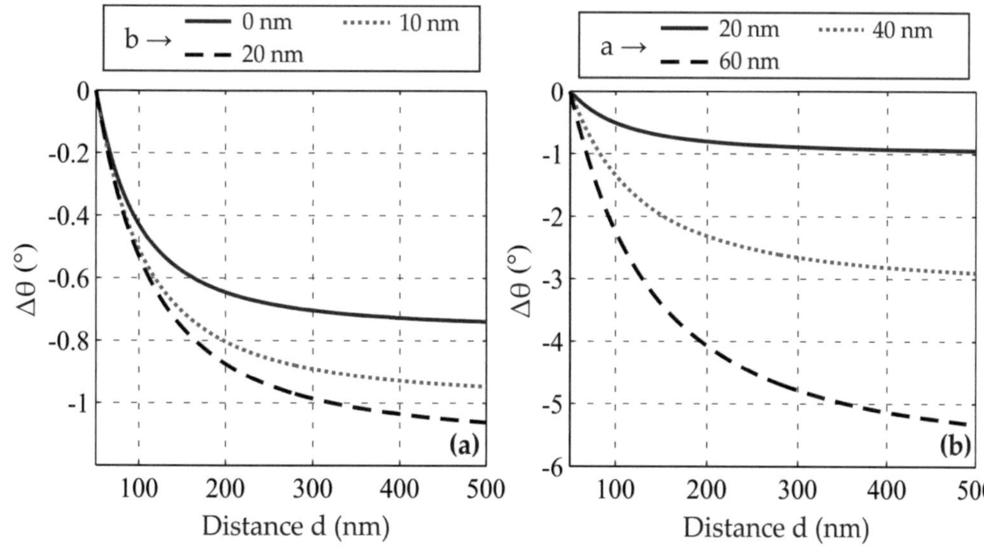

Figure 9. Curves of $\Delta\theta \times d$ for TM2 modes for the values of: (a) dielectric shell thickness $b = 0$, 10, and 20 nm, and the fixed parameter $a = 20$ nm; and (b) radius core $a = 20$, 40, and 60 nm, and the fixed parameter $b = 10$ nm.

values of $\Delta\theta$ in **Figure 9**. In **Figure 9**, the great sensitivity response when CSNps are quite close or for small values of d.

In **Figure 9(a)** we note that the increasing of b results in greater values of $\Delta\theta$, this is a better sensitivity response of the sensor. In **Figure 9(b)**, we can observe that both parameters a and b increase the sensor response because the sensor is more sensitive to variation in the CSNps core radius. This behavior was also observed in [17] and is expected due to the increase of the resonant field interaction with the array when the metal core of the CSNps is bigger. While b increases, it increases the CSNps size, decreasing the field interaction.

3. SPCE configuration

3.1. Functional description

The functioning of an SPR sensor in SPCE configuration (**Figure 10**) is based on the interaction of the field radiated by the immobilized analytes with the sensor structure, composed by a thin metal film deposited on the prism [10]. These interactions generate the SPP wave on the air-gold surface and radiating modes in the prism, that is a high directional emission in a specific SPCE angle and depending on the nanoparticle [26]. The high directional nature of the SPCE emission also increases the efficiency of coupled emission detection [27]. Similar to the sensor in the Kretschmann configuration, here the SPCE configuration is excited by a laser beam operating at the wavelength of 632.8 nm.

In **Figure 10(a)**, the sensor in the SPCE configuration is illustrated. First, a solution with the suspended analytes flows in the microfluidic channel while the target nanoparticles are immobilized on the sensor surface by a specific ligand substance. Then, the solution flow is cut off and drained until only the immobilized CSNps remain to be analyzed. The CSNps

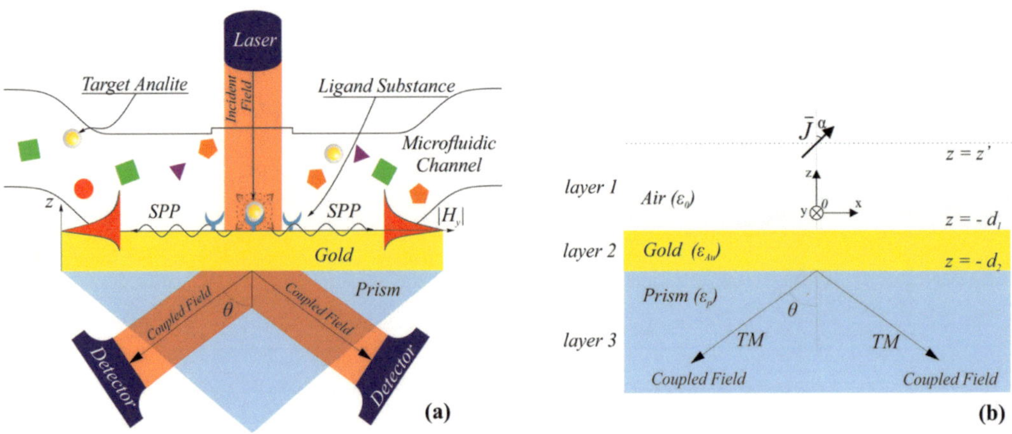

Figure 10. (a) A functional illustration of the analyzed SPCE sensor coupled to a microfluidic channel. (b) The approximate model of the SPCE sensor in (a) by a multilayered planar structure and a resonant dipole.

can be held by the ligand substance at z' or by a dielectric spacer of same thickness with a ligand substance of negligible height. An optical detector evaluates the SPCE angle of the analyte [28].

An approximate model of the sensor in **Figure 10(a)** is presented in **Figure 10(b)**, where the structure is a planar multilayer with three layers: air, thin gold film, and SF4 optical prism, all represented by their respective complex permittivity. The interaction of the laser beam with the analytes and their re-annealing is equivalently modeled by a dipole, which represents the immobilized CSNps and is situated at the height z' in the layer 1 of **Figure 10(b)**.

Although the dipole-type optical emitter is nonpolarized, the coupled field targeting the detector in **Figure 10(b)** is highly polarized in the TM [29]. This occurs because part of the CSNps emission is naturally in the TM polarization and can excite the SPP wave on the air-gold interface, which evanescent wave passes through the thin metallic layer and radiate in the prism as a propagating wave polarized in the TM polarization. Therefore, the SPR sensor in the SPCE configuration can be understood as a reverse functioning of the Kretschmann configuration.

Note the existence of different nanoparticles in the fluidic channel (**Figure 10(a)**); however, only the target nanoparticles are immobilized on the sensor surface. Here, the sensor is analyzed with only immobilized CSNps and the result is a radiating TM field in a specific angle of coupling in the prism that corresponds to this nanoparticle. However, when using different ligand substances, for multichannel evaluation, different coupling angles would be detected, each angle related to a different particle of interest [30].

3.2. Theoretical modeling

For the SPR sensor in SPCE configuration, the SPP wave is created from interactions of the sensor structure and immobilized nanoparticles, which emit radiation and evanescent field when excited by a source. In the SPCE sensor, the nanoparticles on the substrate have dimensions smaller than the excitation wavelength, so they are represented here by infinitesimal dipoles with equivalent dipole moments or by elementary currents given by Eq. (6) [31, 32]:

$$\bar{J} = \bar{J}_{0p}\delta(\bar{r},\bar{r}_p) = \left(\xi_p \overline{E}_p^t\right)\delta(\bar{r},\bar{r}_p) = \xi_p\left[\overline{E}_i(\bar{r}_p) + \overline{E}_r(\bar{r}_p) + \sum_{q=1,\, q\neq p}^{N} \overline{E}_{dip}^q(\bar{r}_p) + \sum_{q=1}^{N} \overline{E}_r^q(\bar{r}_p)\right]\delta(\bar{r},\bar{r}_p) \quad (6)$$

The elementary current of the equivalent dipole is orientated by the laser source. To determine the induced dipole moments of an array of P dipoles on a multilayer structure, one must solve the following system of linear equations for $p,\ q \in \{1,2,\ \ldots,\ P\}$, where $\bar{r}_p$ are the positions of the P dipoles, $\overline{E}_p^t$ is the total external field of excitation on the equivalent dipole, that is the sum of all the fields that arrive in the dipole p, and ξ_p is the polarization constant that depends on the type of element considered (CSNps, biomolecules, QDs) [31].

For the total $\overline{E}_p^t$ in Eq. (6), that has four terms, is considered a linear dependence with the dipole. The first term of $\overline{E}_p^t$ represents the incident field, the second one, the reflection of the field incident on the structure, the third term, the radiation of each dipole, and the fourth term, the reflections in the structure of the field radiated by each dipole.

The total electric field of the dipole defined in Eq. (6), for an arbitrary direction in a homogeneous medium, can be derived from the dyadic Green's functions in Eq. (7) [32]:

$$\widehat{G}^e\left(\overline{r},\overline{r}'\right) = i\omega\mu\left(\widehat{I} + \frac{1}{k^2}\nabla\nabla\right)\overline{\mathrm{J}}\frac{e^{jk\left|\overline{r}-\overline{r}'\right|}}{4\pi\left|\overline{r}-\overline{r}'\right|} \tag{7}$$

where $\widehat{I}$ is a unitary dyad, $\overline{r}$ is the point of observation, and $\overline{r}'$ is the source point. The dipoles irradiated nonpolarized spherical waves; thus, a spherical wave radiated can be expanded as an integral of conical or cylindrical waves in the direction ρ times a plane wave in the z-direction, over all propagation constant $k_\rho = \sqrt{k - k_z}$. It is possible to relate spherical waves with plane waves from the integral Sommerfeld identity in Eq. (8) [33]:

$$\frac{\exp\left(ik\left|\overline{r}-\overline{r}'\right|\right)}{\left|\overline{r}-\overline{r}'\right|} = \frac{i}{2}\int_{-\infty}^{+\infty} dk_\rho k_\rho H_0^2\left(k_\rho\rho\right)\frac{\exp\left(ik_z\left|z-z'\right|\right)}{k_z} \tag{8}$$

The identity in Eq. (8) can be obtained from the solution of the scalar wave equation, obtained first in spherical coordinates, and later in rectangular coordinates using the three-dimensional Fourier transform. For simplicity, only the z-direction component is evaluated, thus, the formalism of a vertical dipole VED is possible to be used. Substituting Eq. (8) in Eq. (7), it can be shown that the field $E_{n,z}$ in an n multiple-layered structure in the presence of a resonant dipole can be represented by Eq. (9) [22]:

$$E_{n,z} = -\frac{\mathrm{J}_z}{4\pi\omega\varepsilon_n}\int_{-\infty}^{+\infty} dk_\rho \frac{k_\rho^3}{2k_{1z}} H_0^2\left(k_\rho\rho\right)A_n^{TM}\left[\exp\left(ik_{n,z}|z|\right) + \tilde{R}_{n,n+1}^{TM}\exp\left(ik_{n,z}(z+2d_n)\right)\right]\exp\left(ik_{1z}z'\right) \tag{9}$$

where A_n^{TM} and $\tilde{R}_{n,n+1}^{TM}$ are the same parameters defined in Eqs. (5) and (3), respectively, over all propagation constantk_ρ. Eq. (9) is here rewritten in terms of the zero-order Bessel function in Eq. (10):

$$E_{n,z} = \frac{i\mathrm{J}_z}{4\pi\omega\varepsilon_n}\left[\frac{\int_0^{+\infty} dk_\rho k_\rho J_0\left(k_\rho\rho\right)\tilde{f}_{n,z}^s\left(k_z,z,z'\right)e^{ik_{n,z}\left|z-z'\right|}}{k_{n,z}} + \int_0^{+\infty} dk_\rho k_\rho J_0\left(k_\rho\rho\right)\frac{\tilde{f}_{n,z}^r\left(k_z,z,z'\right)e^{ik_{n,z}\left|z-z'\right|}}{k_{n,z}}\right] \tag{10}$$

where $\tilde{f}_{n,z}^s\left(k_z,z,z'\right)$ and $\tilde{f}_{n,z}^r\left(k_z,z,z'\right)$ are the spectral functions defined in Eqs. (11) and (12), respectively:

http://dx.doi.org/10.5772/intechopen.70522

$$\tilde{f}^{s}_{n,z}\left(k_z, z, z'\right) = \frac{ik_{n,z}k_\rho^2 A_n^{TM} \exp\left[-i\left(k_{n,z}z - k_{1z}z'\right)\right]}{k_{1z}\exp\left(ik_{n,z}|z - z'|\right)} \tag{11}$$

$$\frac{\tilde{f}^{r}_{n,z}(k_z, z, z') = ik_{n,z}k_\rho^2 A_n^{TM}\tilde{R}^{TM}_{n,n+1}\exp\left[i(k_{n,z}(z + 2d_n) + k_{1z}z')\right]}{k_{1z}\exp\left(ik_{n,z}|z - z'|\right)} \tag{12}$$

where indices *s* and *r* are treated here as a spectral contribution related to incidence and reflection in the multilayer structure. These integrals are extremely complicated to calculate and the solutions do not have closed forms. Some difficulties of the integral representation are highlighted [34, 35]: Low integration kernel convergence, branching points, and branch cuts arise, the appearance of functions of double or multiple values, the choice of a single appropriate Riemann surface, possibility of complex poles, among others. That way, the lack of analytical expressions in a closed form combined with a heavy computational cost associated with the direct integration for this integral, which have low convergence, make direct numerical evaluation an impractical approach to our analysis.

When solving problems involving integrals like Eq. (10), several recent approaches have been proposed [36, 37], all consist of the evaluation of spectral functions using the Sommerfeld Identity with variants of the discrete complex image method (DCIM) as an acceleration tool. Here, we evaluate the integral equations directly from the electric field into a versatile application for the use of DCIM and applies the DCIM directly on the integral field equations.

The DCIM method expands the integral equations of (11) and (12) into a sum of complex terms, that is, it estimates values of complex integrals over an integration path in the complex domain, usually with a range of (0, ∞), by a finite number of samples of the integrand. A solution based on a two-level path is used [38]. We use a sophisticated scheme, where the integrand is approximated by a superposition of complex exponentials, and this approximation is semi-analytical since it is not an exact but approximate solution.

3.3. Modal analysis

In this section, we analyze the SPR sensor of **Figure 10(b)** using the same relative permittivity presented in Section 2.2 for the excitation source wavelength of $\lambda = 632.8$ nm. It is considered that the radiation of the analyte occurs at this same wavelength. The results are presented for the near and far field.

3.3.1. Numerical example

Figure 11(a) shows the real part of the field E_z in the multilayer structure of **Figure 10(b)** and one can observe that the waves radiated by the immobilized nanoparticles induce the surface plasmonic mode in the air-gold interface, whereas in the prism, formed waves are concentrated at specific angles. **Figure 11(b)** shows that the plasmonic mode is excited throughout the air-gold interface as a cylindrical wave symmetrical to the *z*-axis; it is possible to visualize the excited SPPs on the first interface and the rapid fading of the electric field from the source.

Figure 12(a) shows the two-dimensional radiation diagram of the SPCE sensor, where maintaining the operating wavelength at $\lambda = 632.8$ nm makes an evaluation of the intensity of the distant field at different heights: $z' = 20, 50, 100, 150$, and 200 nm. It is observed that the far

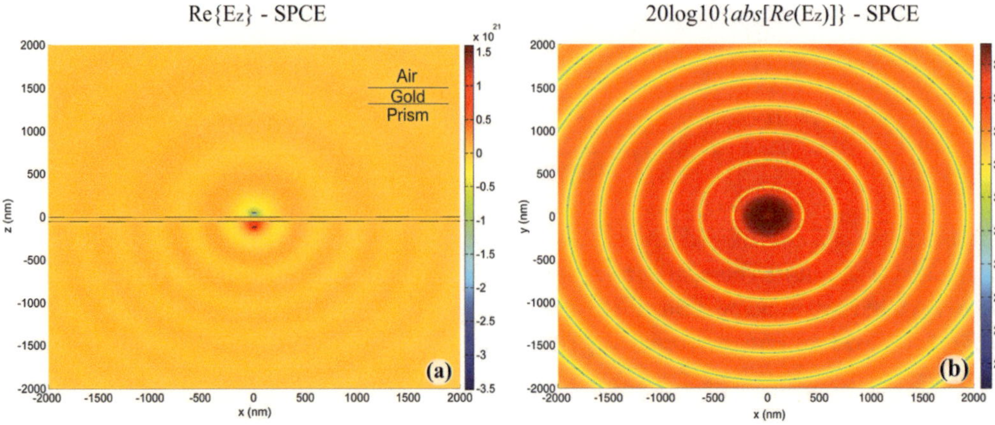

Figure 11. (a) Electric field distribution Re$\{E_z\}$ obtained via DCIM in all three layers air, gold, and prism (SF4). (b) Electric field distribution 20log10{*abs*[Re(E_z)]} obtained in the *xy* plane via DCIM at the air-gold interface.

field strength is greater for height $z' = 20$ nm. Note that the intensity of the lower lobes, which depend on the field coupled in the prism, as well as on the upper lobes that depend on the total field in the first layer, increase in intensity according to the decrease in height z'.

Figure 12(b) shows the far field three-dimensional diagram of the SPCE sensor for the permissiveness values presented above and optimized height for $z' = 20$ nm. The emission coupled to the prism forms a circular cone. Note that the lower lobes have well-directed beams at a very characteristic coupling angle $\theta = 145.2°$, that is, electric field coupled at the angle of $\theta_{SPCE} = 34.8°$.

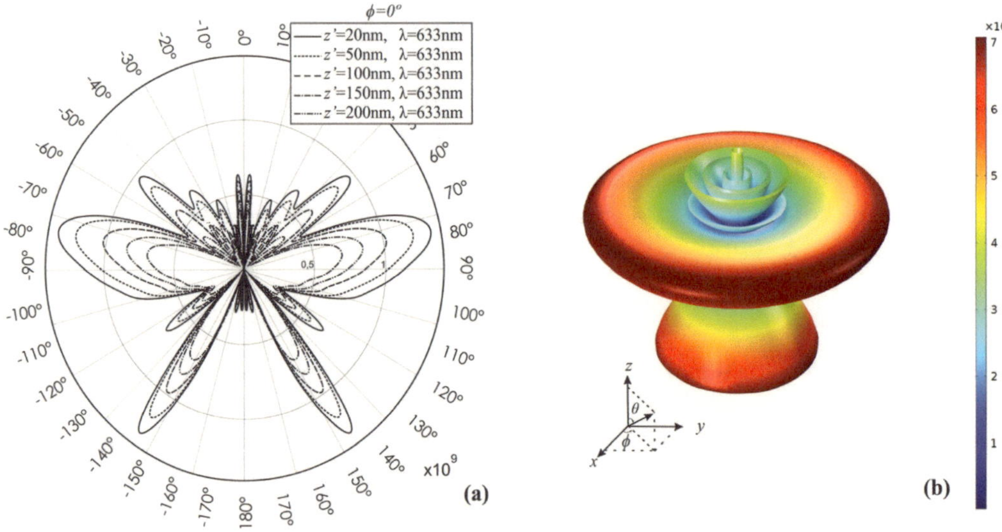

Figure 12. (a) Two-dimensional radiation diagram of the SPCE sensor, evaluation of the intensity of the distant field at different heights: z' = 20, 50, 100, 150, and 200 nm. (b) Far field three-dimensional diagram of the SPCE sensor for optimized height for $z' = 20$ nm. Note that electric field coupled at the angle of $\theta_{SPCE} = 34.8°$.

http://dx.doi.org/10.5772/intechopen.70522

3.4. Reciprocity between SPR sensors in the KR and SPCE configuration

The SPCE sensor is physically the reverse structure of the KR sensor. In this topic, a previous evaluation of the electromagnetic reciprocity between the KR and SPCE configuration sensors is presented. In both configurations, the same materials are used, that is, air, gold, and prism (SF4). The sensor operates on a standard $\lambda = 632.8$ nm wavelength, with a gold layer of 50 nm thickness.

Figure 13 presents results for evaluation of reciprocity between the KR and SPCE sensors with the configurations described above, where **Figure 13(a)** shows the real part of the *Hy* field for a plane wave over the plasma resonance angle $\theta_{SPP} = 36.8°$ in the KR sensor. **Figure 13(b)**

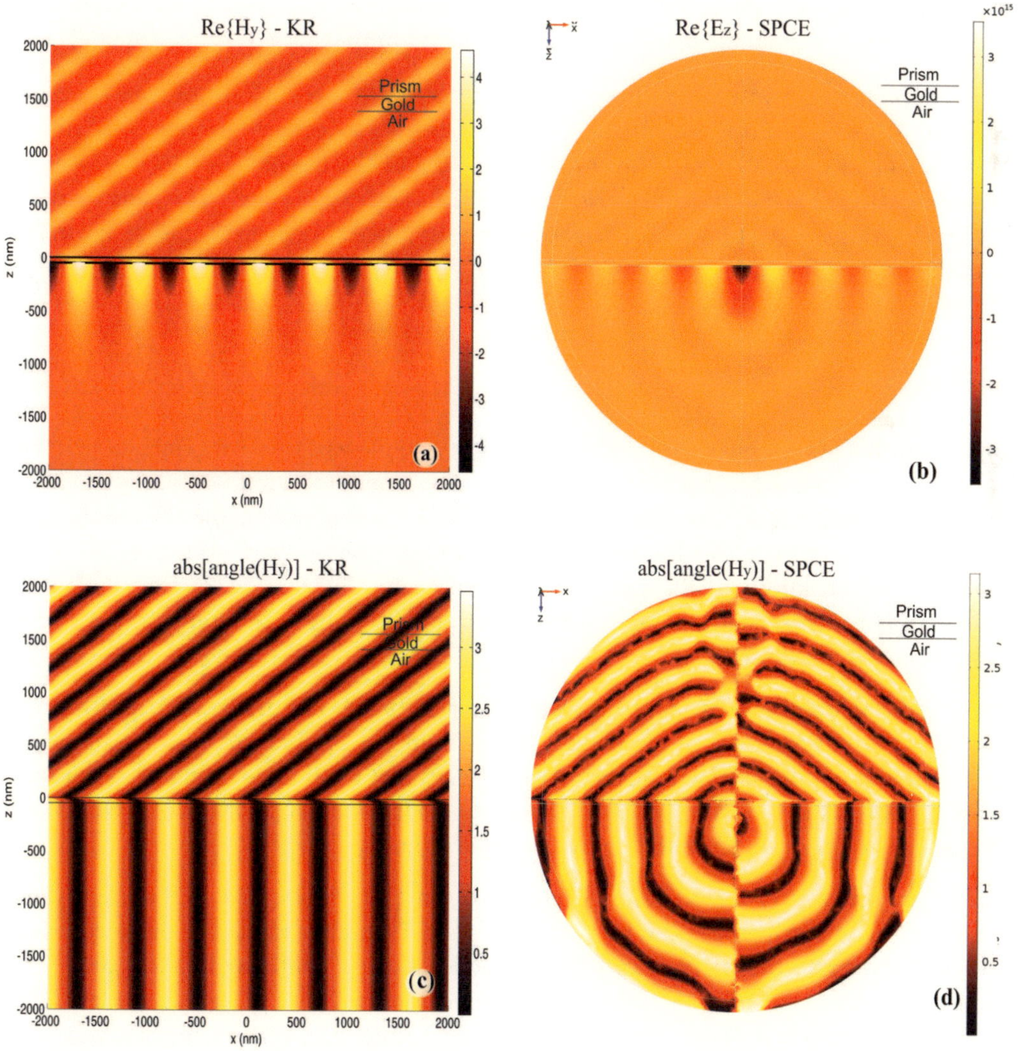

Figure 13. (a) Re(*Hy*) for plane wave incident with $\theta_{SPP} = 36.8163°$obtained analytically for the sensor KR, (b) Re(E_z) for vertical dipole at $z' = 20$ nm for the sensor SPCE, (c) Angle of *Hy* obtained analytically for the KR sensor, (d) Angle of *Hy* obtained in the for the SPCE sensor.

describes the real part E_z field for the SPCE sensor with the source at $z' = 20$ nm. In **Figure 13(b)**, it is possible to observe the formation of surface plasma as a consequence of particle radiation above of the gold layer. Note that the z-axis was inverted only for reciprocal visualization. **Figure 13(c)** represents the phase of Hy obtained analytically for the KR sensor; it is possible to identify the incident phase of the plane wave on the gold layer.

Figure 13(d) illustrates the Hy phase obtained for the SPCE sensor. Note that, in the prism layer, magnetic transverse plane (TM) waves are obtained with the reciprocal phase of the KR sensor phase. The appearance of this polarized wave in the TM mode is explained because part of the emitter's optical emission is naturally in the TM mode and excites an SPP wave on the air/gold surface; then, after the evanescent wave passes through the thin metallic layer, it will radiate in the prism as a polarized propagation wave in TM mode.

So, a coupling angle was set on the SPCE sensor equal to the angle of plasma resonance occurred at the KR sensor, which was actually found $\theta_{SPCE} = 34.8° \approx \theta_{KR} = 36.8°$. It was observed that a gradual increase in the discretization of the meshes used to represent the dipole approximates the coupling angle of the SPCE sensor of the angle of resonance of the KR sensor, which in fact proves its electromagnetic reciprocity between the plasmonic modes of the KR and SPCE sensor.

4. Conclusion

In this article was presented a theoretical analysis of an SPR sensor in Kretschmann (KR) and SPCE configurations, when a periodic array of core-shell nanoparticles (CSNps) is immobilized on the sensor sensitive surface. For the SPR sensor in KR configuration, the CSNps array has approximated by an effective homogeneous layer to treat the resultant structure as a planar multilayer, which improved the computational processing. For the SPCE configuration, the CSNps array has been treated as equivalent dipoles and the study is performed by the discrete complex image method (DCIM).

The approximate model of the KR sensor was validated for low size of the CSNp, parameters a and b, and high distance d in the periodic array from the comparison with the numerical simulations using finite element method. We observed that the increase of the shell thickness tends to depress the validation of this approximate model, such as the metal-core radius, and the parameter d, instead, tends to improve the validation.

The modal analysis of the KR configuration reveled that, besides the SPP surface wave, multiples guide wave modes can be excited, even in TE polarization. The thickness of the SiO_2 layer can alter the order of these guide modes and for configured value guide wave modes of order 1 and 2 was observed. The characteristic field of the guide modes in TM polarization presents a surface wave in the gold-SiO_2 interface, such as the SPP wave. We observed better validation of the approximate model for the TE curves.

There are evidences, such as the higher field intensity in the CSNps array region, that can indicate a greater sensitivity response for the wave modes which the minimum point is closest

http://dx.doi.org/10.5772/intechopen.70522

to the critical ATR angle. As the SiO_2 layer thickness can regulate the minimum point position, this parameter can also improve the sensor sensibility.

The sensitivity analysis revealed that both radius core and shell thickness increase the sensor response, but the radius core presented a greater influence in this behave, showing larger sensitivity to the parameter a. Because the dielectric shell reduces the sensor field interaction with the nanoparticles array, although this increases the size of the CSNps which, in general, intensifies the sensor response. The parameter d always tends to decrease the sensor response due to the reduction of the CSNps concentration in the array.

To develop the modal analysis of the SPCE sensor, we focus in the solution of the field equations for a resonant dipole over a multiple planar structure. The equations are optimized for direct application via DCIM method. The evaluation of the near field is presented and we observe that the waves radiated by the immobilized nanoparticles induce the surface plasmonic mode in the air-gold interface and radiating modes in the prism concentrated at specific angles.

The DCIM method was applied for a general solution of multilayer media using the generalized reflection coefficients. The far field results are presented by numerical simulations performed via finite element method and we observe that in SPCE configuration, the intensity of the lower lobes increases with the decrease in height z′. It is observed that the far field strength is greater for height $z' = 20$ nm.

By the last analysis of the sensor in the SPCE configuration, we demonstrated the reciprocity of the SPP modes in the configurations KR and SPCE. It has been found that the coupling angle of the SPP mode in the SPCE configuration is equal to the angle of maximum coupling of TM0 mode KR configuration.

Author details

Nadson Welkson Pereira de Souza[1,2], Jefferson Souza Costa[1], Rafael Correa dos Santos[1], André Felipe Souza da Cruz[1], Tommaso Del Rosso[3] and Karlo Queiroz da Costa[1]*

*Address all correspondence to: karlo@ufpa.br

1 Department of Electrical Engineering, Federal University of Para, Tucuruí-PA, Brazil

2 Department of Electrical Engineering, Federal University of South and Southeast of Para, Marabá-PA, Brazil

3 Department of Physics, Pontifical Catholic University of Rio de Janeiro, Rua Marques de São Vicente, Rio de Janeiro, Brazil

References

[1] Homola J. Surface plasmon resonance based sensors. In: Springer Series on Chemical Sensors and Biosensors: Methods and Applications. 4th ed. online ed. Springer; 2006. p. 33-44

[2] Wood RW. On a remarkable case of uneven distribution of light in a diffraction grating spectrum. Philosophical Magazine. 1902;**4**:396-402

[3] Wood RW. Diffraction gratings with controlled groove form and abnormal distribution of intensity. Philosophical Magazine. 1912;**23**:310-317

[4] Liedberg B, Nylander C, Lind T. Gas detection by means of surface plasmons. Sensors and Actuators. 1982;**3**:79-88

[5] Nylander C, Liedberg B, Lundstrom I. Surface plasmons resonance for gas detection and biosensing. Sensors and Actuators. 1983;**4**:299-304

[6] Chen Y, Chen JIL, Ginger DS. Plasmonic nanoparticle dimers for optical sensing of DNA in complex media. Journal of the American Chemical Society. 2010;**132**:9600-9601

[7] Feltis BN, Sexton BA, Glenn FL, Best MJ, Wilkins M, Davis TJ. A hand-held surface plasmon resonance biosensor for the detection of ricin and other biological agents. Biosensors and Bioelectronics. 2008;**23**:1131-1136

[8] Lindquist N, Nagpal P, McPeak KM, Norris DJ, Oh SH. Engineering metallic nanostructures for plasmonics and nanophotonics. Reports on Progress in Physics. 2012;**75**(75): 036501

[9] Cooper MA. Optical biosensors in drug discovery. Nature Reviews Drug Discovery. 2002;**1**:515-528

[10] Ignacy G, Lakowicz JR, Malicka J, Zygmunt G. Directional surface plasmon-coupled emission: A new method for high sensitivity detection. Biochemical and Biophysical Research Communications. 2003;**307**:435-439

[11] Cabrini S, Kawata S, editors. Nanofabrication Handbook. CRC Press; 2012

[12] Kaminow IP, Mammel WL, Weber HP. Metal-clad optical waveguides: Analytical and experimental study. Applied Optics. 1974;**13**:396-405

[13] Maier SA. Plasmonics: Fundamentals and Applications. 1st ed. New York: Springer; 2007

[14] Nguyen HH, Park J, Kang S, Kim M. Surface plasmon resonance: A versatile technique for biosensor applications. Sensors. 2015;**15**:10481

[15] Rosso TD, Sánchez JEH, Carvalho RDS, Pandoli O, Cremona M. Accurate and simultaneous measurement of thickness and refractive index of thermally evaporated thin organic films by surface plasmon resonance spectroscopy. Optics Express. 2014;**22**:18914-18923

[16] Mirabet LP. Synthesis, characterization and functionalization of metal and metal oxide nanoparticles. TEM microscopy study. [dissertation]. Barcelona: Universitat Autònoma de Barcelona; 2013

[17] da Costa KQ, Costa JS, Dmitriev V, Rosso TD, Pandoli O, Aucelio RQ. Analysis of surface plasmon resonance sensor coupled to periodic array of gold nanoparticles. In: Microwave and Optoelectronics; Nov 2015; SBMO/IEEE MTT-S International; 2015. pp. 1–5.

[18] Novotny L, Hecht B. Principles of Nano-Optics. 2nd ed. New York: Cambridge; 2012

[19] Choy T. Effective Medium Theory: Principles and Applications. 1st ed. Clarendon Press; 1999

[20] Kong JA. Electromagnetic Wave Theory. Massachusetts: Cambridge; 1998

[21] Ishimaru A. Electromagnetic Wave Propagation, Radiation and Scattering. New Jersey: Prentice Hall; 1991

[22] Chew W. Waves and Fields in Inhomogeneous Media. 1st ed. IEEE Press; 1996

[23] COMSOL Multiphysics 4.2a. COMSOL Inc. Available from: http://www.comsol.com./

[24] Sánchez JEH. Assembly of a surface plasmon resonance (SPR) spectrometer for the characterization of thin organic films [dissertation]. Rio de Janeiro-RJ, Brazil: Pontifícia Universidade Católica; 2013

[25] Costa JS, da Cruz AFS, dos Santos RC, da Costa KQ. Análise de um Sensor Plasmônico Acoplado à Arranjo Periódico de Nanopartículas tipo Core-Shell. In: Sociedade Brasileira de Telecomunicações (SBrT), editor. XXXIV Simpósio Brasileiro de Telecomunicações e Processamento de Sinais (SBrT 2016); Sep. 2016; Santarém-Pará-Brazil. 2016.

[26] Lakowicz JR. Radiative decay engineering: Surface plasmon-coupled directional emission. Analytical Biochemistry. 2004;153-169

[27] Ruckstuhl T, Enderlein J, Seeger S. Highly efficient optical detection of surface-generated fluorescence. Applied Optics. 1999;**38**

[28] Geddes M, Gryczynski L. Directional surface plasmon coupled emission. Journal of Fluorescence. 2004;**14**

[29] Jankowski D, Bojarski P, Kwiek P, Rangełowa JS. Donor–acceptor nonradiative energy transfer mediated by surface plasmons on ultrathin metallic films. Chemical Physics. 2010;**373**:238-242

[30] Lakowicz JR, Malicka J, Matveeva E, Gryczynski I, Gryczynski Z. Plasmonic technology: Novel approach to ultrasensitive immunoassays. Clinical Chemistry. 2005;**51**:1914-1922

[31] Tajdini MM, Mosallaei H. Characterization of Large Array of Plasmonic Nanoparticles on Layered Substrate: Dipole Mode Analysis Integrated with Complex Image Method. Applied EM and Optics Laboratory Northeastern University. Optical Society of America. 2011.

[32] Meng MT. Surface plasmon-coupled emission for applications in biomedical diagnostics. [thesis]. Glasnevin, Dublin, Irlanda: Dublin City University; 2009 Available from: http://doras.dcu.ie/14883/

[33] Kong JA, editor. Electromagnetic Wave Theory. John Wiley and Sons; 1986

[34] Aksun MI, Mittra R. Derivation of closed-form Green's functions for a general microstrip geometry. IEEE Transactions on Microwave Theory and Techniques. 1992;**40**

[35] Messa F, Boix RR, Medina F. Closed-form expressions of multilayered planar Green's functions that account for the continuous spectrum in the far field. IEEE Transactions on Microwave Theory and Techniques. 2008;**56**

[36] Zhuang L, Hu WY. Automatic incorporation of surface wave poles in discrete complex image method. Progress In Electromagnetics Research. 2008;**80**:161-178

[37] Simsek E, Liu J, Liu QH. A spectral integral method (SIM) for layered media. IEEE Transactions on Antennas and Propagation. 2006;**54**

[38] Aksum MI. A robust approach for the derivation of closed-form Green's. IEEE Transactions on Microwave Theory and Techniques. 1996;**44**

Chapter 10

Fano Resonance in High-Permittivity Objects

Xianghong Kong, Lina Qiu and Gaobiao Xiao

Additional information is available at the end of the chapter

http://dx.doi.org/10.5772/intechopen.70518

Abstract

In this chapter, Fano resonances in simple structures with high permittivity such as spheres or core-shell particles are analyzed by Mie theory. The Mie scattering coefficients can be decomposed into slow varying backgrounds and narrow resonances, which cause the Fano resonances in scattered field. For structures of arbitrary shapes, temporal coupled-mode theory is applied to explain the Fano resonances found in the scattering cross section. At last, we analyze the periodic structures by using band diagram, and it shows that the Fano resonances can be viewed as the superposition of the Bloch wave and the Mie scattering wave.

Keywords: Fano resonance, Mie theory, temporal coupled-mode theory, photonic crystal, sensor

1. Introduction

Fano resonance was first discovered in quantum systems to describe the asymmetrically shaped ionization spectral lines of atoms [1]. The asymmetric profile is caused by the interference between a broad background state and a narrow discrete state. The interference phenomenon also exists in electromagnetic system and was first observed by Wood [2]. With the development of metamaterials, Fano resonances have been observed in many classical oscillator systems, such as nonconcentric ring/disk cavities [3], asymmetric split rings, and dolmen structures [4]. Such Fano systems are caused by symmetry breaking of the geometry and are usually consisted of metal and dielectric. Recently, metamaterials composed of high refractive index materials have attracted researchers' attentions since they can enhance efficiency significantly [5]. Fano resonances occur in these metamaterials usually have larger quality factor since metal is replaced by lossless high-permittivity dielectric, which makes Fano curve sharper compared with conventional metamaterials.

In this chapter, we will investigate the Fano resonances in high-permittivity objects theoretically and do the simulations to verify the accuracy of theories, which will provide a guidance for the further study and applications. The chapter is organized as follows:

In the second section, we analyze the Fano resonances in high-permittivity spheres, which are the simplest structures that can be analyzed by applying Mie theory easily.

In the third section, we will use the Mie theory to investigate the Fano resonances in core-shell particles. With more degrees of freedom for design, core-shell structures are more suitable for applications such as sensors.

In the fourth section, a new method called temporal coupled-mode theory (TCMT) is used to explain the Fano resonances found in high permittivity arbitrarily shaped objects. Combined with cylindrical wave expansion (2D) or spherical wave expansion (3D), we can use TCMT to model the Fano resonances in scattering by an arbitrary object.

In the fifth section, we do some numerical simulation on periodic array of cylinders and show that Fano resonances can be observed in transmission spectra as a result of interference of leaky guided modes of cylinders with an incident electromagnetic wave.

In the last section, we will draw a conclusion briefly.

2. Fano resonances in high-permittivity spheres

2.1. Mie theory

Mie scattering was first discovered by Mie in 1908 [6]. In spite of the long history, Mie theory still governs the forefront optical devices such as nanoantennas [7] and metamaterials [8]. It describes the scattering of a plane wave by a homogeneous sphere. The solution takes form of an infinite series of spherical multipole partial waves. For different electromagnetic modes, the positions of resonances which can be calculated by Mie theory are different. Resonance arises when the incident wave reaches an eigenmode frequency and excites localized modes in the sphere.

Let us assume the radius of sphere is a. The relative permittivities and permeabilities of sphere ($r \leq a$) and embedding medium ($r > a$) are (ϵ_1, μ_1) and (ϵ, μ), respectively. The Mie scattering coefficients are [6]:

$$a_n = \frac{\mu_1 x j_n(x)\left[mxj_n(mx)\right]' - \mu m^2 x j_n(mx)\left[xj_n(x)\right]'}{-\mu_1 x h_n^{(1)}(x)\left[mxj_n(mx)\right]' + \mu m^2 x j_n(mx)\left[xh_n^{(1)}(x)\right]'} \tag{1}$$

$$b_n = \frac{\mu x j_n(x)\left[mxj_n(mx)\right]' - \mu_1 x j_n(mx)\left[xj_n(x)\right]'}{-\mu x h_n^{(1)}(x)\left[mxj_n(mx)\right]' + \mu_1 x j_n(mx)\left[xh_n^{(1)}(x)\right]'} \tag{2}$$

where $x = ka$ ($k = \omega/c$ is the wavenumber of incident wave) is the size parameter of the sphere,

http://dx.doi.org/10.5772/intechopen.70518

and $m = \sqrt{\frac{\epsilon_1 \mu_1}{\epsilon \mu}}$ is the relative refractive index. $j_n(x)$ and $h_n^{(1)}(x)$ stand for spherical Bessel functions and Hankel functions of the first kind, respectively.

2.2. Decomposition of Mie scattering coefficients

For simplicity, the embedding medium is considered to be vacuum in the following analysis, so we have $\epsilon = 1$, $\mu = 1$. We assume the relative permeability of sphere to be 1 and the relative permittivity ϵ_1 to be a purely real number, so the relative refractive index $m = \sqrt{\epsilon_1}$. Eq. (1), which represents electric scattering coefficients can be rewritten by

$$a_n = -\frac{[xj_n(x)]'}{\left[xh_n^{(1)}(x)\right]'} + \frac{\frac{i}{\left[xh_n^{(1)}(x)\right]'^2}}{\frac{\epsilon_1 xj_n(\sqrt{\epsilon_1}x)}{\left[\sqrt{\epsilon_1}xj_n(\sqrt{\epsilon_1}x)\right]'} - \frac{xh_n^{(1)}(x)}{\left[xh_n^{(1)}(x)\right]'}} \quad (3)$$

where $-[xj_n(x)]'/\left[xh_n^{(1)}(x)\right]'$ means a slow varying background and $\frac{i}{\left[xh_n^{(1)}(x)\right]'^2} / \left(\frac{\epsilon_1 xj_n(\sqrt{\epsilon_1}x)}{\left[\sqrt{\epsilon_1}xj_n(\sqrt{\epsilon_1}x)\right]'} - \frac{xh_n^{(1)}(x)}{\left[xh_n^{(1)}(x)\right]'}\right)$ means a narrow resonance when high-permittivity dielectric sphere is considered [9].

As shown in **Figure 1**, squared norm of Mie coefficient $|a_1|^2$ is plotted when $\epsilon_1 = 1000$. It can be described by superposition of narrow resonance and slow varying background.

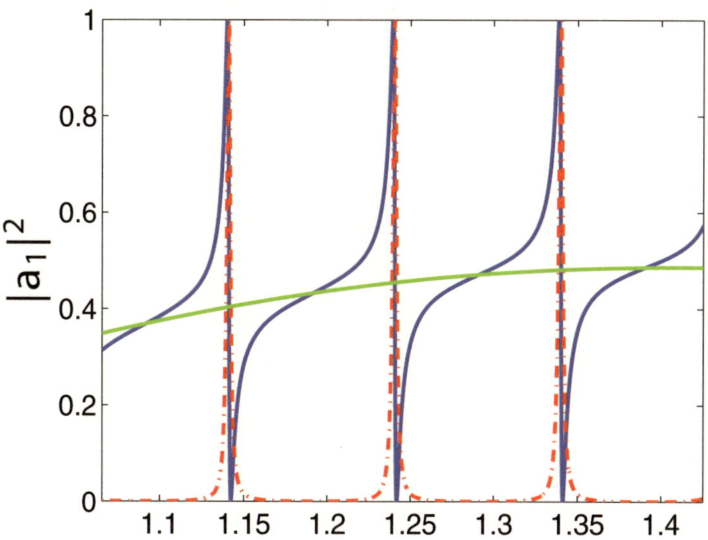

Figure 1. Squared norm of Mie coefficient $|a_1|^2$ (blue curve), slow varying background $\left|\frac{[xj_1(x)]'}{\left[xh_1^{(1)}(x)\right]'}\right|^2$ (green curve), and narrow resonance (red dot-dash line) for a sphere with $\epsilon_1 = 1000$.

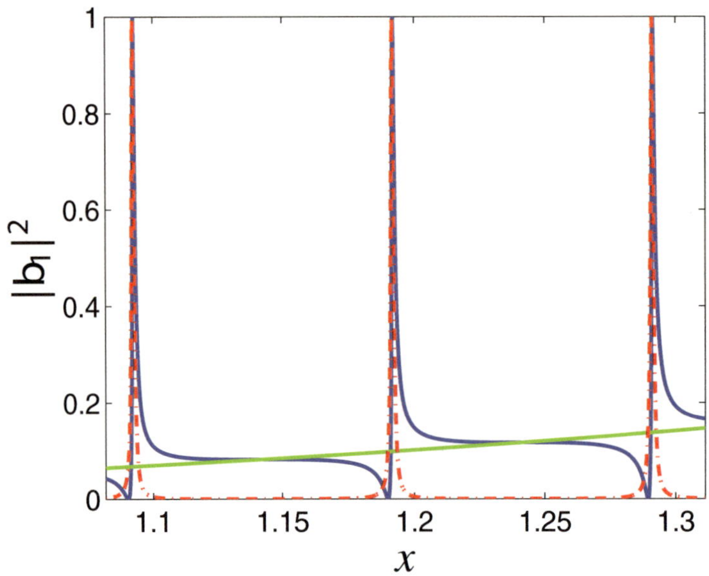

Figure 2. Squared norm of Mie coefficient $|b_1|^2$ (blue curve), slow varying background $\left|\frac{j_1(x)}{h_1^{(1)}(x)}\right|^2$ (green curve), and narrow resonance (red dot-dash line) for a sphere with $\epsilon_1 = 1000$.

Similarly, magnetic scattering coefficient $|b_n|$ can be decomposed into two parts:

$$b_n = -\frac{j_n(x)}{h_n^{(1)}(x)} + \frac{\frac{-i}{\left[xh_n^{(1)}(x)\right]^2}}{\frac{\left[\sqrt{\epsilon_1}xj_n\left(\sqrt{\epsilon_1}x\right)\right]'}{xj_n\left(\sqrt{\epsilon_1}x\right)} - \frac{\left[xh_n^{(1)}(x)\right]'}{xh_n^{(1)}(x)}} \tag{4}$$

where $-j_n(x)/h_n^{(1)}(x)$ represents a slow varying background and $\frac{-i}{\left[xh_n^{(1)}(x)\right]^2}/\left(\frac{\left[\sqrt{\epsilon_1}xj_n\left(\sqrt{\epsilon_1}x\right)\right]'}{xj_n\left(\sqrt{\epsilon_1}x\right)} - \frac{\left[xh_n^{(1)}(x)\right]'}{xh_n^{(1)}(x)}\right)$ represents a narrow resonance (as shown in **Figure 2**).

The slow varying backgrounds are the same as scattering coefficients of PEC spheres.

2.3. Rewrite Mie coefficients in the form of Fano function

Normalized Fano function can be expressed as $\frac{1}{1+q^2}\frac{\left(q+\frac{x-x_0}{\Gamma}\right)^2}{1+\left(\frac{x-x_0}{\Gamma}\right)^2}$, where x_0, Γ, and q represent resonance position, resonance width, and Fano parameter, respectively. Compared with conventional Lorentz resonance, Fano resonance will exhibit asymmetric line shape and usually has sharper resonant curve.

When the permittivity of sphere is high, we can rewrite the Mie coefficients in the form of Fano function. The resonance position, resonance width, and Fano parameter can be achieved by following Eqs. (10):

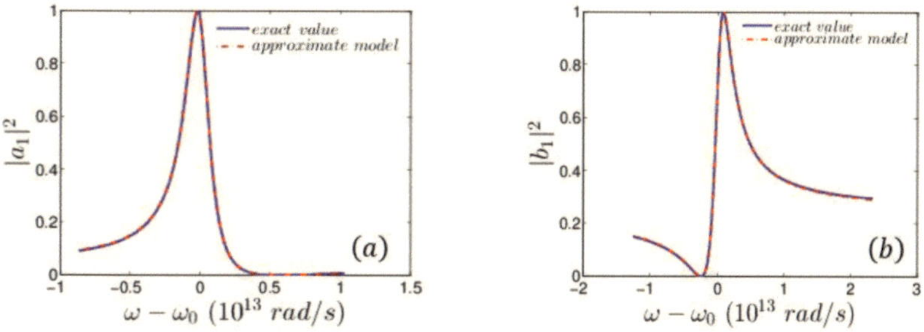

Figure 3. The exact value of Mie scattering coefficients (blue line) and Fano curve predicted by approximate model (red dot-dash line) are shown for (a) electric dipole $|a_1|^2$ when $\omega_0 = 3 \times 10^{15}$ *rad/s* , $a = 64.33$ *nm* , $\epsilon_1 = 1000$ and (b) magnetic dipole $|b_1|^2$ when $\omega_0 = 3 \times 10^{15}$ *rad/s* , $a = 148.97$ *nm* , $\epsilon_1 = 1000$.

$$\frac{\epsilon_1 x_0 j_n(\sqrt{\epsilon_1} x_0)}{[\sqrt{\epsilon_1} x_0 j_n(\sqrt{\epsilon_1} x_0)]'} = Re\left(\frac{x_0 h_n^{(1)}(x_0)}{\left[x_0 h_n^{(1)}(x_0)\right]'}\right) \tag{5}$$

$$q = \frac{[x_0 y_n(x_0)]'}{[x_0 j_n(x_0)]'} sign\left(\frac{Im\left(\frac{x_0 h_n^{(1)}(x_0)}{\left[x_0 h_n^{(1)}(x_0)\right]'}\right)}{\left.\frac{\partial \frac{\epsilon_1 x j_n(\sqrt{\epsilon_1} x)}{[\sqrt{\epsilon_1} x j_n(\sqrt{\epsilon_1} x)]'}}{\partial \omega}\right|_{\omega=\omega_0}}\right) \tag{6}$$

$$\Gamma = \left|\frac{Im\left(\frac{x_0 h_n^{(1)}(x_0)}{\left[x_0 h_n^{(1)}(x_0)\right]'}\right)}{\left.\frac{\partial \frac{\epsilon_1 x j_n(\sqrt{\epsilon_1} x)}{[\sqrt{\epsilon_1} x j_n(\sqrt{\epsilon_1} x)]'}}{\partial \omega}\right|_{\omega=\omega_0}}\right| \tag{7}$$

In Eqs. (5)–(7), Re(x) and Im(x) mean the real and imaginary part of x, $y_n(x)$ represent the spherical Neumann functions, sign(x) denotes the sign function. These equations are the rewrite of electric scattering coefficients. Similarly, magnetic scattering coefficients can also be rewritten in the form of Fano function [10].

As shown in **Figure 3**, the approximate model which can be written in the form of Fano function matches well with the exact Mie scattering coefficient.

3. Fano resonances in core-shell particles with high-permittivity covers

In most researches [11, 12], Fano resonances observed in coated spheres are derived in the Rayleigh limit. However, the approximation may suffer a loss of precision when frequency gets higher. An exact analysis based on Mie theory is proposed to analyze Fano resonances by coated spheres with high-permittivity covers in a precise way [13].

3.1. Theoretical analysis

Let us assume the inner radius of core-shell particle is a_1 and the outer radius is a. The ratio of a_1 and a can be denoted as $\eta = a_1/a$. The relative permittivities and permeabilities of core ($0 < r \leq a_1$), shell ($a_1 < r \leq a$), and embedding medium ($r > a$) are (ϵ_1, μ_1), (ϵ_2, μ_2), and (ϵ_0, μ_0), respectively. The solution of scattering by coated spheres can be described by Mie theory. For simplicity, the embedding medium is considered to be vacuum in the following analysis which means $\epsilon_0 = 1$, $\mu_0 = 1$. Also, we assume $\mu_1 = \mu_2 = 1$, which means both core and shell are nonmagnetic. When the core-shell particles are covered by high-permittivity dielectric shells, we can decompose the scattering coefficients c_n^{TM} and c_n^{TE} into slow varying backgrounds and narrow resonances, which are similar to the high-permittivity spheres. For electric scattering coefficients, we have

$$c_n^{TM} = s_n^{TM} + r_n^{TM} \tag{8}$$

where

$$s_n^{TM} = -\frac{[xj_n(x)]'}{\left[xh_n^{(1)}(x)\right]'} \tag{9}$$

$x = k_0a$ is the size parameter of outer sphere. s_n^{TM} represents the slow varying background and its expression is given in Eq. (9). As we can see, the background is the same as the electric scattering coefficient of a PEC sphere with radius a. r_n^{TM} represents the narrow resonance, and it can be calculated formally by subtracting s_n^{TM} from c_n^{TM}. The expression for narrow resonance can be found in [13].

Similarly, we can decompose magnetic scattering coefficients into two parts:

$$c_n^{TE} = s_n^{TE} + r_n^{TE} \tag{10}$$

where

$$s_n^{TE} = -\frac{j_n(x)}{h_n^{(1)}(x)} \tag{11}$$

As shown in **Figure 4**, the scattering coefficients can be viewed as the cascade of Fano resonances.

3.2. Application of sensors

Due to the sharp resonances near the resonance frequencies, Fano resonances have great potential applications in sensing problems [14–16]. Although some of them may have high sensitivity, the structures which are designed to produce Fano resonances are usually complicated and cannot be analyzed by formula exactly. Because of the simple structure, core-shell particles have the potential to be a great platform for sensing since they can be fabricated easily. In fact, core-shell particles consisted of metal and dielectric, can exhibit Fano resonances due to the

http://dx.doi.org/10.5772/intechopen.70518

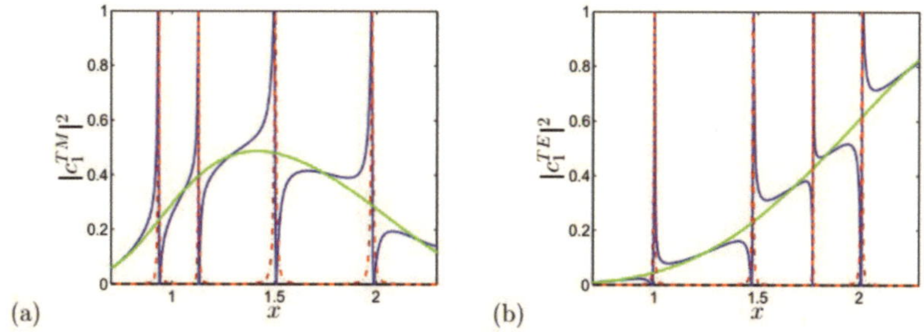

Figure 4. Squared norm of Mie coefficient (a) $|c_1^{TM}|^2$ (blue curve), slow varying background $\left|\frac{[xj_1(x)]'}{[xh_1^{(1)}(x)]'}\right|^2$ (green curve) and narrow resonances (red dot-dash line) (b) $|c_1^{TE}|^2$ (blue curve), slow varying background $\left|\frac{j_1(x)}{h_1^{(1)}(x)}\right|^2$ (green curve), and narrow resonances (red dot-dash line) for a core-shell particle with $a = 100\ nm$, $a_1 = 80\ nm$, $\epsilon_1 = 10$, $\epsilon_2 = 1000$.

hybridization between the plasmon resonances of the core and shell [17, 18]. However, the loss in metal may flatten the shape of Fano curve, which affects the sensitivity of Fano resonance sensor. Hence, we use lossless high-permittivity dielectric to replace the metal.

For the high-permittivity shell sensors, we can fill the core with unknown materials. The permittivity of the unknown material can be varied continuously such as liquid solvents. By detecting the scattering field over a discrete set of frequencies near Fano resonance position, we can achieve the permittivity we want with high accuracy.

The sensitivity for Fano resonance sensing can be examined by comparing the changes in the scattering coefficients between core-shell structure with high-permittivity shell and homogeneous sphere when the permittivity of material changes. We can define the sensitivity as an analogy to [14]

$$S_n^{TM} = \lim_{\Delta\epsilon \to 0} \frac{\Delta|c_n^{TM}|^2}{\Delta\epsilon} \tag{12}$$

As shown in **Figure 5(a)**, the difference of sensitivity between core-shell structure and sphere is plotted. As for the core-shell structure, the relative permittivity of core is increased by $\Delta\epsilon_1 = 0.1$. The maximum value of $\Delta|c_1^{TM}|^2$ occurs at $x = 0.49495$ (located by a vertical blue line) when $0.493 \leq x \leq 0.496$, which is $\Delta|c_1^{TM}|^2 = 0.1971$. In order to make a comparison with Fano resonance sensor, we figure out the scattering coefficients of a sphere with different permittivities as given for the core in core-shell structure. As shown in **Figure 5(b)**, the maximum value $\Delta|c_1^{TM}|^2 = 4.2259 \times 10^{-5}$ occurs at $x = 0.496$, which shows that the Fano resonance sensor offers a high sensitivity.

Since Fano resonances of high permittivity core-shell particles mentioned above only exist in scattering coefficients of multipole partial waves, it is difficult to achieve these coefficients separately by measuring the electromagnetic field distribution around the scatterers. In fact, by choosing operating frequency range properly, we can achieve the scattering coefficient of a

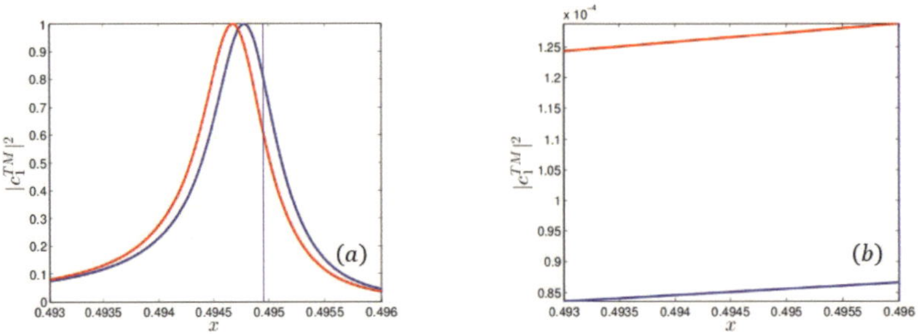

Figure 5. (a) $|c_1^{TM}|^2$ as a function of x for core-shell structure with different core permittivities $\epsilon_1 = 1.4$ (blue line) and $\epsilon_1 = 1.5$ (red line), high-permittivity shell $\epsilon_2 = 1000$, $\eta = 0.8$. (b) $|c_1^{TM}|^2$ as a function of x for sphere structure with $\epsilon_1 = 1.4$ (blue line) and $\epsilon_1 = 1.5$ (red line).

single partial wave ($c_1^{TM}$for example) without filtering out from the total electromagnetic fields. When size parameter x is small, most of the scattering coefficients c_n^{TM}, c_n^{TE} are zero, except for several coefficients with small n. We choose the first resonance frequency of c_1^{TM} as the operating frequency. Since Fano resonance is usually sharp and narrow, we find that overlap between different modes can be avoided if the frequency range narrows down. To explain it explicitly, we define the scattering cross section as [19]

$$Q_{sca} = \frac{2}{x^2} \sum_{n=1}^{\infty} (2n+1)\left(\left|c_n^{TM}\right|^2 + \left|c_n^{TE}\right|^2\right) \tag{13}$$

The contribution of c_1^{TM} to Eq. (13) can be defined as

$$Q_1^{TM} = \frac{6}{x^2} \left|c_1^{TM}\right|^2 \tag{14}$$

As shown in **Figure 6**, when we choose the first Fano resonance position of c_1^{TM} as the operating frequency, we find that the scattering cross section of multipole partial waves is the same as the scattering cross section of c_1^{TM} for a narrow frequency range. As shown in **Figure 6(a)**, Fano resonance can also be observed in scattering cross section which can be used to sense the permittivity.

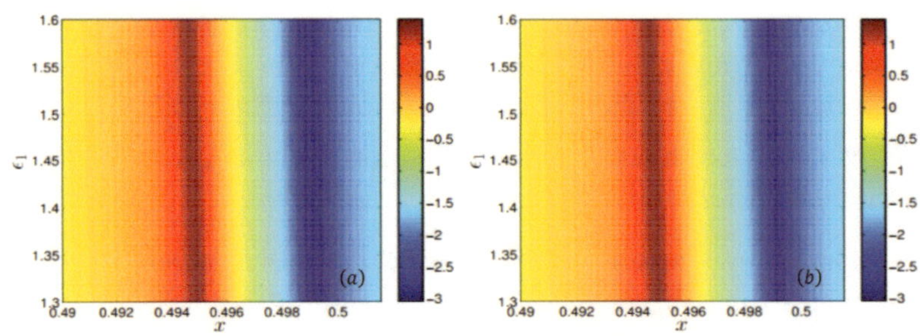

Figure 6. For the core-shell particle with high-permittivity shell $\epsilon_2 = 1000$, $\eta = 0.8$, (a) $log_{10}(Q_{sca})$ is plotted as a function of x and ϵ_1 when the summation in Eq. (13) is truncated to $n = 5$. (b) $log_{10}\left(Q_1^{TM}\right)$ (calculated by Eq. (14)) is plotted as a function of x and ϵ_1.

http://dx.doi.org/10.5772/intechopen.70518

The asymmetry parameter ⟨cosΘ⟩ is defined as the average cosine of the scattering angle Θ. For a spherical particle, the asymmetry parameter can be calculated by [20]

$$\begin{aligned}\langle \cos\Theta\rangle = & \frac{4\pi}{x^2 Q_{sca}} Re\left(\sum_{n=1}^{\infty}\left(\frac{n(n+2)}{n+1}\left[c_n^{TM}\left(c_{n+1}^{TM}\right)^* + c_n^{TE}\left(c_{n+1}^{TE}\right)^*\right]\right)\right) \\ & + \frac{4\pi}{x^2 Q_{sca}} Re\left(\sum_{n=1}^{\infty}\left(\frac{2n+1}{n(n+1)}\left[c_n^{TM}\left(c_n^{TE}\right)^*\right]\right)\right)\end{aligned} \tag{15}$$

The asymmetry parameter is positive if the particle scatters more light toward the forward direction while it is negative if more light is scattered toward the backscattering direction.

As shown in **Figure 7**, the width between maximum and minimum for a fixed core permittivity ϵ_1 narrows down compared with scattering cross section which is shown in **Figure 6**. With the increase of size parameter x, the asymmetry parameter reaches its maximum and decreases sharply to its minimum.

To check the average scattering direction changes from front to back, we use numerical simulation software COMSOL 5.0 to simulate the scattering of a plane wave by a core-shell

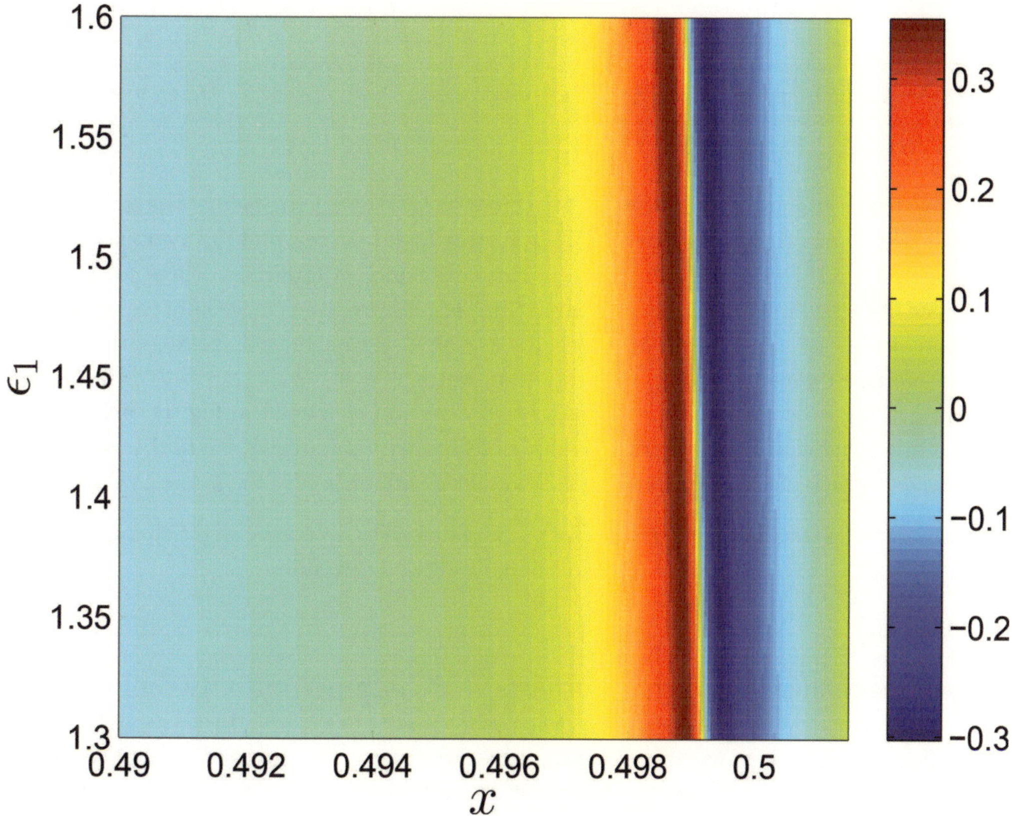

Figure 7. The asymmetry parameters for high-permittivity shell particles with $\epsilon_2 = 1000$, $\eta = 0.8$.

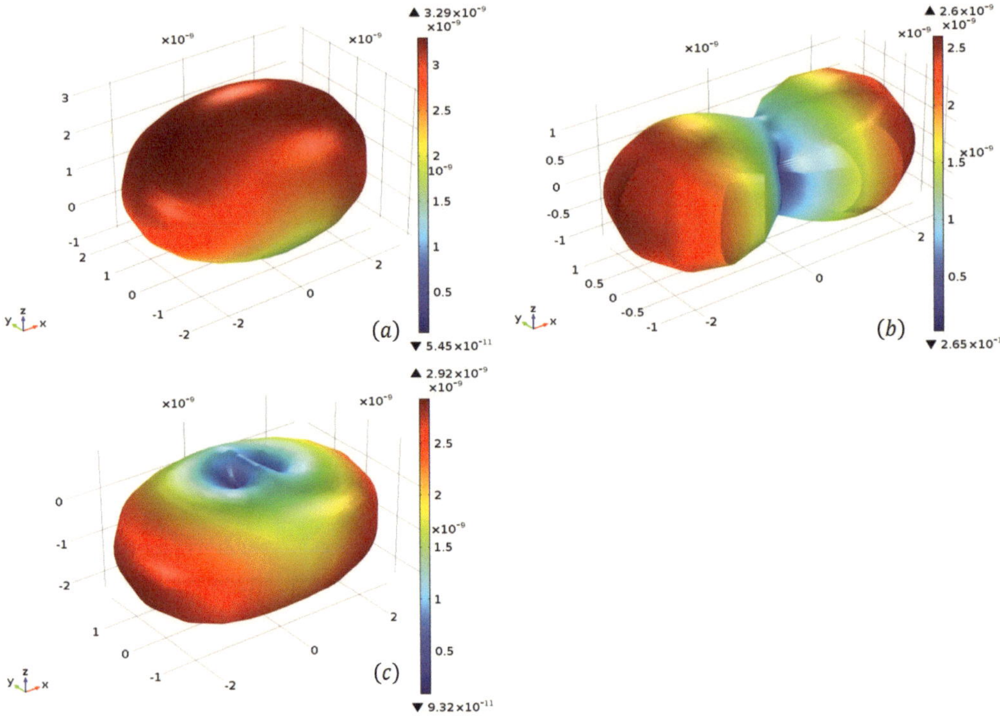

Figure 8. Scattering pattern of a high-permittivity shell particle with $\epsilon_2 = 1000$, $\epsilon_1 = 1.5$, $\eta = 0.8$, $a = 100\ nm$ when (a) $x = 0.498672$ (b) $x = 0.498980$ (c) $x = 0.499276$.

structure. The incident wave travels in the +z-direction and the electric field is oriented in the x-direction. Draw a horizontal line at $\epsilon_1 = 1.5$ in **Figure 7**, and we can find the average cosine of the scattering angle $\langle cos\Theta \rangle$ has a lineshape of Fano resonance as a function of size parameter x. As shown in **Figure 8(a)**, the high-permittivity shell structure scatters more light to the forward direction at $x = 0.498672$. When x increases, the asymmetry parameter decreases sharply from positive value to negative value. The minimum value is achieved at $x = 0.499276$ and the scattering wave is concentrated in the backward direction as shown in **Figure 8(c)**. Among the maximum value and the minimum value of asymmetry parameter, we find $\langle cos\Theta \rangle \approx 0$ at $x = 0.498980$, which means the scattering is symmetric with respect to the plane $z = 0$.

Hence, Fano resonances in core-shell particles can be used to detect the slight changes of core permittivity since they are sensitive in both magnitude and direction.

4. Fano resonances in arbitrary objects with high-permittivity dielectric

When the structure gets more complicated, the Mie theory is no longer valid for the solution of scattering field. We have to use the temporal coupled-mode theory (TCMT) to replace Mie theory when investigating the Fano resonances in arbitrary objects with high-permittivity dielectric.

4.1. Temporal coupled-mode theory

The temporal coupled-mode theory provides a useful general framework to study the interaction of a resonance with external waves. It has been well developed when dealing with particles that have cylindrical or spherical shapes [21]. In [22], TCMT has been generalized to analysis the scattering of arbitrary shape structures. The temporal coupled-mode equations can be expressed as [23]

$$\begin{cases} \dfrac{dA}{dt} = \left(-i\omega_0 - \dfrac{1}{\tau}\right)A + \boldsymbol{\kappa}^T \boldsymbol{s}^+ \\ \boldsymbol{s}^- = \boldsymbol{B}\boldsymbol{s}^+ + A\boldsymbol{d} \end{cases} \tag{16}$$

In Eq. (16), $|A|^2$ corresponds to the energy inside the resonator. $\boldsymbol{s}^+$ and $\boldsymbol{s}^-$ represent incoming waves and outgoing waves, respectively. They couple directly by the resonant mode A is coupled with the outgoing waves $\boldsymbol{s}^-$ through $\boldsymbol{d}$ and is excited by the incoming waves $\boldsymbol{s}^+$ through $\boldsymbol{\kappa}^T$. ω_0 is the resonance frequency and $\frac{1}{\tau}$ is the external leakage rate.

There exists some constrains between **B**, **d**, and **κ**, which are imposed by energy conservation and time-reversal invariance [22]. The constrain conditions are

$$\begin{cases} |\boldsymbol{d}|^2 = \dfrac{2}{\tau} \\ \boldsymbol{\kappa}^T \boldsymbol{d}^* = \dfrac{2}{\tau} \\ \boldsymbol{B}\boldsymbol{d}^* + \boldsymbol{d} = 0 \end{cases} \tag{17}$$

For a 2D arbitrary object, we can expand scattering field into cylindrical waves

$$H_{sca} = \sum_{m=-\infty}^{\infty} H_0 \, a_m \, H_{|m|}^{(1)}(k\rho) e^{im\theta} \tag{18}$$

$\boldsymbol{s}^+$ and $\boldsymbol{s}^-$ in Eq. (16) can be viewed as coefficients of input wave and outgoing wave on the basis of cylindrical waves.

The incident plane wave can also be expanded into cylindrical waves

$$e^{ikr} = \sum_{m=-\infty}^{\infty} i^{|m|} e^{-i\theta_0 m} \left(\frac{H_{|m|}^{(1)}(k\rho) + H_{|m|}^{(2)}(k\rho)}{2} \right) \times e^{i\theta m} \tag{19}$$

where θ_0 is the incident angle. Combined with cylindrical wave expansion, we can use TCMT to describe the Fano resonances in arbitrary objects with high-permittivity dielectric.

4.2. Numerical simulation

The method we determine the coefficients in Eq. (16) is similar to the method described in [22]. Firstly, we use eigenmode analysis in COMSOL 5.0 to figure out the resonance frequency ω_0 and the external leakage rate $\frac{1}{\tau}$. Secondly, being different from the method in [22] where they set $\boldsymbol{B} = \boldsymbol{I}$, we calculate the $\boldsymbol{B}$ through the simulation results of the arbitrary object covered by PEC illuminated by the plane wave. Since the slow varying background of high-permittivity sphere is the PEC sphere as mentioned above, it is intuitive to assume the slow varying background of arbitrary object which is described by $\boldsymbol{B}$ is the same as the object covered by PEC. Thirdly, combined with the field distribution of eigenmode simulation and background scattering matrix $\boldsymbol{B}$, we can figure out the resonant radiation coefficients $\boldsymbol{d}$. At last, $\boldsymbol{\kappa}$ can be solved through constrain conditions in Eq. (17).

Once the coefficients in Eq. (16) are determined, we can use the TCMT to predict the scattering fields by different incident frequencies and incident angles.

As shown in **Figure 9**, the relative permittivity of rounded-corner triangle is 600. The structure has a resonance frequency of $\omega_0 = 0.12172\omega_p$ and the leakage rate is $\frac{1}{\tau} = 1.1771 \times 10^{-4}$, which can be figured out by COMSOL. We use Matlab to set the temporal coupled-mode model as shown in Eq. (16). By comparing with the simulation results of COMSOL at different incident frequencies (near the resonance frequency) and incident angles, we can prove the validity of TCMT.

When a TM wave impinges on the scatterer, the scattering cross section can be defined as

$$C_{sct} = \frac{P_{sct}}{I_0} \tag{20}$$

where P_{sct} is the rate at which energy is scattered across the circle far away from the scatterer and $I_0 = \frac{1}{2}\sqrt{\frac{\mu_0}{\epsilon_0}}|H_0|^2$ is the intensity of the incident plane wave. For TCMT, the scattering cross section can be calculated by [22]

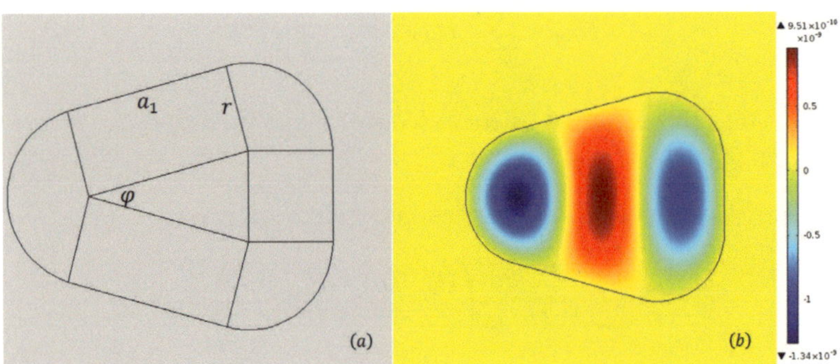

Figure 9. (a) Rounded-corner triangle with $r = 0.15\lambda_p, a_1 = 0.3\lambda_p, \phi = \frac{\pi}{6}, \epsilon = 600$. (b) The real part of H_z for the eigenmode analysis at the frequency $\omega_0 = 0.12172\omega_p$.

http://dx.doi.org/10.5772/intechopen.70518

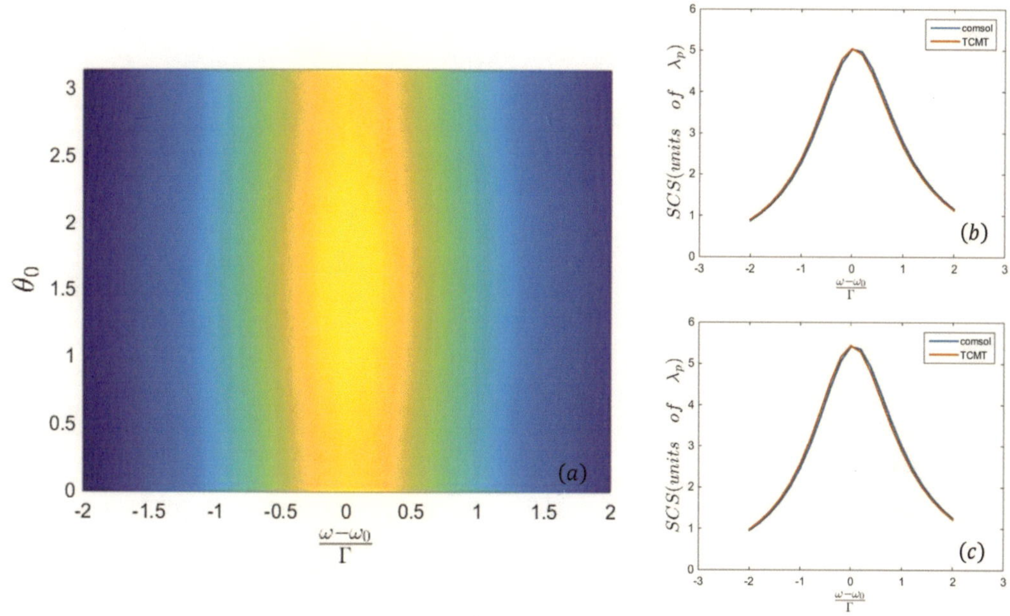

Figure 10. (a) Scattering cross section predicted by TCMT as a function of incident frequency ω and incident angle θ_0. Comparison between TCMT and COMSOL simulation results for different incident angles $\theta_0 = 0$ (b) and $\theta_0 = \frac{\pi}{2}$ (c).

$$C_{sct} = \frac{((S-I)s^+)^\dagger (S-I)s^+}{I_0} \tag{21}$$

where

$$\mathbf{S} = \mathbf{B} + \frac{d\boldsymbol{\kappa}^T}{i\omega_0 - i\omega + \frac{1}{\tau}} \tag{22}$$

As shown in **Figure 10(b, c)**, for different incident angles, scattering cross sections predicted by TCMT match well with the results simulated by COMSOL.

As shown in **Figure 11**, the green line represents the assumption in [22] that the background scattering matrix $\boldsymbol{B}$ can be set to $\boldsymbol{I}$ while the blue line represents the assumption that $\boldsymbol{B}$ is achieved through the simulation results of the arbitrary object covered by PEC. As we can see, when the incident frequency equals to the resonance frequency, both the green line and blue line (TCMT models with different parameters) can match the simulation results. However, when the incident frequency deviates from the resonance frequency, our TCMT model shows a better accuracy compared with the TCMT model in [22], which indicates that the background scattering matrix $\boldsymbol{B}$ cannot be set to $\boldsymbol{I}$ easily when the permittivity of object is high.

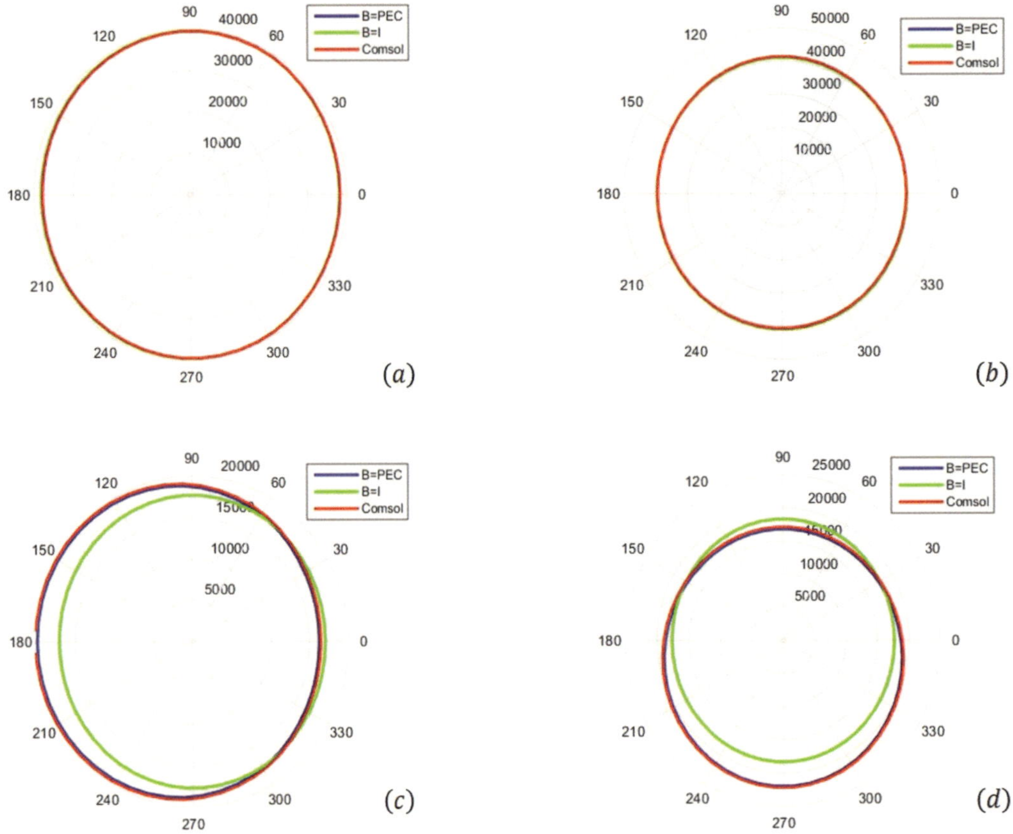

Figure 11. The far-field amplitude of the scattering field with different incident frequencies and angles of (a) $\omega = \omega_0$, $\theta_0 = 0$, (b) $\omega = \omega_0, \theta_0 = \frac{\pi}{2}$, (c) $\omega = \omega_0 + 2\Gamma$, $\theta_0 = 0$, (d) $\omega = \omega_0 + 2\Gamma, \theta_0 = \frac{\pi}{2}$.

5. Fano resonances in periodic structures

Fano resonances have been widely observed in the various periodic structures [24, 25]. The theory of Fano resonance in periodic structures is well developed. In [26], temporal coupled-mode theory is applied to analysis the transmission spectra of photonic crystal slab. According to the TCMT, the Fano resonances existed in transmission spectra are the result of the coupling of leaky mode to the external waves. Recently, the experimental discovery of Fano resonances involving interference between Mie scattering and Bragg scattering is studied in [27]. By comparing the disordered system with the periodic structure, they conclude the sharp resonances in periodic structure are caused by the Bloch waves. In order to study the interference between Mie scattering and Bragg scattering theoretically, the inverse dispersion method is proposed to calculate the photonic band diagram and distinguish unambiguously between Bragg and Mie gaps in the spectra [28]. The method reduces Maxwell's equations to a problem with the eigenvalue k while ω is considered to be a real parameter. It is not so intuitive since conventional approach will reduce the Maxwell's equations to standard eigenproblem for the frequency [29]. In [30], the author shows that the frequencies of observed Fano resonances existed in a linear array of dielectric cylinders coincide with the position of narrow frequency bands found in the spectra of corresponding two-dimensional photonic crystals. Inspired by [28, 30], we figure out the eigenfrequency of the photonic crystal slab and compare with the band diagram of two-dimensional photonic crystal. We are surprised to find that the occurrence of Fano resonances in photonic crystal slab can be predicted by the band diagram of photonic crystal.

5.1. Transmission spectra of the photonic Crystal slab

The structure of photonic crystal slab is shown in **Figure 12**. We assume the dielectric cylinders are parallel to the z axis. When TE waves with different angles incident on the slab, we can calculate the transmission coefficients and plot them in **Figure 13**.

As shown in **Figure 13**, Fano resonances with narrow resonance width can be observed. The permittivity of photonic crystal slab does not need to be as high as single cylinder in order to achieve same quality factor and such materials may be easily found in nature.

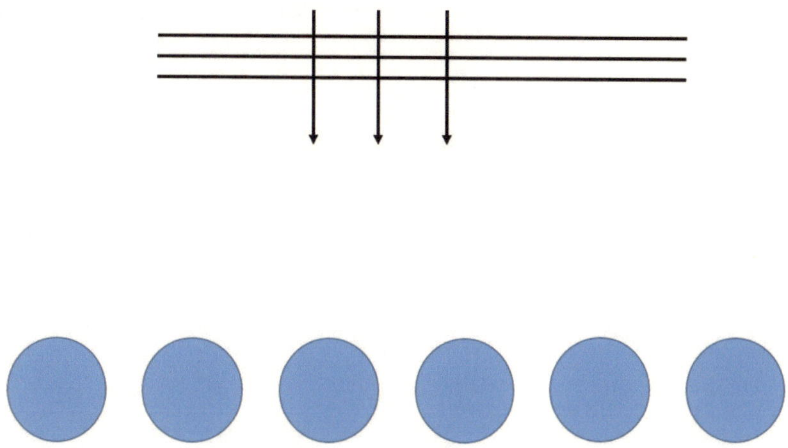

Figure 12. Photonic crystal slab with radius of cylinders $r = 0.4a$ (a is the period of the slab), $\epsilon = 12$, $\mu = 1$.

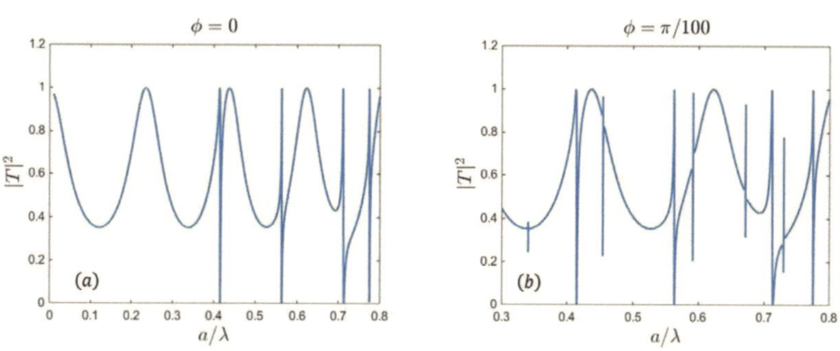

Figure 13. Transmission coefficient of plane wave incident on the photonic crystal slab. The incident angles are (a) $\varphi = 0$ and (b) $\varphi = \frac{\pi}{100}$.

Let us assume the Fano resonances in **Figure 13** satisfy the Fano function $\frac{1}{1+q^2}\frac{\left(q+\frac{\omega-\omega_0}{\Gamma}\right)^2}{1+\left(\frac{\omega-\omega_0}{\Gamma}\right)^2}$. Firstly, we use eigenmode analysis in COMSOL to figure out the eigenfrequency of photonic crystal slab. The real and imaginary part of eigenfrequency represent the resonance frequency ω_0 and resonance width Γ respectively. Secondly, we use the fitting method in Matlab to get the optimal Fano parameter q in Fano function. Thirdly, with given ω_0, Γ, and q, we can plot the Fano function with respect to frequency ω. As shown in **Figure 14**, the Fano curve matches well with the transmission coefficient simulated by COMSOL. The horizontal ordinate is chosen as a/λ for convenience, which is proportional to frequency ω.

5.2. Band diagram of photonic Crystal

The photonic crystal slab is periodic in only one direction while two-dimensional photonic crystal is periodic in two directions. For a photonic crystal as shown in **Figure 15**, the band diagram for $k_y = 0, 0 \le k_x \le \frac{\pi}{a}$ is plotted in **Figure 16(a)**. The eigenfrequencies of the photonic crystal are real while the eigenfrequencies of the photonic crystal slab are complex due to the existence of radiation loss. Hence, only the real parts of eigenfrequencies are plotted as shown in **Figure 16(b)**. By comparing the resonance frequencies shown in **Figure 13** and eigenfrequencies in **Figure 16(b)**, we can conclude that the occurrence of Fano resonances in transmission spectra of photonic crystal slab can be predicted by the real parts of the eigenfrequencies of the system. In addition, for the Fano resonances, which are observed in **Figure 13(b)** but cannot be observed in **Figure 13(a)**, they all have the eigenfrequencies with $Q \to \infty$. Hence, the resonance widths tend to zero and the resonances cannot be observed.

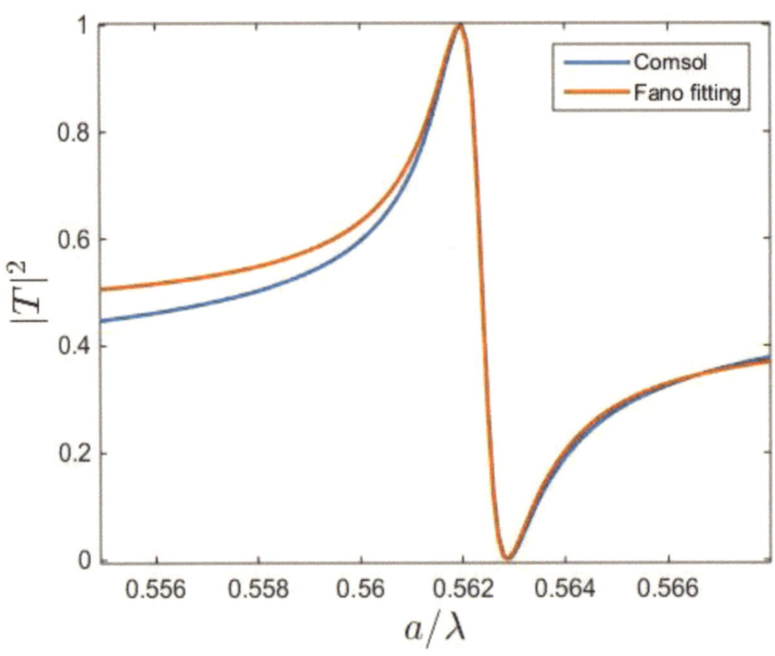

Figure 14. The Fano curve and the simulation result of photonic crystal slab are plotted.

http://dx.doi.org/10.5772/intechopen.70518

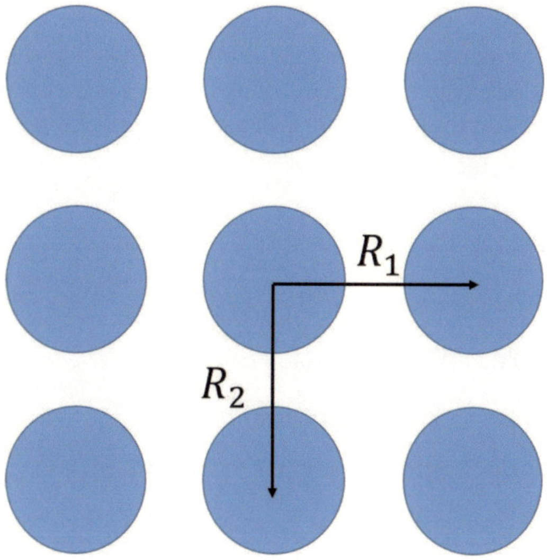

Figure 15. Photonic crystal with radius of cylinders $r = 0.4a$ ($R_1 = R_2 = a$), $\epsilon = 12$, $\mu = 1$.

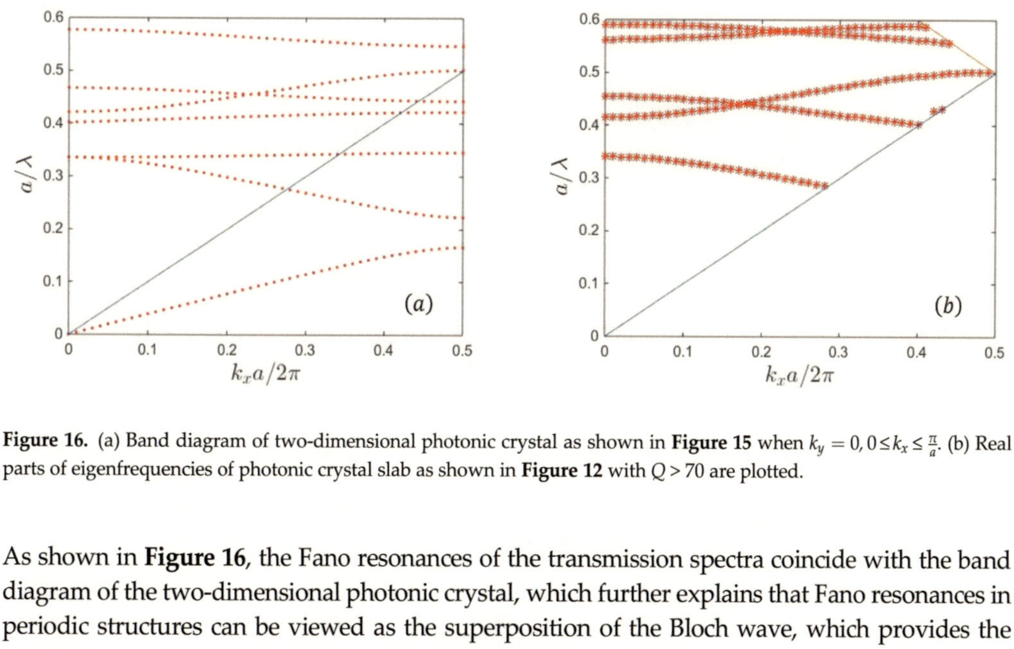

Figure 16. (a) Band diagram of two-dimensional photonic crystal as shown in **Figure 15** when $k_y = 0, 0 \le k_x \le \frac{\pi}{a}$. (b) Real parts of eigenfrequencies of photonic crystal slab as shown in **Figure 12** with $Q > 70$ are plotted.

As shown in **Figure 16**, the Fano resonances of the transmission spectra coincide with the band diagram of the two-dimensional photonic crystal, which further explains that Fano resonances in periodic structures can be viewed as the superposition of the Bloch wave, which provides the narrow resonances and the Mie scattering wave which provides the slow varying background.

6. Conclusion

In this chapter, we have presented various structures with high permittivity, which have Fano resonances, such as spheres, core-shell particles, arbitrary shape objects, and periodic structures. For each structure, different theoretical methods together with numerical analysis have been presented. Compared with conventional Fano resonances observed in structures consisted of metal and dielectric, high-permittivity structures can enhance the quality factor significantly, which may open up new opportunities for applications such as sensors, switches, and permittivity measuring technique.

Author details

Xianghong Kong*, Lina Qiu and Gaobiao Xiao

*Address all correspondence to: klovek@sjtu.edu.cn

Department of Electronic Engineering, Shanghai Jiao Tong University, China

References

[1] Fano U. Effects of configuration interaction on intensities and phase shifts. Physical Review. 1961;**124**:1866-1878. DOI: 10.1103/PhysRev.124.1866

[2] Wood RW. On a remarkable case of uneven distribution of light in a diffraction grating Spectrum. Proceedings of the Physical Society of London. 1902;**18**:269-275. DOI: 10.1088/1478-7814/18/1/325

[3] Hao F, Sonnefraud Y, Van Dorpe P, Maier SA, Halas NJ, Nordlander P. Symmetry breaking in plasmonic nanocavities: Subradiant LSPR sensing and a tunable Fano resonance. Nano Letters. 2008;**8**(11):3983-3988. DOI: 10.1021/nl802509r

[4] Luk'yanchuk B, Zheludev NI, Maier SA, Halas NJ, Nordlander P, Giessen H, et al. The Fano resonance in plasmonic nanostructures and metamaterials. Nature Materials. 2010;**9**(9):707-715

[5] Jahani S, Jacob Z. All-dielectric metamaterials. Nature Nanotechnology. 2016;**11**(1):23-26

[6] Bohren CF, Huffman DR. Absorption and scattering by a sphere. In: Absorption and Scattering of Light by Small Particles; 2004. p. 82-129

[7] Foteinopoulou S, Vigneron JP, Vandenbem C. Optical near-field excitations on plasmonic nanoparticle-based structures. Optics Express. 2007;**15**:4253-4267

[8] Yannopapas V, Moroz A. Negative refractive index metamaterials from inherently non-magnetic materials for deep infrared to terahertz frequency ranges. Journal of Physics. Condensed Matter. 2005;**17**:3717-3734

[9] Rybin MV, Filonov DS, Belov PA, Kivshar YS, Limonov MF. Switching from visibility to invisibility via Fano resonances: Theory and experiment. Scientific Reports. 2015;**5**:8774

[10] Kong X, Xiao G. Fano resonance in high-permittivity dielectric spheres. Journal of the Optical Society of America. A. 2016;**33**(4):707-711

[11] Arruda TJ, Martinez AS, Pinheiro FA. Unconventional Fano effect and off-resonance field enhancement in plasmonic coated spheres. Physical Review A. 2013;**87**:043841

[12] Arruda TJ, Martinez AS, Pinheiro FA. Tunable multiple Fano resonances in magnetic single-layered core-shell particles. Physical Review A. 2015;**92**:023835

[13] Kong X., Xiao G. Fano resonances in core-shell particles with high permittivity covers. In: Progress in Electromagnetic Research Symposium (PIERS); IEEE. 2016. pp. 1715–1719.

[14] Zheng G, Chen Y, Bu L, Xu L, Su W. Waveguide-coupled surface phonon resonance sensors with super-resolution in the mid-infrared region. Optics Letters. 2016;**41**(7):1582-1585

[15] Rahmani M, Luk'yanchuk B, Hong M. Fano resonance in novel plasmonic nanostructures. Laser and Photonics Reviews. 2013;**7**(3):329-349

[16] Zafar R, Salim M. Enhanced figure of merit in Fano resonance-based plasmonic refractive index sensor. IEEE Sensors Journal. 2015;**15**(11):6313-6317

[17] Chen H, Shao L, Man YC, Zhao C, Wang J, Yang B. Fano resonance in (gold core)-(dielectric shell) nanostructures without symmetry breaking. Small. 2012;**8**(10):1503-1509

[18] Raschke G, Brogl S, Susha A, Rogach A, Klar T, Feldmann J, et al. Gold nanoshells improve single nanoparticle molecular sensors. Nano Letters. 2004;**4**(10):1853-1857

[19] Fan X, Zheng W, Singh DJ. Light scattering and surface plasmons on small spherical particles. Light: Science & Applications. 2014;**3**(6):e179

[20] Mishchenko MI, Travis LD, Lacis AA. Scattering, Absorption, and Emission of Light by Small Particles. Cambridge, England: Cambridge University Press; 2002. p. 486

[21] Ruan Z, Fan S. Temporal coupled-mode theory for Fano resonance in light scattering by a Sinble obstacle. Journal of Physical Chemistry C. 2010;**114**(16):7324-7329

[22] Ruan Z, Fan S. Temporal coupled-mode theory for light scattering by an arbitrarily shaped object supporting a single resonance. Physical Review A. 2012;**85**:043828

[23] Haus HA. Waves and Fields in Optoelectronics. Upper Saddle River, New Jersey: Prentice-Hall; 1984. p. 748

[24] Kanskar M, Paddon P, Pacradouni V, Morin R, Busch A, Young JF, et al. Observation of leaky slab modes in an air-bridged semiconductor waveguide with a two-dimensional photonic lattice. Applied Physics Letters. 1997;**70**:1438-1440

[25] Astratov VN, Culshaw IS, Stevenson RM, Whittaker DM, Skolnick MS, Kraus TF, et al. Resonant coupling of near-infrared radiation to photonic band structure waveguides. Journal of Lightwave Technology. 1999;**17**:2050-2057

[26] Fan S, Suh W, Joannopoulos JD. Temporal coupled-mode theory for the Fano resonance in optical resonators. Journal of the Optical Society of America. A. 2003;**20**(3):569-572

[27] Rybin MV, Khanikaev AB, Inoue M, Samusev KB, Steel MJ, Yushin G, et al. Fano resonance between Mie and Bragg scattering in photonic crystals. Physical Review Letters. 2009;**103**(2):023901

[28] Rybin MV, Limonov MF. Inverse dispersion method for calculation of complex photonic band diagram and PT symmetry. Physical Review B. 2016;**93**:165132

[29] Joannopoulos JD, Johnson SG, Winn JN, Meade RD. Photonic Crystals: Molding the Flow of Light. 2nd ed. Princeton: Princeton University Press; 2008

[30] Markos P. Fano resonances and band structure of two-dimensional photonic structures. Physical Review A. 2015;**92**:043814

Chapter 11

Laser-Induced Fano Resonance in Condensed Matter Physics

Ken-ichi Hino, Yohei Watanabe,
Nobuya Maeshima and Muneaki Hase

Additional information is available at the end of the chapter

http://dx.doi.org/10.5772/intechopen.70524

Abstract

Recent development of laser technology toward the realization of high-power laser has opened up a new research area exploring various fascinating phenomena governed by strongly photoexcited electronic states in diverse fields of science. In this chapter, we review the laser-induced Fano resonance (FR) in condensed matter systems, which is one of the representative resonance effects successfully exposed by strong laser field. The FR of concern sharply differs from FR effects commonly observed in conventional quantum systems where FR is caused by a weak external perturbation in a stationary system in the following two aspects. One is that the present FR is a transient phenomenon caused by nonequilibrium photoexcited states. The other is that this is induced by an optically nonlinear process. Here, we introduce two physical processes causing such transient and optically nonlinear FR in condensed matter, followed by highlighting anomalous effects inherent in it. The first is a Floquet exciton realized in semiconductor superlattices driven by a strong continuous-wave laser, and the second is the coherent phonon induced by an ultrashort pulse laser in bulk crystals.

Keywords: laser, Fano resonance, photodressed states, exciton, dynamic localization, Floquet theorem, coherent phonon, ultrafast phenomena, polaronic quasiparticle

1. Introduction

In quantum systems where discrete levels are embedded in energetically degenerate continuum states, resonance phenomenon is likely manifested, that is, characteristic of asymmetric spectral profiles consisting of both a peak and a dip. This is known as Fano resonance (FR) [1]; this is also termed as either Feshbach resonance or many-channel resonance. FR is one of the common and fundamental concepts in diverse fields of physics and chemistry; FR processes are observed, for instance, in strongly interacting Bose-Einstein condensates in an ultracold atomic system [2–4], superexcited states of molecules [5], a semiconductor quantum dot in an Aharonov-Bohm interferometer [6], an electronic transition near Weyl points strongly coupled with an infrared-active phonon in a Weyl semimetal [7].

In particular, within the restriction just to the FR processes triggered by laser irradiation, these may be classified in terms of the three categories as shown in **Table 1**. The first category is regarding whether a process is a linear one or a nonlinear one with respect to an order of a laser-matter interaction, as categorized as (a1) and (a2), respectively. For instance, the former is a photoabsorption process [8–11], and the latter is a multiphoton process [12–16]. The second category is regarding whether the process results from a built-in interaction between the discrete level and continuum that is intrinsic to a material itself or from a coupling induced extrinsically by a laser, as categorized as (b1) and (b2), respectively. For instance, the former is the interaction of an electron with a longitudinal optical (LO)-phonon in incoherent Raman scattering [17–22], and the electron-electron interaction brings about autoionization and the Auger process [23]. The latter FR process is known as a laser-induced continuum structure [2–4, 24]. The third category is regarding whether the process is a (quasi)stationary one or a transient one, as categorized as (c1) and (c2), respectively. In other words, this is whether (quasi)time-independent or time-dependent. For instance, the former is induced by a continuous-wave (cw) laser (monochromatic laser) [15, 16, 25, 26], and the latter is by a short pulsed laser [27–31].

It is stressed that for the FR categorized as (a2), its physical characters—such as asymmetry in spectral profile, spectral intensity, resonance position, and spectral width—are controllable in a quantum-mechanic manner by tuning various laser parameters. Thus, it is expected that underlying physics is enriched by intriguing effects inherent in this sort of FR. This differs from most of FR processes observed thus far because of being simply classified as (a1)-(b1)-(c1).

Currently, new research areas have been opened up owing to the progress of laser technology toward the realization of sophisticated high-power light sources. In particular, in the field of condensed matter physics, the development of high-intensity terahertz (THz) wave enables us to explore a photodressed quantum state in which a temporally periodic interaction of THz wave with matter is renormalized in the original quantum state in a nonperturbative manner [32–34]. Such an anomalous state is termed as a Floquet state because of ensuring the Floquet theorem [35]. Further, the development of ultrashort pulse laser—with its temporal width being of an

Category	Characteristic			
Optical process	(a1)	Linear (perturbative)	(a2)	Nonlinear (nonperturbative)
Interaction causing FR	(b1)	Intrinsic (built-in)	(b2)	Extrinsic (external)
Light source	(c1)	Monochromatic, continuous wave (stationary/quasistationary)	(c2)	Pulsed (transient)

Table 1. Classification of FR into three categories.

order of 10 femtosecond (fs)—enables us to explore ultrafast transitory phenomena governed by strongly photoexcited electronic states. Bearing in mind such current situations, here, we focus exclusively on the laser-induced FR effects realized in the following two physical systems. One is a Floquet exciton formed in semiconductor superlattices (SLs) driven by a strong THz wave, and the other is a coherent phonon (CP) generated by ultrashort pulse laser in bulk crystals. In the light of **Table 1**, the FR effects of concern sharply differ from those commonly observed in conventional quantum systems classified as (a1)-(b1)-(c1) in the following aspects. Both of the Floquet exciton and the CP are induced by optically nonlinear processes, and hence the significant quantum controls of FR are feasible by means of tuning the respective applied light sources. Further, the Floquet exciton forms manifolds of quasistationary states with quasienergy as a constant of motion, where the FR is mediated by the ac-Zener tunneling caused by the THz wave. Hence, this is classified as (a2)-(b2)-(c1) and is herein termed as dynamic FR (DFR). On the other hand, the CP is a transient phenomenon caused by the built-in interaction of an LO-phonon with nonequilibrium photoexcited carriers. Hence, this is classified as (a2)-(b1)-(c2) and is herein termed as transient FR (TFR).

Below, we survey the present research backgrounds of DFR and TFR in brief. In both cases, the applied electric field of pumping laser is represented as $F(t) = F_0(t)\cos(\omega t)$ with an envelope function $F_0(t)$ at time t and the center frequency ω.

To begin with the DFR, this is closely related with the photodressed miniband formation [36]. Here, the cw laser with a constant amplitude $F_0(t) \equiv F_{ac}$ gives rise to a nonlinear optical interaction with electron to result in a photodressed miniband with effective width $\Delta_{eff} = \Delta_0 |J_0(x)|$, where Δ_0 and $J_0(x)$ represent the width of the original SL miniband and the zeroth-order Bessel function of the first kind with $x = eF_{ac}d/\hbar\omega$, respectively, and e, d, and $\hbar$ represent the elementary charge, a lattice constant of the SLs, and Planck's constant divided by 2π, respectively. Each photodressed miniband forms a sequence of photon sidebands arrayed at equidistant energy intervals of $\hbar\omega$ following the Floquet theorem. The DFR is caused by the interaction due to the ac-Zener tunneling between photon sidebands pertaining to different sequences, and this is coherently controlled by tuning F_{ac} and ω. In particular, it is expected that an anomalous effect attributed to dynamic localization (DL) on DFR is revealed on the occasion that all of the photodressed minibands collapse by tuning x to ensure $J_0(x) = 0$ [36–38]. The DL was first observed in electron-doped semiconductor SLs driven by a THz wave [39]. In addition, this has also been observed in diverse physical systems such as a cold atomic gas in one-dimensional optical lattices [40], a Bose-Einstein condensate [41], and light in curved waveguide arrays [42–44].

As regards the TFR, this was observed in a lightly n-doped Si crystal immediately after carriers were excited by an ultrashort laser pulse [45], where the speculation was made that the observed FR would show the evidence of the birth of a polaronic-quasiparticle (PQ) likely formed in a strongly interacting carrier-LO-phonon system in a moment [46]. The TFR of concern has been observed exclusively in this system and semimetals/metals such as Bi and Zn [47, 48] till now, however, not observed in p-doped Si and GaAs crystals [49, 50]. Thus far, there are a number of theoretical studies regarding these experimental findings. The time-dependent Schrödinger equation in the system of GaAs was calculated to show the asymmetric shape featuring FR spectra, though apparently opposed to existing experimental results, as mentioned above [51]. Further, the classical Fano oscillator model was presented based on the Fano-Anderson Hamiltonian [52, 53], and the close comparison of the experimental results of the CP signals of Bi was made with the time signal obtained by taking the Fourier transform of Fano's spectral formula into a temporal region [48]. Recently, the authors have constructed a fully quantum-mechanical model based on the PQ picture in a unified manner on an equal footing between both of polar and nonpolar semiconductors such as undoped GaAs and undoped Si [31]. Here, it has been shown that the TFR is manifested in a flash only before the carrier relaxation time (~100 fs) in undoped Si, whereas this is absent from GaAs.

Acronyms used in the text and the corresponding meanings are summarized in **Table 2**. The remainder of this chapter is organized as follows. In Section 2, the theoretical framework is described, where the models of the DFR and TFR are presented separately in Sections 2.1 and 2.2, respectively. The results and discussion are given in Section 3, and the conclusion with summary is given in Section 4. Atomic units (a.u.) are used throughout unless otherwise stated.

Acronyms	Meanings
CP	Coherent phonon
cw	Continuous wave
DFR	Dynamic FR
DL	Dynamic localization
FR	Fano resonance
fs	Femtosecond
LO	Longitudinal optical
PQ	Polaronic-quasiparticle
SL	Superlattice
TFR	Transient FR
THz	Terahertz

Table 2. Summary of acronyms used in text in alphabetical order and corresponding meanings.

2. Theoretical framework

2.1. Theoretical model for DFR in the photodressed exciton

2.1.1. Optical absorption spectra

The total Hamiltonian $\hat{H}^{(DFR)}(t)$ concerned comprises a SL Hamiltonian consisting of field-free Hamiltonians of the conduction (c) and valence (v) bands, a Coulomb interaction between electrons, an intersubband interaction caused by the driving laser $F(t)$ polarized in the direction of crystal growth (the z-axis), and an interband interaction caused by the probe laser $f(t) = f_p \cos(\omega_p t)$ with the center frequency ω_p and the constant amplitude f_p; it is assumed that $F_{ac} \gg f_p$ and

$\omega \ll \omega_p$. The microscopic polarization defined as $p_{\lambda\lambda' \boldsymbol{k}_\parallel}(t) \equiv \left\langle a^{(v)\dagger}_{\lambda \boldsymbol{k}_\parallel} a^{(c)}_{\lambda' \boldsymbol{k}_\parallel} \right\rangle$ is examined to shed light on the detail of DFR of the Floquet exciton; $\langle O \rangle$ represents the expectation value of the operator O. Here, $\lambda^{(\prime)} = (b^{(\prime)}, l^{(\prime)})$, which represents the lump of the SL miniband index $b^{(\prime)}$ and the SL site $l^{(\prime)}$. In addition, $\boldsymbol{k}_\parallel$ represents the in-plane momentum associated with the relative motion of the pair of c band and v band electrons, where the in-plane is defined as the plane normal to the z-axis; hereafter, the relative position conjugate to $\boldsymbol{k}_\parallel$ is represented as $\boldsymbol{\rho}$. Further, $a^{(s)\dagger}_{\lambda \boldsymbol{k}_\parallel} \left(a^{(s)}_{\lambda \boldsymbol{k}_\parallel} \right)$ represents the creation (annihilation) operator of the electron with λ and $\boldsymbol{k}_\parallel$ in band s.

The equation of motion for the microscopic polarization is given by the semiconductor Bloch equation

$$i\left(\frac{d}{dt} + \frac{1}{T_2}\right) p_{\lambda\lambda' \boldsymbol{k}_\parallel}(t) = \left\langle [a^{(v)\dagger}_{\lambda \boldsymbol{k}_\parallel} a^{(c)}_{\lambda \boldsymbol{k}_\parallel}, \widehat{H}^{(DFR)}(t)] \right\rangle \tag{1}$$

with T_2 dephasing time. For the practical purpose of tackling the multichannel scattering problem of exciton, it is convenient to transform it into the equation for $\overline{p}(\boldsymbol{\rho}, z_v, z_c, t)$ defined in the real-space representation as

$$\overline{p}(\boldsymbol{\rho}, z_v, z_c, t) = e^{i\omega_p t} \sum_{\lambda, \lambda'} \int d\boldsymbol{k}_\parallel e^{i\boldsymbol{k}_\parallel \cdot \boldsymbol{\rho}} \langle z_v | \lambda \rangle p_{\lambda\lambda' \boldsymbol{k}_\parallel}(t) \langle \lambda' | z_c \rangle, \tag{2}$$

where $\langle \lambda | z \rangle$ represents the Wannier function at position $z - ld$ in SL miniband b. The resulting equation becomes

$$\begin{aligned} &i\left(\frac{d}{dt} + \frac{1}{T_2} - i\omega_p\right) \overline{p}(\boldsymbol{\rho}, z_v, z_c, t) + (2\pi)^2 e^{i\omega_p t} f_0^{(+)}(t) d_0^{(vc)} \delta(\boldsymbol{\rho}) \delta(z_v - z_c) \\ &= \int dz \left[\overline{p}(\boldsymbol{\rho}, z_v, z, t) H^{(c)}_{\mathrm{TB}}(z, z_c, t) - H^{(v)}_{\mathrm{TB}}(z_v, z, t) \overline{p}(\boldsymbol{\rho}, z, z_c, t) \right] + \mathcal{H}(\boldsymbol{\rho}, z_v, z_c) \overline{p}(\boldsymbol{\rho}, z_v, z_c, t), \end{aligned} \tag{3}$$

where the rotating wave approximation is employed by replacing $f(t)$ by $f_0^{(+)}(t) \equiv \left(f_p/2\right) e^{-i\omega_p t}$ and $d_0^{(vc)}$ represents the interband dipole moment of a bulk material. Here, the Hamiltonian $\mathcal{H}(\boldsymbol{\rho}, z_v, z_c)$ for the in-plane motion is given by $\mathcal{H}(\boldsymbol{\rho}, z_v, z_c) = -\nabla^2_\rho / 2m_\parallel + V(r)$, where $m_\parallel$ and $V(r) = -1/(\varepsilon_0 r)$ represent an in-plane reduced mass and the Coulomb interaction, respectively, with $r = \sqrt{\rho^2 + (z_v - z_c)^2}$ and ε_0 the dielectricity of vacuum. The nearest-neighbor tight-binding Hamiltonian of the laser-driven SLs is given by $H^{(s)}_{\mathrm{TB}}(z, z', t) \equiv \left\langle z | \widehat{H}^{(s)}_{\mathrm{TB}}(t) | z' \right\rangle$, where

$$\widehat{H}^{(s)}_{\mathrm{TB}}(t) = \sum_{\lambda=(b,l)} \left[\varepsilon^{(s)}_{0b} |\lambda\rangle\langle\lambda| + \frac{(-1)^{b+\sigma^s}}{4} \Delta^{(s)}_b (|l, b\rangle\langle l+1, b| + |l+1, b\rangle\langle l, b|) \right] - F(t) \sum_{\lambda\lambda'} |\lambda\rangle Z^{(s)}_{\lambda\lambda'} \langle \lambda' |, \tag{4}$$

and $\varepsilon^{(s)}_{0b}$ and $\Delta^{(s)}_b$ represent the band center and the band width of b, respectively, with $\sigma^s = 0$ (for $s = c$) and 1 (for $s = v$). The last term of Eq. (4) represents the dipole interaction induced by the driving laser $F(t)$ with $Z^{(s)}_{\lambda,\lambda'}$ as a dipole matrix element. It should be noted that the off-diagonal contribution of $Z^{(s)}_{\lambda,\lambda'}$ with $b \neq b'$ induces the ac-Zener tunneling, which plays a significant role of quantum control of DFR, as shown below.

The concerned function $\overline{p}(\boldsymbol{\rho}, z_v, z_c, t)$ can be expressed in terms of the complete set of the Floquet wave functions $\{\psi_{E\beta}(\boldsymbol{\rho}, z_v, z_c, t)\}$, that is,

$$\overline{p}(\boldsymbol{\rho}, z_v, z_c, t) = \int dE \sum_{\beta} a_{E\beta}\, \psi_{E\beta}(\boldsymbol{\rho}, z_v, z_c, t) \tag{5}$$

with $a_{E\beta}$ as an expansion coefficient. Here, the Floquet wave function ensures the following homogeneous equation associated with the inhomogeneous equation of Eq. (3) as

$$i\left(\frac{d}{dt} + E\right)\psi_{E\beta}(\boldsymbol{\rho}, z_v, z_c, t) = \int dz\left[\psi_{E\beta}(\boldsymbol{\rho}, z_v, z, t) H^{(c)}_{\mathrm{TB}}(z, z_c, t) - H^{(v)}_{\mathrm{TB}}(z_v, z, t)\psi_{E\beta}\boldsymbol{\rho}, z, z_c, t)\right]$$

$$+\mathcal{H}(\boldsymbol{\rho}, z_v, z_c)\psi_{E\beta}(\boldsymbol{\rho}, z_v, z_c, t), \tag{6}$$

where the temporally periodic boundary condition $\psi_{E\beta}(\boldsymbol{\rho}, z_v, z_c, t) = \psi_{E\beta}(\boldsymbol{\rho}, z_v, z_c, t+T)$ is imposed on it with E and $T = 2\pi/\omega$ as quasienergy and the time period of the driving laser field, respectively. Equation (6) is the Wannier equation of the Floquet exciton of concern. It should be noted that this is cast into the multichannel scattering equations and the Floquet state of $\psi_{E\beta}(\boldsymbol{\rho}, z_v, z_c, t)$ forms a continuum spectrum designated by both E and β with β representing the label of an open channel. Such a multichannel feature is introduced by the strong driving laser $F(t)$ that closely couples an excitonic-bound state with continua; more detail of the multichannel scattering problem is described in Section 2.1.2. The expansion coefficient $a_{E\beta}$ is readily obtained by inserting Eq. (5) into Eq. (3) in view of Eq. (6) as

$$a_{E\beta} = \frac{(2\pi)^2 d_0^{(vc)}\left(f_p/2\right)}{(E - \omega_p - i\gamma)T}\int_0^T dt' \overline{\psi}_{E\beta}(t'), \tag{7}$$

where $\overline{\psi}_{E\beta}(t) = \int dz \psi_{E\beta}(\mathbf{0}, z, z, t)$ and $\gamma = 1/T_2$.

Since the macroscopic polarization is given by $P(t) = \sum_{\lambda,\lambda'} \int d\boldsymbol{k}_{\|} d_0^{(vc)*} p_{\lambda\lambda'\boldsymbol{k}_{\|}}(t)$, the linear optical susceptibility $\chi(t)$ with respect to the weak probe laser $f(t)$ is cast into [54]

$$\chi(t) = \frac{\left|d_0^{(vc)}\right|^2}{\varepsilon_0}\int dE \sum_{\beta} \frac{O_{E\beta}(t)}{E - \omega_p - i\gamma}, \tag{8}$$

where $O_{E\beta}(t) = \left[\overline{\psi}_{E\beta}(t)/T\right] \int_0^T dt' \overline{\psi}^*_{E\beta}(t')$. Taking the Fourier transform of $\chi(t) \equiv \sum_j e^{ij\omega t} \chi_j(\omega_p; \omega)$, leads to the expression of the absorption coefficient to be calculated as

$$\alpha(\omega_p;\omega) = \frac{\omega_p}{C}\sum_j \mathrm{Im}\,\chi_j(\omega_p;\omega) \tag{9}$$

with C the speed of light; $\chi_{j\neq 0}(\omega_p;\omega)$ vanishes in the limit of $F_{ac}\rightarrow 0$.

2.1.2. Multichannel scattering problem

The absorption coefficient of Eq. (9) is obtained by evaluating a set of the wave functions, $\{\psi_{E\beta}(\boldsymbol{\rho}, z_v, z_c, t)\}$. To do this, first, the wave function is expanded as

$$\psi_{E\beta}(\boldsymbol{\rho}, z_v, z_c, t) = \sum_\mu \Phi_\mu(z_v, z_c, t)F_{\mu\beta}(\rho), \tag{10}$$

where $\rho = |\boldsymbol{\rho}|$, and just the contribution of the *s*-angular-momentum component is incorporated because of little effects from higher-order components. Here, $\Phi_\mu(z_v, z_c, t)$ is the real-space representation of the Floquet state $|\Phi_\mu\rangle$, that is, $\Phi_\mu(z_v, z_c, t) = \langle z_v, z_c | \Phi_\mu\rangle$, satisfying $\left(\widehat{H}_{\mathrm{TB}} - i\partial/\partial t\right)$ $|\Phi_\mu\rangle = \mathcal{E}_\mu|\Phi_\beta\rangle$, where $\widehat{H}_{\mathrm{TB}} \equiv \widehat{H}^{(c)}_{\mathrm{TB}} + \widehat{H}^{(v)}_{\mathrm{TB}}$ and $\mathcal{E}_\mu$ is the μth quasienergy. The index μ is considered as the approximate quantum number $\mu \approx [\overline{\mu}, k]$ with $\overline{\mu} \equiv [b_c, b_v, n_p]$ as a photon sideband index, where b_c and b_v are SL miniband indexes belonging to the *c*- and *v*-bands, respectively, and k and n_p represent the Bloch momentum of the joint miniband of (b_c, b_v) and the number of photons relevant to absorption and emission, respectively. The quantum number $\overline{\mu}$ becomes a set of the good quantum numbers with F_{ac} decreasing, while k always remains conserved because of the spatial periodicity in the laser-driven SLs of concern. In view of Eq. (10), Eq. (6) is recast into the coupled equations for the radial wave function $F_{\nu\beta}(\rho)$, that is,

$$\sum_\mu L_{\mu\nu}F_{\nu\beta}(\rho) = EF_{\mu\beta}(\rho), \tag{11}$$

where $L_{\mu\nu}$ is an operator given by $L_{\mu\nu} = \delta_{\mu\nu}[-(2m_{\|})^1(d^2/d\rho^2+\rho^1 d/d\rho)+\mathcal{E}_\mu]+V_{\mu\nu}(\rho)$ and $V_{\mu\nu}(\rho)$ is a Coulomb matrix element defined as $V_{\mu\nu}(\rho) = T^{-1}\int_0^T dt \int dz_v \int dz_c \Phi^*_\mu(z_v, z_c, t)V(\rho, z_v, z_c)$ $\Phi_\nu(z_v, z_c, t)$.

The Floquet exciton in the laser-driven SL system pertains to the multichannel scattering problem, because $V_{\mu\nu}(\rho)$ vanishes at $\rho \gg 1$. Actually, for a given E, the channel μ satisfying $E > \mathcal{E}_\mu$ is an open channel, while the channel μ satisfying $E < \mathcal{E}_\mu$ is a closed channel. Thus, the label μ of $F_{\mu\beta}$ plays the role of the scattering channel. On the other hand, the label β means the number of independent solutions satisfying Eq. (11). Here, there are same number of independent solutions as open channels, since as many scattering boundary conditions are imposed on $F_{\mu\beta}$ at $\rho \gg 1$; while evanescent boundary conditions that $F_{\mu\beta}$ vanishes at $\rho \gg 1$ are imposed on closed channels. Eq. (11) can be numerically evaluated by virtue of the R-matrix propagation method, which is a sophisticated formalism providing a stable numerical algorithm with extremely high accuracy [55].

It is expected that the DFR of concern is caused by a coupling between photon sidebands mediated by ac-Zener tunneling, as mentioned in Section 1. To see this situation in more detail,

Figure 1 shows the interacting two photon sidebands $\overline{\mu}$ and $\overline{\mu}' \equiv \left[b_c', b_v', n_p'\right]$, where the discrete Floquet excitonic state is supported by the photon sideband $\overline{\mu}$, and this is also embedded in the continuum of the alternative photon sideband $\overline{\mu}'$. It is likely that the DFR occurs due to a close coupling between these photon sidebands, and, eventually, the exciton state decays into the continuum state pertaining to $\overline{\mu}'$. In fact, it is noted that the Coulomb interaction incorporated in Eq. (6) also gives rise to FR. Defining the difference between the photon numbers of both photon sidebands, namely, $\Delta n_p = |n_p - n_p'|$, the ac-Zener tunneling is featured by $\Delta n_p \neq 0$, while the Coulomb coupling is by $\Delta n_p = 0$. The spectral profile and intensity of FR in the former can be even more effectively controlled than in the latter by modulating the laser parameters F_{ac} and ω, since the degree of magnitude of ac-Zener tunneling depends exclusively on both of the external parameters, differing from the Coulomb interaction. In the region of F_{ac} weak enough to suppress the ac-Zener tunneling, the FR is dominated by the Coulomb coupling, similarly to the conventional FR observed in the original SLs without laser irradiation [56].

2.2. Theoretical model for TFR in the CP generation

2.2.1. *Introduction of polaronic quasiparticle operators*

The total Hamiltonian $\widehat{H}^{(TFR)}$ of concern is given by $\widehat{H}^{(TFR)} = \widehat{H}_e + \widehat{H}'(t) + \widehat{H}_p + \widehat{H}_{e-p}$. Here, $\widehat{H}_e$ represents an electron Hamiltonian including an interelectronic Coulomb potential, where a two-band model is employed that consists of the energetically lowest c band and the energetically highest valence v band, and a creation (annihilation) operator of electron with band index b and Bloch momentum $\boldsymbol{k}$ is represented as $a_{b\boldsymbol{k}}^{\dagger}\,(a_{b\boldsymbol{k}})$. $\widehat{H}_p$ represents an LO-phonon

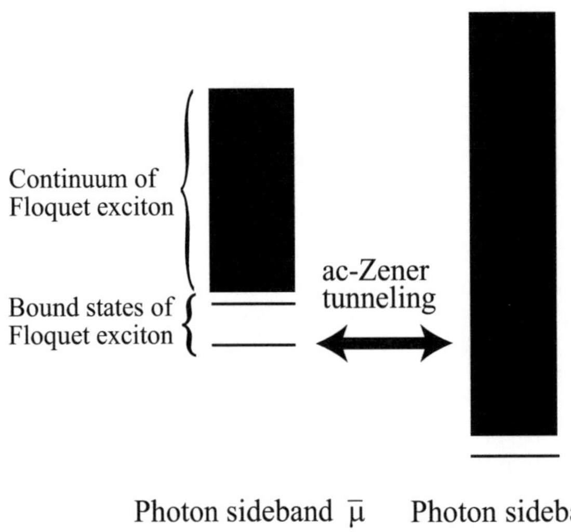

Figure 1. Schematic diagram of the DFR formation in the Floquet excitonic system. This shows the coupling mechanism that a bound state supported by the sideband $\overline{\mu}$ interacts with a continuum state belonging to the sideband $\overline{\mu}'$ by the ac-Zener tunneling to result in the Fano decay (from Ref. [15] with partial modification).

Hamiltonian, where a creation (annihilation) operator of LO-phonon with an energy dispersion $\omega_q^{(LO)}$ at momentum $\boldsymbol{q}$ is represented as $c_q^\dagger\,(c_q)$. Further, $\widehat{H}'(t)$ and $\widehat{H}_{e-p}$ represent interaction Hamiltonians of electron with the pump pulse and the LO-phonon, respectively. These are given as follows:

$$\widehat{H}'(t) = -\frac{1}{2}\sum_{b,\,b'(\neq b),\,\boldsymbol{k}}\left[\Omega_{bb'}(t)\,a_{b\boldsymbol{k}}^\dagger a_{b'\boldsymbol{k}} + \Omega_{bb'}^*(t)\,a_{b'\boldsymbol{k}}^\dagger a_{b\boldsymbol{k}}\right], \tag{12}$$

where $\Omega_{bb'}(t) = d_{bb'}F(t)$ with $d_{bb'}$ an electric dipole moment between b and b' bands, and

$$\widehat{H}_{e-p} = \sum_{b,\,\boldsymbol{q},\,\boldsymbol{k}}\left(g_{b\boldsymbol{q}}c_{\boldsymbol{q}}a_{b\boldsymbol{k}+\boldsymbol{q}}^\dagger a_{b\boldsymbol{k}} + g_{b\boldsymbol{q}}^* c_{\boldsymbol{q}}^\dagger a_{b\boldsymbol{k}}^\dagger a_{b\boldsymbol{k}+\boldsymbol{q}}\right), \tag{13}$$

where $g_{b\boldsymbol{q}}$ is a coupling constant of b band electron with the LO-phonon. Here, let the envelope of $F(t)$ be of squared shape just for the sake of simplicity, that is, $F_0(t) = \mathcal{F}_0\theta(t+\tau_L/2)\theta(\tau_L/2-t)$ with $\mathcal{F}_0$ constant, where temporal width τ_L is of the order of a couple of 10 fs at most, satisfying $\tau_L \ll 2\pi/\omega_q^{(LO)}$.

The equation of motion of a composite operator $A_{\boldsymbol{q}}^\dagger(\boldsymbol{k}bb') \equiv a_{b,\,\boldsymbol{k}+\boldsymbol{q}}^\dagger a_{b'\boldsymbol{k}}$ is considered below, where this represents an induced carrier density with spatial anisotropy determined by $\boldsymbol{q}$; $|\boldsymbol{q}|$ is finite, though quite small, that is, $\boldsymbol{q}\neq\boldsymbol{0}$. It is convenient to remove from this equation high-frequency contributions by means of the rotating-wave approximation [57] by replacing $A_{\boldsymbol{q}}^\dagger(\boldsymbol{k}bb')$ by $e^{i\overline{\omega}_{bb'}t}\overline{A}_{\boldsymbol{q}}^\dagger(\boldsymbol{k}bb')$, where $\overline{\omega}_{cv} = \omega$, $\overline{\omega}_{vc} = -\omega$, and $\overline{\omega}_{bb} = 0$. Thus, the resulting equation of motion is as follows:

$$\begin{aligned}-i\left(\frac{d}{dt}+\frac{1}{T_{\boldsymbol{q}}(\boldsymbol{k}bb')}\right)\overline{A}_{\boldsymbol{q}}^\dagger(\boldsymbol{k}bb') &= \left[\widehat{\mathcal{H}}_e(t),\overline{A}_{\boldsymbol{q}}^\dagger(\boldsymbol{k}bb')\right] - \overline{A}_{\boldsymbol{q}}^\dagger(\boldsymbol{k}bb')\overline{\omega}_{bb'} + \left[\widehat{H}_{e-p},\overline{A}_{\boldsymbol{q}}^\dagger(\boldsymbol{k}bb')\right] \\ &\approx \sum_{\tilde{\boldsymbol{k}}\tilde{b}\tilde{b}'}\overline{A}_{\boldsymbol{q}}^\dagger\left(\tilde{\boldsymbol{k}}\tilde{b}\tilde{b}'\right)\overline{Z}_{\boldsymbol{q}}\left(\tilde{\boldsymbol{k}}\tilde{b}\tilde{b}',\boldsymbol{k}bb'\right) + \left[\widehat{H}_{e-p},\overline{A}_{\boldsymbol{q}}^\dagger(\boldsymbol{k}bb')\right],\end{aligned} \tag{14}$$

where the total electronic Hamiltonian is defined as $\widehat{\mathcal{H}}_e(t) = \widehat{H}_e + \widehat{H}'(t)$, the first commutator in the right-hand side of the first equality is evaluated by making a factorization approximation, and $T_{\boldsymbol{q}}(\boldsymbol{k}bb')$ represents a phenomenological relaxation time constant relevant to $A_{\boldsymbol{q}}^\dagger(\boldsymbol{k}bb')$. Further, $\overline{Z}_{\boldsymbol{q}}$ represents a non-Hermitian matrix, which is a slowly varying function in time, since rapidly time-varying contributions are removed owing to the above rotating-wave approximation, aside from the discontinuity at $t = \pm\tau_L/2$.

Bearing in mind this situation, we tackle left and right eigenvalue problems of $\overline{Z}_{\boldsymbol{q}}$ [58], described by $U_{\boldsymbol{q}}^{L\dagger}\overline{Z}_{\boldsymbol{q}} = \mathcal{E}_{\boldsymbol{q}}U_{\boldsymbol{q}}^{L\dagger}$ and $\overline{Z}_{\boldsymbol{q}}U_{\boldsymbol{q}}^R = U_{\boldsymbol{q}}^R\mathcal{E}_{\boldsymbol{q}}$, respectively, in terms of an adiabatic-eigenvalue diagonal matrix $\mathcal{E}_{\boldsymbol{q}}$ and the associated biorthogonal set of adiabatic eigenvectors $\left\{U_{\boldsymbol{q}}^L, U_{\boldsymbol{q}}^R\right\}$ with time t fixed as a parameter. The orthogonality relation and the completeness are read as $U_q^{L\dagger}U_q^R = 1$ and $U_q^R U_q^{L\dagger} = 1$, respectively [58]. Given the relation $\overline{Z}_q = U_q^R \mathcal{E}_q U_q^{L\dagger}$, Eq. (14) is recast into the form of adiabatic coupled equations:

$$-i\frac{dB_{q\alpha}^{\dagger}}{dt} = B_{q\alpha}^{\dagger}\mathcal{E}_{q\alpha} + i\sum_{\alpha'} B_{q\alpha'}^{\dagger}\mathcal{W}_{q\alpha'\alpha} + \left[\widehat{H}_{e-p}, B_{q\alpha}^{\dagger}\right], \tag{15}$$

where the operator $B_{q\alpha}^{\dagger}$ is defined as $B_{q\alpha}^{\dagger} \equiv \overline{A}_q^{\dagger} U_{q\alpha}^R$, $\mathcal{W}_{q\alpha'\alpha} \equiv \left[dU_{q\alpha'}^{L\dagger}/dt\right]U_{q\alpha'}^R + U_{q\alpha'}^{L\dagger}T_q^{-1}U_{q\alpha'}^R$ and $\mathcal{E}_{q\alpha}(t)$ is complex adiabatic energy at time t associated with the operator $B_{q\alpha}^{\dagger}(t)$ thus introduced. Hereafter, this operator is termed as a creation operator of quasiboson, and the corresponding annihilation operator is defined as $B_{q\alpha} \equiv U_{q\alpha}^{R\dagger}\overline{A}_q$. The set of eigenstates $\{\alpha\}$ is composed of continuum states represented as β with eigenenergy $\mathcal{E}_{q\beta}$ and a single discrete energy state represented as α_1 with eigenenergy $\mathcal{E}_{q\alpha_1}$, that is, $\{\alpha\}$ = $(\{\beta\},\alpha_1)$; the state β corresponds to electron-hole continuum arising from interband transitions, and the state α_1 corresponds to a plasmon-like mode. It is noted that the relation of $\left\langle \left[B_{q\alpha}(t), B_{q'\alpha'}^{\dagger}(t)\right]\right\rangle = \delta_{qq'}\delta_{\alpha\alpha'}$ is assumed, though $B_{q\alpha}(t)$ and $B_{q\alpha}^{\dagger}(t)$ do not satisfy the equal-time commutation relations for a real boson, where $\widehat{X}$ means an expectation value of operator $\widehat{X}$ with respect to the ground state; the validity of the criterion of this relation is discussed in detail in Ref. [31].

Eq. (13) is rewritten as $\widehat{H}_{e-p} = \sum_{q,\alpha}\left(c_q B_{q\alpha}^{\dagger}M_{q\alpha} + M_{q\alpha}^{*}B_{q\alpha}c_q^{\dagger}\right)$ with an effective coupling between quasiboson and LO-phonon as $M_{q\alpha} = \sum_{kb} g_{bq} U_{q\alpha}^{L\dagger}(\boldsymbol{k}bb)$. Thus, the commutator in Eq. (15) is approximately evaluated as $\left[\widehat{H}_{e-p}, B_{q\alpha}^{\dagger}\right] \approx M_{q\alpha}^{'*}c_q^{\dagger}$, though $M_{q\alpha}^{'*} \neq M_{q\alpha}^{*}$. On the other hand, the equation of motion of the LO-phonon is described by $-idc_q^{\dagger}/dt = \sum_{\alpha} B_{q\alpha}^{\dagger}M_{q\alpha} + c_q^{\dagger}\omega_q^{(LO)}$. Both of the equations of motion for $B_q^{\dagger}$ and $c_q^{\dagger}$ are integrated into a single equation in terms of matrix notations as follows:

$$-i\frac{d}{dt}\left(B_q^{\dagger}, c_q^{\dagger}\right) \approx \left(B_q^{\dagger}, c_q^{\dagger}\right)h_q + \left(iB_q^{\dagger}\mathcal{W}_q, 0\right). \tag{16}$$

Here, the non-Hermitian matrix $h_q \equiv \{h_{q\gamma\gamma'}\}$ given by $h_q = \begin{pmatrix} \mathcal{E}_q & M_q \\ M_q^{'\dagger} & \omega_q^{(LO)} \end{pmatrix}$ is introduced with $\gamma, \gamma' = 1 \sim N+2$, where N represents the number of electron-hole (discretized) continua, namely, $\beta = 1 \sim N$, aside from two discrete states attributed to a plasmon-like mode and an LO-phonon mode designated by α_1 and α_2, respectively: $\{\gamma\}$ = $(\{\beta\},\alpha_1,\alpha_2)$. In the system of the TFR of concern, the case is exclusively examined that both of the discrete levels of α_1 and α_2 are embedded into the continua $\{\beta\}$. Thus, the following coupled equations for the multichannel scattering problem are taken account of

$$\sum_{\gamma'} h_{q\gamma\gamma'}V_{q\gamma'\beta}^R = V_{q\gamma\beta}^R\mathcal{E}_{q\beta}, \tag{17}$$

where $V^R_{q\beta} = \left\{ V^R_{q\gamma\beta} \right\}$ is the right vector representing the solution for given energy $\mathcal{E}_{q\beta}$; similarly to Eq. (11) for the DFR, the indices of γ and β play the roles of a scattering channel and the number of independent solutions, respectively. Eq. (17) provides the theoretical basis on which both of LO-phonon and plasmon-like modes are brought into connection with the CP dynamics on an equal footing. In terms of this vector, a set of N-independent operators, $F^\dagger_{q\beta}\ (\beta = 1 \sim N)$, is defined as

$$F^\dagger_{q\beta} = \sum_{\beta'} B^\dagger_{q\beta'} V^R_{q\beta'\beta} + B^\dagger_{q\alpha_1} V^R_{q\alpha_1\beta} + c^\dagger_q V^R_{q\alpha_2\beta}. \tag{18}$$

In addition, the left vector $V^{L\dagger}_{q\beta} = \left\{ V^{L\dagger}_{q\beta\gamma} \right\}$ associated with $V^R_{q\beta}$ is introduced to ensure the inverse relations $B^\dagger_{q\alpha} = F^\dagger_q V^{L\dagger}_{q\alpha}$ and $c^\dagger_q = F^\dagger_q V^{L\dagger}_{q\alpha_2}$, where $V^{L\dagger}_q V^R_q = 1$ and $V^R_q V^{L\dagger}_q = 1$. Hereafter, the operator $F^\dagger_{q\beta}(t)$ thus introduced is termed as a creation operator of PQ, and then the corresponding annihilation operator is $F_{q\beta}(t)$; these are not bosonic operators. The bosonization scheme for the PQ operators is similar to that for the quasiboson operators, where the PQ ground state is given by the direct product of the ground states of quasiboson and LO-phonon and $\mathcal{E}_{q\beta}(t)$ is read as the single-PQ adiabatic energy at time t with mode $q\beta$.

Given Eq. (18), Eq. (16) becomes adiabatic coupled equations for $F^\dagger_q$:

$$-i\frac{d}{dt}F^\dagger_{q\beta} \approx F^\dagger_{q\beta}\mathcal{E}_{q\beta} + i\sum_{\beta'} F^\dagger_{q\beta'}\mathcal{I}_{q\beta'\beta}, \tag{19}$$

where $\mathcal{I}_q = \left(dV^{L\dagger}_q/dt \right) V^R_q + V^{L\dagger}_q \mathcal{W}_q V^R_q$. In terms of F_q and $F^\dagger_q$, the associated retarded Green function is given by [59]

$$G^R_{q\beta\beta'}(t,t') = -i\theta(t-t')\left\langle \left[F_{q\beta}(t), F^\dagger_{q\beta'}(t') \right] \right\rangle. \tag{20}$$

2.2.2. *Transient induced photoemission spectra*

A weak external potential $f_q(t)$ additionally introduced in the transient and nonequilibrium system of concern induces an electron density $n^{(ind)}_q(t)$ given by

$$n^{(ind)}_q(t) = \frac{1}{4\pi V}\int_{-\infty}^{t} dt' \chi^{(t)}_q(t,t') f_q(t'), \tag{21}$$

based on the linear response theory [59, 60] with V the volume of crystal. It is noted that $n^{(ind)}_q(t)$ is nonlinear with respect to the pump field. Here, $\chi^{(t)}_q(t,t')$ represents the retarded longitudinal susceptibility that depends on passage of t and the relative time $\tau = t - t'$, differing from equilibrium systems depending solely on τ. Introducing a retarded longitudinal susceptibility due to the electron-induced interaction and that of an LO-phonon-induced interaction represented as $\chi_q(t,t')$ and $\chi'_q(t,t')$, respectively, $\chi_q^{(t)}(t,t')$ is given by [59]

$$\chi_q^{(t)}(t,t') = \chi_q(t,t') + \chi'_q(t,t'). \tag{22}$$

Let $f_q(t')$ be assumed to be $f_q(t') = f_{q0}\,\delta(t' - t_p)$ in the present system; f_{q0} is independent of t', and t_p represents the time at which $f_q(t')$ probes transient dynamics of concern. Thus, it is seen that $\chi_q^{(t)}(t,t_p)$ reveals the way of alteration of $n_q^{(ind)}(t)$ after t_p, since Eq. (21) becomes $n_q^{(ind)}(t) = f_{q0}\chi_q^{(t)}(t,t_p)\theta(t-t_p)/4\pi V$.

In terms of $\chi_q^{(t)}(t_p+\tau,t_p)$, the dielectric function $\varepsilon_q(t_p+\tau,t_p)$ is readily obtained, and by taking the Fourier transform of it as $\tilde{\varepsilon}_q(t_p;\omega_p) = \int_0^\infty d\tau e^{-i\omega_p\tau}\varepsilon_q(t_p+\tau,t_p)$, this leads to a transient absorption coefficient $\alpha_q(t_p;\omega_p)$ at time t_p. This is given by $\alpha_q(t_p;\omega_p) = \omega A_q(t_p;\omega_p)/n(t_p;\omega_p)C$, where $A_q(t_p;\omega_p) = \mathrm{Im}\,\tilde{\varepsilon}_q(t_p;\omega_p)$ and $n(t_p;\omega_p)$ represents the index of refraction. It is remarked that according to the definition of the sign of ω_p made above, transient photoemission spectra, where $A_q(t_p;\omega_p)<0$, peak at positive ω_p, while transient photoabsorption spectra, where $A_q(t_p;\omega_p)>0$, peak at negative ω_p. For the sake of the later convenience, the transient induced photoemission spectra are defined as $\overline{A}_q(t_p;\omega_p) = -A_q(t_p;\omega_p)$.

Based on the PQ model developed in Section 2.2.1, $\chi_q(t,t')$ and $\chi'_q(t,t')$ can be explicitly expressed in terms of the retarded Green function given by Eq. (20). Here, the obtained results are shown below; for more detail, consult Ref. [31]:

$$\chi_q^*(t,t') = \frac{4\pi}{V}\sum_{\alpha\alpha'\beta\beta'} N_{q\alpha}^{L*}(t)V_{q\alpha\beta}^{L}(t)G_{q\beta\beta'}^{R}(t,t')V_{q\beta'\alpha'}^{L\dagger}(t')N_{q\alpha'}^{L}(t'), \tag{23}$$

where $N_{q\alpha}^{L} = \sum_{kb}U_{q\alpha}^{L\dagger}(\mathbf{k}bb)$, and this is equivalent to a normalization constant of the left vector $U_{q\alpha}^{L\dagger}$:

$$\chi'_q(t,t') = \frac{4\pi}{V}\left|g'_q\right|^2\left[\overline{D}'^{R}_{q}(t,t') + \left[\overline{D}'^{R}_{-q}(t,t')\right]^*\right], \tag{24}$$

where g'_q is a constant in proportion to $(g_{cq}+g_{vq})/2$ and

$$\overline{D}'^{R}_{q}(t,t') = \sum_{\beta\beta'} V_{q\alpha_2\beta}^{L}(t)G_{q\beta\beta'}^{R}(t,t')V_{q\beta'\alpha_2}^{L\dagger}(t'). \tag{25}$$

Finally, the TFR dynamics caused by the CP generation is mentioned based on the PQ picture. As shown in **Figure 2**, the LO-phonon state α_2 is embedded in the quasiboson state β, and the effective coupling between both states induces the formation of transient PQ FR state. This composite state is deexcited into the PQ ground state via two paths: one is the transient photoemission from α_2, and the other is from β. It is likely that these two paths interfere to

http://dx.doi.org/10.5772/intechopen.70524

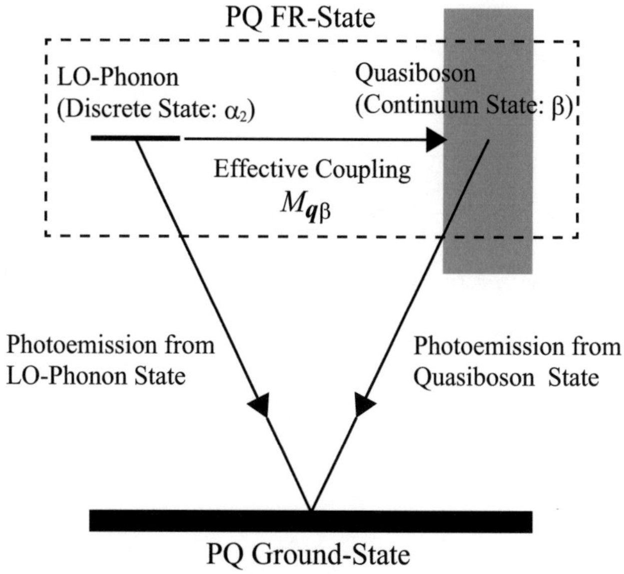

Figure 2. Schematic diagram of the TFR dynamics based on the PQ picture, where the LO-phonon state α_2 is embedded in the quasiboson state β. The PQ FR state composed of α_2 and β is deexcited by induced photoemission process. For more detail, consult the text (from Ref. [31] with partial modification).

give rise to asymmetry in spectra. It is remarked that the contribution from the plasmon-like mode α_1 is omitted because of a negligibly smaller effect on the TFR.

3. Results and discussion

3.1. DFR in the photodressed exciton

For the calculations of DFR spectra, the semiconductor SLs of $GaAs/Ga_{0.75}Al_{0.25}As$ are employed with 35/11 monolayers (ML) for the well and barrier thickness, where 1 ML = 2.83 Å. Here, 14 photon sidebands of $\overline{\mu} = [1, 1, -3 \sim 3]$ and $[2,1,-3{\sim}3]$ are incorporated by setting ω to 91 meV; this equals to the difference between the centers of the joint minibands of (1,1) and (2,1). Other photon sidebands are neglected for the sake of simplicity.

First of all, the calculated quasienergy bands $\{\mathcal{E}_\mu\}$ as a function of F_{ac} are shown in **Figure 3** to illustrate the effect of ac-Zener coupling. The two photon sidebands labeled by $\mu_1 = [1,1,0,k]$ and $\mu_2 = [2,1,-1,k]$ are mixed by the coupling induced by the driving laser $F(t)$. With the increase of F_{ac}, the quasienergy bands are branched into two distinct photon sidebands, termed as the upper sideband μ_+ and the lower sideband μ_-, where both labels of μ_1 and μ_2 are no longer good quantum numbers, aside from k. It is noted that both of μ_+ and μ_- form dynamic localization showing band collapse around two points $F_{ac} = F_{DL1} \equiv 170$ kV/cm and $F_{DL2} \equiv 395$ kV/cm. **Figure 4** shows the absorption spectra $\alpha(\omega_p;\omega)$ obtained by solving Eq. (9) in the range of F_{ac} from 10 to 450 kV/cm. Asymmetric spectral profiles characteristic of DFR are discerned at the arrowed positions of ω_p when $F_{ac} \geq 150$ kV/cm, where all peaks are followed by dips. These

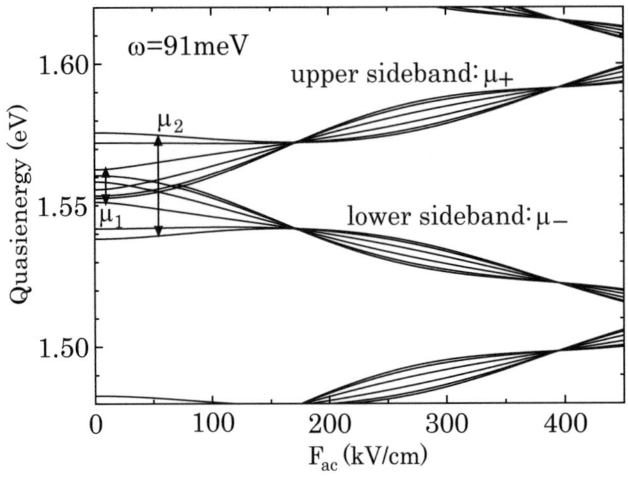

Figure 3. The quasienergy $\mathscr{E}_\mu$ as a function of F_{ac}. ω is set to 91 meV. The vertical double arrows represent the original SL miniband widths corresponding to the photon sidebands of μ_1 and μ_2 (from Ref. [16] with partial modification).

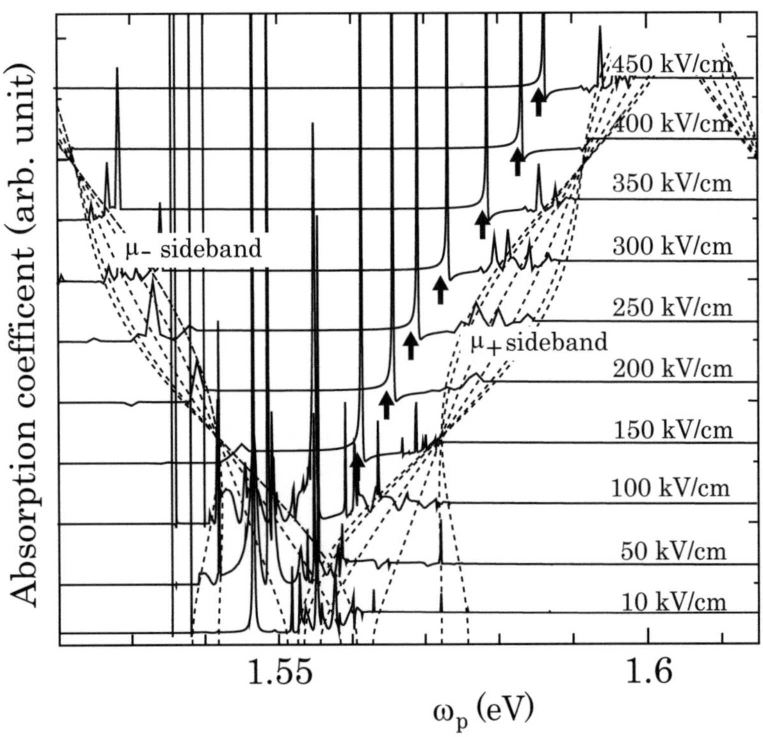

Figure 4. Absorption spectra $\alpha(\omega_p;\omega)$ as a function of ω_p for F_{ac}=50–450 (kV/cm) with ω= 91 meV. A series of the arrowed spectral profiles are examined exclusively in the text. The quasienergies shown in **Figure 3** are also plotted (dotted lines) (from Ref. [16] with partial modification).

http://dx.doi.org/10.5772/intechopen.70524

peaks are located just below the upper sideband μ_+, thereby being blue shifted. Consulting **Figure 1**, the DFR is dominantly formed by the interaction between one open channel μ_- and one closed channel μ_+.

To deepen the understanding of the DFR exciton, its characteristic quantities determining the spectral profiles are extracted from $\alpha(\varepsilon)\equiv\alpha(\omega_p;\omega)$ arrowed in **Figure 4** by being fitted to Fano's Formula [1]:

$$\alpha(\varepsilon) = \alpha_0 \frac{\left(\varepsilon + q^{(F)}\right)^2}{\varepsilon^2 + 1}, \tag{26}$$

in the vicinity of an excitonic resonance quasienergy $\mathcal{E}_{ex}$, where $\varepsilon = 2(\omega_p - \mathcal{E}_{ex})/\Gamma$ with the spectral width Γ and the asymmetry parameter (Fano's q-parameter) $q^{(F)}<0$. **Figure 5(a)** shows the evaluated values of $|1/q^{(F)}|$ and Γ as a function of F_{ac}, while **Figure 5(b)** shows the peak intensity $\alpha(0) = \alpha_0[q^{(F)}]^2 \equiv \alpha_{max}$ and background spectra $\alpha(\pm\infty) = \alpha_0$ as a function of F_{ac}. It is seen that these functions are affected pronouncedly by F_{ac}; in particular, extrema are formed around $F_{ac} = F_{DL1}$. It is remarked that with the decrease in $|1/q^{(F)}|$ and Γ, the DFR state becomes a pure bound state. In addition, there still exist faint extrema around $F_{ac} = F_{DL2}$ in the concerned quantities except Γ. Therefore, the DL is considered to fulfill a special role of the quantum control of photodressed excitonic states.

For the purpose of confirming such an effect of DL and the pronounced F_{ac} dependence of related quantities on the excitonic DFR, one evaluates the transition probability between the photon sidebands of μ_1 and μ_2 due to the ac-Zener coupling; this value is represented as $M(F_{ac})$ as a function of F_{ac}. This corresponds to the degree of mixing between these two photon

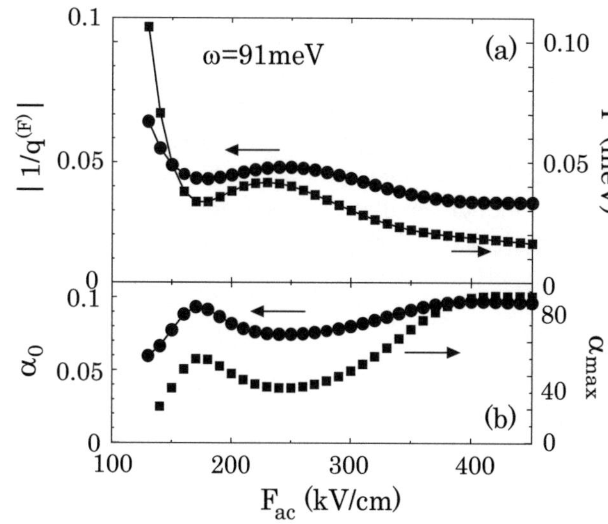

Figure 5. The DFR-related quantities as a function of F_{ac} with the fixed value of ω=91 meV. The calculated results represented by the filled symbols are connected by the solid lines in order to aid the presentation. (a) $|1/q^{(F)}|$ and Γ and (b) α_0 and α_{max} (from Ref. [16] with partial modification).

sidebands. $M(F_{ac})$ is readily obtained by solving the associated coupled equations between μ_1 and μ_2 in an approximate manner of neglecting contributions from all other photon sidebands [16]. Given $\Delta\varepsilon$ and v as the difference of ac-Zener-free quasienergies between μ_1 and μ_2, and the matrix element of the ac-Zener coupling between them, respectively, $M(F_{ac})$ is provided as

$$M(F_{ac}) = \left[\sin\left(\phi/2\right)\right]^2 = \frac{1}{2}\left(1 - \frac{1}{\sqrt{1+z^2}}\right), \tag{27}$$

where $z \equiv \tan\phi = 2|v|/|\Delta\varepsilon|$. With $x = F_{ac}d/\omega$, v and $\Delta\varepsilon$ are evaluated as $v \propto x$ and $\Delta\varepsilon \propto \cos(kd)J_0(x)$, respectively; see Section 1. Thus, one has $z = 2x/\eta J_0(x)$ where η is a proportional constant between $\Delta\varepsilon$ and v. According to Eq. (27), for finite values of η, with the increase of F_{ac}, $M(F_{ac})$ increases from 0 to 1/2 in an oscillating manner; for more detail of the shape of $M(F_{ac})$ for several values of η, consult Ref. [16].

The alteration pattern of $M(F_{ac})$ with respect to F_{ac} looks somewhat similar to the shapes of the DFR-related functions shown in **Figure 5**. In particular, it is noted that $M(F_{ac})$ has extrema at zeros of $J_0(x)$, which just correspond to DL concerned here; that is, $M(F_{ac})$ shows extrema at $F_{ac} = F_{DL1}$ and F_{DL2}. In fact, $M(F_{ac})$ shows a clear extremum at $F_{ac} = F_{DL1}$, while the second extremum at $F_{ac} = F_{DL2}$ is not obviously discernible. This is understood by the behavior that the oscillating component incorporated in $J_0(x)$ is overwhelmed by the ac-Zener coupling v for large x. Therefore, it is concluded that the characteristic F_{ac} dependence of the functions of $|1/q^{(F)}|$, Γ, α_{max}, and α_0 is attributed to the competition between the ac-Zener effect and the band width of the free electron-hole pair states in the vicinity of the DL positions.

Finally, one mentions in brief the ω dependence of the physical quantities $|1/q^{(F)}|$ and Γ at F_{ac}=180 kV/cm in the vicinity of $F_{ac} = F_{DL1}$. As shown in **Figure 6(a)**, $|1/q^{(F)}|$ decreases sharply with the increase in ω, while Γ is maximized around $\omega = 91$ meV at which the centers of two photon sidebands μ_1 and μ_2 coincide. The tendency of $|1/q^{(F)}|$ is in harmony with the ω dependence of the ratio of d_c to d_o, namely, $r_d = d_c/d_o$, as shown in **Figure 6(b)**, where d_c and d_o represent a dipole-transition matrix from the ground state to the closed channel μ_+ and that to the open channel μ_-, respectively. Actually, r_d is in proportion to $q^{(F)}$ [16]. Such alteration of r_d is interpreted on the basis of the anticrossing formation of photon sidebands of μ_+ and μ_- due to the Autler-Townes effect, though not discussed here; for more detail, consult Ref. [16]. Thus, it seems that comparing **Figure 6(a)** with **Figure 5(a)**, the q parameter is even more controllable by changing ω than by F_{ac}.

3.2. TFR in the CP generation

For the calculations of TFR spectra of undoped Si and undoped GaAs, the associated materials parameters employed in the present study are shown in Ref. [31], while parameters of a square-shaped pulse laser employed are as follows. For undoped Si and undoped GaAs, detuning with reference to energy band gap $\Delta\omega$=82 and 73 meV, respectively, temporal width τ_L=15 fs, pulse area A_L=0.12π and 0.20π, respectively, and the maximum excited electron density N_{ex}^0=6.31$\times 10^{17}$ and 5.30$\times 10^{17}$cm^3, respectively; by $\Delta\omega$>0, it is meant that opaque interband transitions with real excited carriers are examined. Further, two time constants of

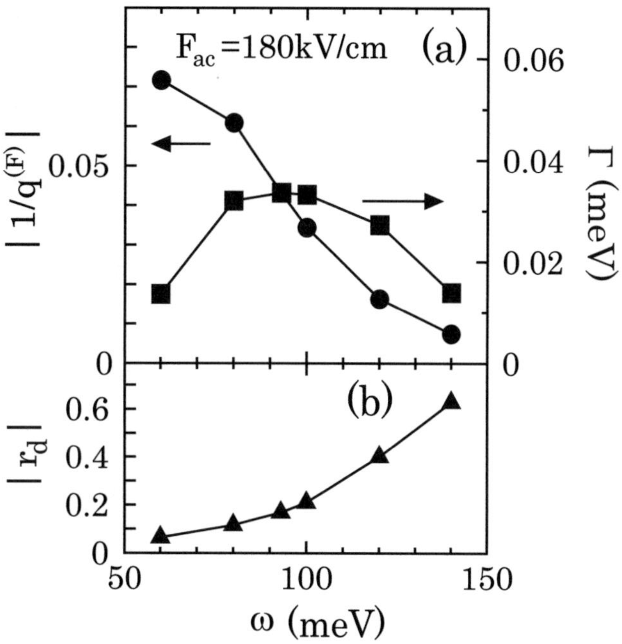

Figure 6. The DFR-related quantities as a function of ω with the fixed value of F_{ac}=180 meV. The calculated results represented by the filled symbols are connected by the solid lines in order to aid the presentation. (a) $|1/q^{(F)}|$ and Γ and (b) $|r_d|$ (from Ref. [16] with partial modification).

T_{12} and T_{q12} are introduced, which represent phenomenological damping time constants of induced carrier density with isotropic momentum distribution and anisotropic momentum distribution with $\boldsymbol{q}$, respectively. The temporal region $t<T_{12}$ is termed as the early-time region during which a great number of carriers still stay in excited states, and the quantum processes govern the CP dynamics; T_{q12} is approximately equal to $T_q(\boldsymbol{k}bb')$ introduced in Eq. (14). On the other hand, the temporal region $t\gtrsim T_{12}$ is termed as the classical region. For the present calculations, T_{q12} and T_{12} are set equal to 20 and 90 fs, respectively. As regards experimental estimates of these time constants for Si, T_{q12} and T_{12} extracted from the CP measurements in Ref. [45] are 16 and 100 fs, respectively, at $N_{ex}^0 = 4 \times 10^{19}\,\mathrm{cm}^{-3}$.

Transient induced photoemission spectra $\overline{A}_q(t_p, \omega_p)$ defined in Section 2.2.2 show the change of excited electronic structure due to the pump field at probe time t_p, and this is crucial to understand the TFR accompanied by CP generation. The total retarded longitudinal susceptibility consists of the dynamically screened Coulomb interaction induced by electron and the LO-phonon-induced interaction. That is, $\tilde{\chi}_q^{(t)}(t_p;\omega_p) = \tilde{\chi}_q(t_p;\omega_p) + \tilde{\chi}_q'(t_p;\omega_p)$, where this is a Fourier transform of Eq. (22) with respect to τ into the ω_p domain. In the small transferred momentum $\boldsymbol{q}$ limit, $\tilde{\chi}_q(t_p;\omega_p)$ is proportional to $|\boldsymbol{q}|^2$, while $\tilde{\chi}_q'(t_p;\omega_p)$ is proportional to $|\boldsymbol{q}|^2$ for the Fröhlich interaction exclusively for a polar crystal such as GaAs and to $|\boldsymbol{q}|^4$ for the deformation potential interaction. This difference is attributed to the fact that the Fröhlich interaction is of long range, and the deformation potential interaction is of short range. It is noted that in a nonpolar crystal such as Si, a dipole transition for lattice absorption vanishes in the limit of $\boldsymbol{q} = \mathbf{0}$ because of the presence of spatial inversion symmetry [61].

In **Figures 7** and **8**, $\overline{A}_q(t_p;\omega_p)$ of Si and GaAs as a function of ω_p is shown, respectively, by solid lines at t_p equal to $t_1 \equiv 15$, $t_2 \equiv 65$ and $t_3 \equiv 100$ fs, where the separate contributions from $\tilde{\chi}_q(t_p;\omega_p)$ and $\tilde{\chi}'_q(t_p;\omega_p)$ are also shown by chain and dashed lines, respectively. $\tilde{\chi}_q(t_p;\omega_p)$ and $\tilde{\chi}'_q(t_p;\omega_p)$ are mostly governed by the plasmon-like mode α_1 and the LO-phonon mode α_2, respectively. In both figures, it is seen that just $\tilde{\chi}'_q(t_p;\omega_p)$ contributes to the formation of spectral peaks and becomes dominant over $\tilde{\chi}_q(t_p;\omega_p)$ in the classical region.

Figure 7(a) shows $\overline{A}_q(t_p;\omega_p)$ of Si at $t_p = t_1 < T_{\mathbf{q}12}$, where the obtained continuum spectra are governed by the contribution from $\tilde{\chi}_q(t_p;\omega_p)$, whereas the contribution from $\tilde{\chi}'_q(t_p;\omega_p)$ is negligibly small due to the proportion of it to $|\boldsymbol{q}|^4$. In Figure 7(b) at $t_p = t_2$ with $T_{q12} < t_p < T_{12}$, the contributions from $\tilde{\chi}_q(t_p;\omega_p)$ are damped to be comparable to those from $\tilde{\chi}'_q(t_p;\omega_p)$. It is noted that asymmetric spectra characteristic of FR are manifested with a dip followed by a peak. This is in sharp contrast with a symmetric Lorentzian profile shown in **Figure 7(c)** at $t_p = t_3 > T_{12}$. As regards $\overline{A}_q(t_p;\omega_p)$ of GaAs, it is shown in **Figure 8(a)** that at $t_p = t_1$, a pronounced peak due to the α_2 mode, is superimposed with a continuum background composed of $\tilde{\chi}_q(t_p;\omega_p)$ and $\tilde{\chi}'_q(t_p;\omega_p)$ with comparable order, since both are in proportion to $|\boldsymbol{q}|^2$. The spectra at $t_p = t_2$ shown in **Figure 8(b)** are dominated by $\tilde{\chi}'_q(t_p;\omega_p)$, differing a lot from those shown in **Figure 7(b)** of Si. The spectra at $t_p = t_3$ in **Figure 8(c)** are similar to those in **Figure 7(c)**.

The origin of the manifestation of TFR in Si shown in **Figure 7(b)** is examined below. The principal difference between Si and GaAs observed here is attributed just to the effective coupling $M_{q\beta}$ between quasiboson and LO-phonon aside from less significant difference in other material parameters; this appears in the matrix h_q introduced in Eq. (16), and the approximation of $M_q \approx M'_q$ is employed here. The primitive coupling constant $g_{b\mathbf{q}}$ incorporated in $M_{q\beta}$ consists of g^D_{bq} and g^F_{bq} representing the coupling constants due to a phenomenological

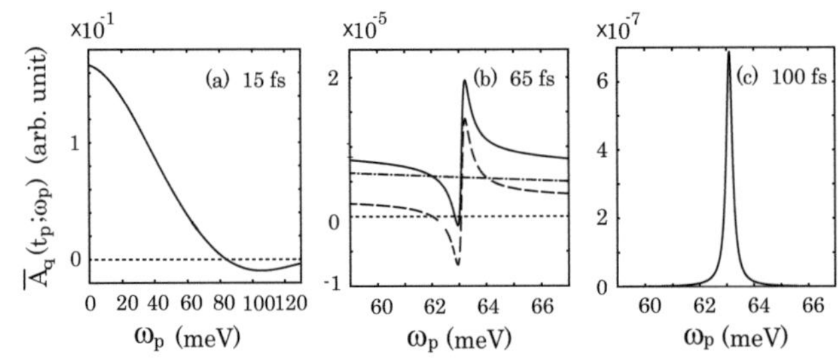

Figure 7. $\overline{A}_q(t_p,\omega_p)$ of undoped Si (solid line) as a function of ω_p at t_p equal to (a) 15 fs, (b) 65 fs, and (c) 100 fs. Separate contributions to the spectra from $\tilde{\chi}_q(t_p;\omega_p)$ and $\tilde{\chi}'_q(t_p;\omega_p)$ are also shown by chain and dashed lines, respectively. $\overline{A}_q(t_p;\omega)$ is reckoned from structureless background due to electron-hole continuum states β that are almost constant in the ω_p region concerned. The widths of the spectral peaks are determined by a phenomenological damping constant T_{ph} of LO-phonon due to lattice anharmonicity: $2/T_{ph}$=0.27 meV (from Ref. [31] with partial modification).

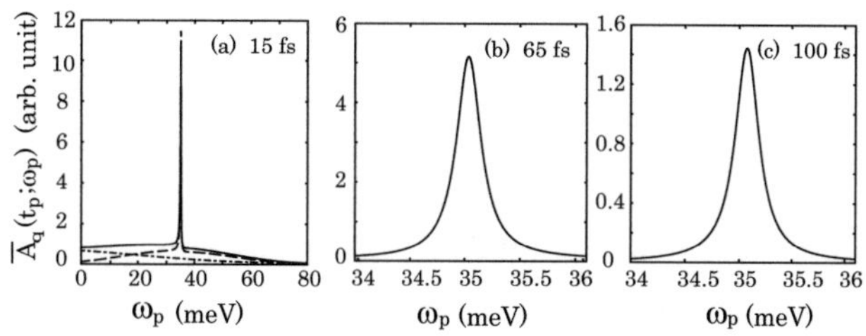

Figure 8. The same as **Figure 7** but for undoped GaAs (from Ref. [31] with partial modification).

LO-phonon deformation potential interaction and the Fröhlich interaction, respectively, that is, $g_{bq} = g_{bq}^D + g_{bq}^F$. Here, g_{bq}^D is real and approximately independent of q, while g_{bq}^F is pure imaginary and $|g_{bq}^F| \propto |q|^{-1}$ [61]. In a nonpolar crystal such as Si, $g_{bq} = g_{bq}^D$, whereas in a polar or partially ionic crystal such as GaAs, g_{bq}^F is much dominant to g_{bq}^D, namely, $g_{bq} \approx g_{bq}^F$. Actually, the phase of $M_{q\beta}$ is almost determined by that of g_{bq}, since a residual factor defining $M_{q\beta}$ is almost considered real. Thus, $M_{q\beta}$ is a complex number given by $M_{q\beta} = |M_{q\beta}|e^{i\phi_{q\beta}}$ in general; $\phi_{q\beta} = 0, \pi$ for Si, while $\phi_{q\beta} = \pm\pi/2$ for GaAs.

Next, discussion is made on how such difference of $M_{q\beta}$ affects the spectral profile of $\overline{A}_q(t_p;\omega_p)$ based on the PQ picture depicted in **Figure 2**. It is seen that there are two transition paths for the process: one is a direct path mediated by an optical transition matrix $D_{q\alpha_2}^{(r)}$ from LO-phonon state α_2 to the PQ ground state, and the other is a two-step resonant path mediated by $M_{q\beta}$ from α_2 to quasiboson state β, followed by a deexcited process mediated by an optical transition matrix $D_{q\alpha_2}^{(c)}$ from β to the PQ ground state. Accordingly, owing to Shore's model [62], the induced photoemission spectra in the proximity of $\omega_p \approx \omega_q^{(LO)}$ is read as

$$\overline{A}_q(t_p;\omega_p) \approx \mathcal{C}_{q\beta} + \frac{\mathcal{A}_{q\alpha_2}\left(\omega_p - \omega_q^{(LO)}\right) + \mathcal{B}_{q\alpha_2}\Gamma_{q\alpha_2}/2}{\left(\omega_p - \omega_q^{(LO)}\right)^2 + \left(\Gamma_{q\alpha_2}/2\right)^2}, \tag{28}$$

where a set of Shore's spectral parameters of $\mathcal{A}_{q\alpha_2}$, $\mathcal{B}_{q\alpha_2}$, and $\mathcal{C}_{q\beta}$ are provided by

$$\begin{cases} \mathcal{A}_{q\alpha_2} = 2|D_{q\beta}^{(c)}||D_{q\alpha_2}^{(r)}||M_{q\beta}|\cos\phi_{q\beta} \\ \mathcal{B}_{q\alpha_2} = -2|D_{q\beta}^{(c)}||D_{q\alpha_2}^{(r)}||M_{q\beta}|\sin\phi_{q\beta} + \left|D_{q\alpha_2}^{(r)}\right|^2 |M_{q\beta}|^2/(\Gamma_{q\alpha_2}/2) \\ \mathcal{C}_{q\beta} = \left|D_{q\beta}^{(c)}\right|^2 \end{cases} \tag{29}$$

and the natural spectral width is represented by $\Gamma_{q\alpha_2} = 2\pi\rho_{q\alpha_2}|M_{q\alpha_2}|^2$; $\rho_{q\alpha_2}$ and $M_{q\alpha_2}$ are the density of state of quasiboson and the coupling matrix at $\mathcal{E}_{q\beta} = \omega_q^{(LO)}$, respectively. The associated Fano's q parameter is determined in terms of Shore's parameters as $q_{q\alpha_2}(t_p) = r_{q\alpha_2}(t_p) + \sigma_{q\alpha_2}(t_p)\sqrt{\left[r_{q\alpha_2}(t_p)\right]^2 + 1}$ with $r_{q\alpha_2}(t_p) = \mathcal{B}_{q\alpha_2}/\mathcal{A}_{q\alpha_2}$ and $\sigma_{q\alpha_2}(t_p) = \mathcal{A}_{q\alpha_2}/|\mathcal{A}_{q\alpha_2}|$.

An asymmetric spectral profile is exclusively determined by $\mathcal{A}_{q\alpha_2}$. It is seen that $\mathcal{A}_{q\alpha_2}(t_p)$ vanishes for $\phi_{q\beta} = \pm\pi/2$ and $\overline{A}_q(t_p;\omega_p)$ becomes of symmetric shape with $|q_{q\alpha_2}(t_p)|$ infinite. The spectral profile of GaAs shown in **Figure 8(b)** corresponds to this case. For $\phi_{q\beta}\neq\pm\pi/2$, both $\mathcal{A}_{q\alpha_2}(t_p)$ and $\mathcal{B}_{q\alpha_2}(t_p)$ are finite, and $\overline{A}_q(t_p;\omega)$ becomes of asymmetric shape with $|q_{q\alpha_2}(t_p)|$ finite. The spectral profile of Si shown in **Figure 7(b)** corresponds to this case, where $\phi_{q\beta}\approx 0,\pi$. For **Figures 7(c)** and **8(c)**, since $D^{(c)}_{q\overline{\alpha}}$ and $|M_{q\beta}|$ become negligibly small, $\overline{A}_q(t_p;\omega)$ is governed by the second term of the expression of $\mathcal{B}_{q\alpha_2}(t_p)$, and this becomes symmetric with $\Gamma_{q\alpha_2}\approx 0$. To conclude, the effective coupling $M_{q\beta}$ around $\mathcal{E}_{q\beta}\approx\omega_q^{(LO)}$ plays the crucial role of the manifestation of TFR, and the asymmetry of profile is mostly determined by $\phi_{q\beta}$ as long as $|M_{q\beta}|$ is still large.

Finally, the manifestation of TFR of Si is discussed from the viewpoint of the allocation of time constants T_{q12} and T_{12}, where one sets $T_{q12}<T_{12}$. This is an important issue for deepening the understanding of TFR. As shown in Figure 7(b), in the region of $T_{q12}\lesssim t_p<T_{12}$, the asymmetric spectral profile bursts into view from the structureless continuum $\tilde{\chi}_q(t_p;\omega)$. Actually, in the early-time region of $t_p<T_{12}$, the excited carrier density is still populated enough around the energy region of $\mathcal{E}_{q\beta}\approx\omega_q^{(LO)}$ to couple strongly with LO-phonon, while the effect of $\tilde{\chi}_q(t_p;\omega)$ is much suppressed in the region of $T_{q12}\lesssim t_p$. As regards a different allocation of these time constants, for instance, $T_{q12}\sim T_{12}$, the TFR profile is no longer observed in the region of $t_p<T_{12}$, since this is covered with still dominant contributions from $\tilde{\chi}_q(t_p;\omega)$, and the effect of $M_{q\beta}$ becomes too small to cause TFR in the region of $t_p\approx T_{12}$. Therefore, the allocation of time constants such as $T_{q12}<T_{12}$ is a necessary condition for realizing the TFR of Si in $\overline{A}_q(t_p;\omega)$; otherwise this never appears.

4. Conclusion

Transient and optically nonlinear FR in condensed matter is examined here, which differs from conventional FR processes caused by a weak external perturbation in a stationary system. In particular, the following two FR processes are discussed: one is the DFR of Floquet exciton realized in semiconductor superlattices driven by a strong cw laser, and the other is the TFR accompanied by the CP generated by an ultrashort pulse laser in bulk crystals of undoped Si and undoped GaAs.

It is shown that the physical quantities relevant to the DFR spectra can be controlled by modulating F_{ac} and ω. In particular, the quantities as a function of F_{ac} take the extrema due to the ac-Zener coupling between the photon sidebands of μ_1 and μ_2, when F_{ac} is suitably adjusted to satisfy the DL condition. Further, the strong ω dependence is explained on the basis of the Autler-Townes effect forming the anticrossing between these two photon sidebands. It is remarked that the spectral width shown in **Figures 5** and **6** seems too small to be confirmed by experiments. Actually, in the present calculations, the Coulomb many-body effect is neglected. At least at the Hartree-Fock level, the vertex correction to the Rabi energy would make the net ac-Zener coupling stronger to result in such a great DFR width that experimental measurement would be accessible.

As regards the TFR spectra, the PQ model succeeds in demonstrating the appearance of asymmetric spectral profile in Si in a flash, whereas the profile observed in GaAs remains symmetric; the obtained results are in harmony with the existing experimental ones [45]. The difference between Si and GaAs is attributed to the phase factor of the effective coupling $M_{q\beta}(t_p)$. To conclude, it is found that in order to realize the TFR in the CP dynamics, the following conditions are to be fulfilled simultaneously. First, the coupling of an LO-phonon with an electron-hole continuum is conducted by the LO-phonon deformation potential interaction rather than by the Fröhlich interaction. Second, photoexcited carriers are populated enough around the energy region $\mathcal{E}_{q\beta} \approx \omega_q^{(LO)}$ in the early-time region $T_{q12}<t_p<T_{12}$ with $T_{q12} \ll T_{12}$.

Acknowledgements

This work was supported by JSPS KAKENHI Grants No. JP21104504, JP23540360, and JP15K05121.

Author details

Ken-ichi Hino[1,2]*, Yohei Watanabe[3], Nobuya Maeshima[2] and Muneaki Hase[1]

*Address all correspondence to: hino@ims.tsukuba.ac.jp

1 Faculty of Pure and Applied Sciences, University of Tsukuba, Tsukuba, Japan

2 Center for Computational Science, University of Tsukuba, Tsukuba, Japan

3 Graduate School of Pure and Applied Sciences, University of Tsukuba, Tsukuba, Japan

References

[1] Fano U. Physics Review. 1961;**124**. DOI: 10.1103/PhysRev.124.1866

[2] Bohn JL, Julienne PS. Physical Review A. 1997;**56**. DOI: 10.1103/PhysRevA.56.1486

[3] Ciurylo R, Tiesinga E, Julienne PS. Physical Review A. 2006;**74**. DOI: 10.1103/PhysRevA.74.022710

[4] Enomoto K, Kasa K, Kitagawa M, Takahashi Y. Physical Review Letters. 2008;**101**. DOI: 10.1103/PhysRevLett.101.203201

[5] Kouchi K. Superexcited states of molecules. In: Hiraoka K, editor. Fundamentals of Mass Spectrometry. Berlin: Springer-Verlag; 2013. DOI: 10.1007/978-1-4614-7233-9. Chap. 5.

[6] Kobayashi K, Aikawa H, Katsumoto S, Iye Y. Physical Review Letters. 2002;**88**. DOI: 10.1103/PhysRevLett.88.256806

[7] Xu B, Dai YM, Zhao LX, Wang K, Yang R, Zhang W, Liu JY, Xiao H, Chen GF, Trugman SA, Zhu J-X, Taylor AJ, Yarotski DA, Prasankumar RP, Qiu XG. Nature Communications. 2017;**8**. DOI: 10.1038/ncomms14933

[8] Bar-Ad S, Kner P, Marquezini MV, Mukamel S, Chemla DS. Physical Review Letters. 1997;**78**. DOI: 10.1103/PhysRevLett.78.1363

[9] Hino K. Physical Review B. 2000;**62**. DOI: 10.1103/PhysRevB.62.R10626

[10] Hino K. Physical Review B. 2001;**64**. DOI: 10.1103/PhysRevB.64.075318

[11] Hino K, Toshima N. Physical Review B. 2005;**71**. DOI: 10.1103/PhysRevB.71.205326

[12] Chu S-I, Telnov DA. Physics Reports. 2004;**390**. DOI: 10.1016/j.physrep.2003.10.001

[13] Potvliege RM, Shakeshaft R. In: Gavrila M, editor. Atoms in Intense Laser Fields. New York: Academic Press; 1992. p. 373

[14] Kroner M, Govorov AO, Remi S, Biedermann B, Seidl S, Badolato A, Petroff PM, Zhang W, Barbour R, Gerardot BD, Warburton RJ, Karrai K. Nature. 2008;**451**. DOI: 10.1038/nature06506

[15] Maeshima N, Hino K. Physical Review B. 2012;**85**. DOI: 10.1103/PhysRevB.85.205305

[16] Maeshima N, Yamada K, Hino K. Journal of Physics: Condensed Matter. 2013;**25**. DOI: 10.1088/0953-8984/25/43/435801

[17] Russell JP. Applied Physics Letters. 1965;**6**:223. DOI: 10.1063/1.1754144

[18] Parker JH Jr, Feldman DW, Ashkin M. Physics Review. 1967;**155**. DOI: 10.1103/PhysRev.155.712

[19] Hart TR, Aggarwal RL, Lax B. Physical Review B. 1970;**1**. DOI: 10.1103/PhysRevB.1.638

[20] Cerdeira F, Cardona M. Physical Review B. 1972;**5**. DOI: 10.1103/PhysRevB.5.1440

[21] Belitsky VI, Cantarero A, Cardona M, Trallero-Giner C, Pavlov ST. Journal of Physics: Condensed Matter. 1997;**9**. DOI: 10.1088/0953-8984/9/27/022

[22] Jin K-J, Zhang J, Chen Z-H, Yang G-Z, Chen ZH, Shi XH, Shen SC. Physical Review B. 2001;**64**. DOI: 10.1103/PhysRevB.64.205203

[23] Fano U, Rau ARP. Atomic Collisions and Spectra. Orlando: Academic Press, Inc; 1986 Chaps. 7–11

[24] Knight PL, Lauder MA, Dalton BJ. Physics Reports. 1990;**190**. DOI: 10.1016/0370-1573(90)90089-K

[25] Kukuu A, Amano T, Karasawa T, Maeshima N, Hino K. Physical Review B. 2010;**82**. DOI: 10.1103/PhysRevB.82.115315

[26] Nemoto Y, Hino K, Maeshima N. Physical Review B. 2013;**87**. DOI: 10.1103/PhysRevB.87.205305

[27] Meier T, Schulze A, Thomas P, Vaupel H. Physical Review B. 1995;**51**. DOI: 10.1103/PhysRevB.51.13977

[28] Siegner U, Mycek M-A, Glutsch S, Chemla DS. Physical Review Letters. 1995;**74**. DOI: 10.1103/PhysRevLett.74.470

[29] Siegner U, Mycek M-A, Glutsch S, Chemla DS. Physical Review B. 1995;**51**. DOI: 10.1103/PhysRevB.51.4953

[30] Hino K, Goto K, Toshima N. Physical Review B. 2004;**69**. DOI: 10.1103/PhysRevB.69.035322

[31] Watanabe Y, Hino K. M, Hase, and N. Maeshima. Physical Review B. 2017;**95**. DOI: 10.1103/PhysRevB.95.014301

[32] Bartal B, Kozma IZ, Stepanov AG, Almási G, Kuhl J, Riedle E, Hebling J. Applied Physics B: Lasers and Optics. 2007;**86**. DOI: 10.1007/s00340-006-2512-7

[33] Karpowicz N, Dai J, Lu X, Chen Y, Yamaguchi M, Zhang L, Zhang C, Price-Gallagher M, Fletcher C, Mamer O, Lesimple A, Johnson K. Applied Physics Letters. 2008;**92**. DOI: 10.1063/1.2828709

[34] Hirori H, Doi A, Blanchard F, Tanaka K. Applied Physics Letters. 2011;**98**. DOI: 10.1063/1.3560062

[35] Shirley J. Physics Review. 1965;**138**. DOI: 10.1103/PhysRev.138.B979

[36] Holthaus M. Physical Review Letters. 1992;**69**. DOI: 10.1103/PhysRevLett.69.351

[37] Grifoni M, Hänggi P. Physics Reports. 1998;**304**. DOI: 10.1016/S0370-1573(98)00022-2

[38] Dunlap DH, Kenkre VM. Physical Review B. 1986;**34**. DOI: 10.1103/PhysRevB.34.3625

[39] Keay BJ, Zeuner S, Allen SJ Jr.,Maranowski KD,Gossard AC, Bhattacharya U, and Rodwell MJW, Physical Review Letters **75**, (1995). DOI: 10.1103/PhysRevLett.75.4102

[40] Madison KW, Fischer MC, Diener RB, Niu Q, Raizen MG. Physical Review Letters. 1998;**81**. DOI: 10.1103/PhysRevLett.81.5093

[41] Eckardt A, Holthaus M, Lignier H, Al Z, Ciampini D, Morsch O, Arimondo E. Physical Review A. 2009;**79**. DOI: 10.1103/PhysRevA.79.013611

[42] Longhi S, Marangomi M, Lobino M, Ramponi R, Laporta P, Cianci E, Foglietti V. Physical Review Letters. 2006;**96**. DOI: 10.1103/PhysRevLett.96.243901

[43] Della Valle G, Ornigotti M, Cianci E, Foglietti V, Laporta P, Longhi S. Physical Review Letters. 2007;**98**. DOI: 10.1103/PhysRevLett.98.263601

[44] Szameit A, Garanovich IL, Heinrich M, Sukhorukov AA, Dreisow F, Pertsch T, Nolte S, Tünnermann A, Longhi S, Kivshar YS. Physical Review Letters. 2010;**104**. DOI: 10.1103/PhysRevLett.104.223903

[45] Hase M, Kitajima M, Constantinescu AM, Petek H. Nature (London). 2003;**426**:51. DOI: 10.1038/nature02044

[46] Gaal P, Kuehn W, Reimann K, Woerner M, Elsaesser T, Hey R. Nature. 2007;**450**. DOI: 10.1038/nature06399

[47] Hase M, Ishioka K, Demsar J, Ushida K, Kitajima M. Physical Review B. 2005;**71**. DOI: 10.1103/PhysRevB.71.184301

[48] Misochko OV, Lebedeva MV. JETP **120**; 2015. DOI: 10.1134/S1063776115020168

[49] Misochko OV. JETP. 2001;**92**. DOI: 10.1134/1.1354682

[50] Kato K, Ishizawa A, Oguri K, Tateno K, Tawara T, Gotoh H, Kitajima M, Nakano H. Japanese Journal of Applied Physics. 2009;**48**. DOI: 10.1143/JJAP.48.100205

[51] Lee JD, Inoue J, Hase M. Physical Review Letters. 2006;**97**. DOI: 10.1103/PhysRevLett.97.157405

[52] Mahan GD. Many-Particle Physics. New York: Plenum; 1981. DOI: 10.1007/978-1-4757-5714-9. Chaps. 4 and 5.

[53] Riffe DM. Physical Review B. 2011;**84**. DOI: 10.1103/PhysRevB.84.064308

[54] Hino K, Tong XM, Toshima N. Physical Review B. 2008;**77**. DOI: 10.1103/PhysRevB.77.045322

[55] Maeshima N, Hino K. Computer Physics Communications. 2012;**183**. DOI: 10.1016/j.cpc.2011.07.022

[56] Glutsch S, Bechstedt F. Physical Review B. 1999;**60**. DOI: 10.1103/PhysRevB.60.16584

[57] Meystre P, Sargent M III. Elements of Quantum Optics. 3rd ed. Berlin: Springer-Verlarg; 1999. DOI: 10.1007/978-3-540-74211-1. Chaps. 3 and 15

[58] Morse PM, Feshbach H. Methods of Theoretical Physics. New York: McGraw-Hill; 1953 Chap. 7

[59] Schäfer W, Wegener M. Semiconductor Optics and Transport Phenomena. Berlin: Springer-Verlag; 2002. DOI: 10.1007/978-3-662-04663-0. Chaps. 2, 10, and 11.

[60] Fetter AL, Walecka JD. Quantum Theory of Many-Particle Systems. New York: McGraw-Hill, Inc.; 1971 Chaps. 3–5

[61] Yu PY, Cardona M. Fundamentals of Semiconductors. 4th ed. Berlin: Springer-Verlarg; 2010. DOI: 10.1007/978-3-642-00710-1. Chaps. 3 and 7

[62] Shore BW. Reviews of Modern Physics. 1967;**39**. DOI: 10.1103/RevModPhys.39.439

Chapter 12

Resonances and Exceptional Broadcasting Conditions

Juan Manuel Velázquez-Arcos,
Alejandro Pérez-Ricardez,
Ricardo Teodoro Páez-Hernández and
Jaime Granados-Samaniego

Additional information is available at the end of the chapter

http://dx.doi.org/10.5772/intechopen.70520

Abstract

Many technologies have been developed to improve the quality of broadcasting, but persist with the problem that avoids the continuity of communications when the physical conditions of the media change. However, loss of signal propagation cannot be avoided because the refractive index of propagation media changes at the same time as magnetization, electromagnetic potential and other local parameters. That is, there is neither a device nor theories that take into account the effect of the sign of the refractive index under the broadcasting process. Simultaneously with the change of refractive index, conventional waves may find travel conditions inaccessible to the desired destination. In this chapter, we proposed that a sudden change in conditions is due to a resonant behavior of the media naturally described by a homogeneous integral equation of Fredholm. In addition, we propose a method to avoid the loss of the signal due to drastic changes in the broadcasting regime.

Keywords: resonances, broadcasting, evanescent waves, communications, negative refraction index

1. Introduction

As we mentioned in the abstract, we propose the behavior of the electromagnetic waves propagating media—a model that consists in the division of the space in several portions and layers that eventually are considered as a superposition of thin layers of plasma. We must underline that only when exceptional conditions locally prevail in a particular portion of space, we can suppose the existence of these plasma layers. When an alternation of unmagnetized and magnetized layers occurs, we can observe that for some intervals of the magnetization and electric potential values, the refraction index of the set of alternating plasma layers becomes negative. That is, we have left-hand material conditions as we have called them. Because

Xiang-kun Kong et al. [1] found experimentally that they could change the refraction index sign as they wanted in a succession of magnetized and unmagnetized plasma layers (which they called as a plasma sandwich), we have applied the plasma sandwich model (PSM) to our proposal. The reason they assume for the change in the refraction index sign is very different to the explanation we present nowadays. Xiang-Kun Kong et al. [1], suppose is the coupling between the electromagnetic polarized waves and the evanescent waves. Instead this reasoning, we have shown in several papers that the homogeneous integral equation of Fredholm (HFE) and its Fourier transform (THFE) give us a simple reason, that is, the brake of confinement of the evanescent waves that turn to be traveling waves. In addition, last explanation is accompanied by the properties of the resonant solutions of the HFE and THFE equations. One of the most important resonance properties is the orthogonality that allows the possibility to send signals with little loss. Another important property of the resonances is the fact that the resonances cannot live on the original sites where the evanescent waves lived. The generation of propagation modes from the evanescent ones is due to a resonant behavior mechanism. We also preserved the term precursor for the evanescent waves that become traveling waves. With this definition, the traveling resonant waves cannot live where the precursors lived. One of the advantages of the PSM is the fact that we can model the resonant broadcasting regime from a little set of PSM parameters. Also, instead of the formalism employed by Xiang-kun Kong et al. [1], we used our own formalism, the vector matrix formalism (VMF) [2–4]. The most decisive variables are the electrical potential and the magnetization intensities.

2. Resonances and the Fredholm's eigenvalue

We remember that we can represent the broadcasting process through a Fourier transform of a generalized inhomogeneous Fredholm equation (TGIFE) [5,6] for the electric and magnetic fields that is by the Fourier transform of the equation:

$$E^m{}_j(t)=E^m{}_j^{(\circ)}(t)+\int_{-\infty}^{\infty} K_{j,k}^{m,n(\circ)}(t,t')E_k^n(t')dt' \tag{1}$$

In Eq. (1), $E_j^m(\omega)$ represents any of the electric or the magnetic field components and the kernel is

$$K_{j,k}^{m,n(\circ)}(t,T')=G^{m,n(\circ)}(\mathbf{r}_j,t;\mathbf{r}_k,t')A_k^{m,n} \tag{2}$$

In Eq. (2), $G^{m,n(\circ)}(\mathbf{r}_j,t;\mathbf{r}_k,t')$ is the free Green function and $A_k^{m,n}$ is the interaction.

Then, the Fourier transform of Eq. (1) can be written as [2–4]:

$$\mathbf{f}^{m(\circ)}(\omega)=\left[\mathbf{1}-\mathbf{K}^{(\circ)}(\omega)\right]_n^m\mathbf{f}^n(\omega) \tag{3}$$

In Eq. (3), $f^n(\omega)$ represents the Fourier transform of any of the electric or magnetic fields due to the source $f^{m(\circ)}(\omega)$, but as we can see from Eq. (1), both are vectors whose components are also

http://dx.doi.org/10.5772/intechopen.70520

vectors. Eq. (3) is an example of which we have called the vector-matrix formalism that avoids a more complicated treatment in terms of integral equations.

From Eq. (3), we can input the condition for the existence of a resonance, which implies that the source term vanishes; in other words, we are imposing the left-hand material conditions [5–10]. Simultaneously, for a purpose of mathematical clearance, we let the discrete indexes J and K in Eq. (1) take continuum values, so we now have a spatial dependence on $\mathbf{r}$ and $\mathbf{r}'$; so Eq. (3) with the left term equal to zero yields

$$w_R^m(\mathbf{r};\omega)=\eta_R(\omega)\int_0^\infty K_n^{m(\circ)}(\omega;\mathbf{r},\mathbf{r}')w_R^n(\mathbf{r}';\omega)dr' \tag{4}$$

In this equation, $w_R^m(\mathbf{r};\omega)$ are the resonances, and we have introduced the Fredholm eigenvalue [2] $\eta_R(\omega)$, corresponding to the R resonance. The introduced parameter $\eta_R(\omega)$ allows for asking about nontrivial solutions for Eq. (4) by means of Fredholm theory of integral equations. We have shown that the structure of $\eta_R(\omega)$ can be chosen in the same way as a phase factor [6]:

$$\eta_R(\omega)=e^{ih(\omega_R)} \tag{5}$$

For the resonant frequency ω_R, where in general it is given as:

$$\omega_R=K_R-i\Lambda_R \tag{6}$$

So, we must ask for $h(\omega_R)$ to be a real number even when refraction index can be complex and dependent of arbitrary magnetization or ionization conditions.

Resonances analyzed in the present chapter are electromagnetic traveling waves that comes from the so-called precursors or evanescent waves, but they share very close mathematical properties with the quantum mechanics resonances; i.e., it fulfills the following theorem we have tested elsewhere [4]:

Theorem I

Suppose that $\mathbf{w}_l(\omega)$ and $\mathbf{w}_u(\omega)$ are solutions of Eq. (4), then,

$$\mathbf{w}_l^t(\omega)\mathbf{A}\mathbf{w}_u(\omega)\left[\lambda_u^{-1}-\lambda_l^{-1}\right]=0 \tag{7}$$

We must remember the relation:

$$w_R^m(\mathbf{r};\omega)\rightarrow \mathbf{w}_R(\omega) \tag{8}$$

On the other hand, the resonances $w_{\mathbf{R}}^m(\mathbf{r};\omega)$ comply with the important orthogonality condition between the eigenvalue function $\frac{1}{\eta_R(\omega)}$ and resonance on the site of a punctual antenna located at $\mathbf{r}_A$ [5]:

$$\frac{1}{\eta_R(\omega)}w_R^m(\mathbf{r}_A;\omega)=0 \tag{9}$$

That implies that the resonances vanish on the sites of the antennas that generate this precursor signals, but we underline that not on the sites that generates the precursor signals of distinct resonances $w_R^m(\mathbf{r};\omega)$.

3. The VMF formalism

Now, we can return to our discrete proposal where we can put the parameters appeared in the PSM [2, 3, 5] into the VMF model [2–4]. To this end, let us recall that Eq. (1) can be written as:

$$\left[\mathbf{1}-\eta_R(\omega)\mathbf{K}^{(\circ)}(\omega)\right]_n^m \mathbf{w}_R^n(\omega)=0 \tag{10}$$

where the kernel $\mathbf{K}^{(\circ)}(\omega)$ is the product of the free Green function $\mathbf{G}^{(\circ)}(\omega)$ with the interaction $\mathbf{A}$ so explicitly,

$$\left[\mathbf{1}-\eta_R(\omega)\mathbf{G}^{(\circ)}(\omega)\mathbf{A}\right]_n^m \mathbf{w}_R^n(\omega)=0 \tag{11}$$

Now, we can find the resonant frequencies in an academic example. To this end, we choose a convenient kernel $\mathbf{K}^{(\circ)}(\omega)$; for simplicity, we do not take into account the three components of the electromagnetic field. Supposing that we only have one component of the field, but we have two emitting antennas, a possible kernel is [2]:

$$\mathbf{K}^{(\circ)}(\omega)=\begin{pmatrix} \dfrac{\sin(\omega-\omega_p)\delta}{(\omega-\omega_p)\delta} & -i\dfrac{\cos(\omega-\omega_p)\delta}{(\omega-\omega_p)\delta} \\ i\dfrac{\cos(\omega-\omega_p)\delta}{(\omega-\omega_p)\delta} & \dfrac{\sin(\omega-\omega_p)\delta}{(\omega-\omega_p)\delta} \end{pmatrix} \tag{12}$$

In Eq. (12), we have introduced the PSM parameter δ. This parameter is defined as:

$$\delta=\kappa \mathrm{d}_M \tag{13}$$

where d_M is the average thickness of the plasma-magnetized layer involved in the change of sign of the refraction index; κ is the wave number of an incident beam interacting with the electric and magnetic fields in a way that the whole kernel is expressed in Eq. (12). The parameter ω_p is an average value for the plasma frequency over the referred layer and can be expressed in terms of the local electron concentration in the layer as:

$$\omega_p=\frac{1}{2\pi}\left(\frac{Ne^2}{m\varepsilon_0}\right)^{\frac{1}{2}} \tag{14}$$

In Eq. (14), ε_0 is the permittivity of vacuum, N is the electron concentration and e is the electronic charge.

Different broadcasting regimes occur when these parameters change, that is the refraction index sign changes. The PSM also considers a dynamical condition in the sense that we have a series of sets of iterated layers changing with time in a random manner and therefore with different effects for distinct frequencies.

Let us remember that the equation we must solve is Eq. (10) where the kernel is

$$\mathbf{K}^{n(\circ)}_{i,j,m}(\omega)=\begin{cases} 0 \;\; if\; i=j \\ A^{n,m}_{j} G^{n,m(\circ)}_{\omega}(\mathbf{r}_i,\mathbf{r}_j) \;\; if \;\; i\neq j \end{cases} \tag{15}$$

The conditions for resonances are that Fredholm's determinant for Eq. (10) equals zero and that Fredholm's eigenvalue equals to one Eq. (16).

These last two conditions allow us to obtain the resonant frequencies for the system constituted by these two antennas but dependent on the parameters of plasma sandwich model. As their similar quantum mechanics case, the wave number or the resonant frequency has an imaginary part; that is, a resonant frequency can be represented by a complex frequency:

$$\omega = K - i\Lambda \tag{16}$$

The transformation of the evanescent waves into traveling waves is a consequence of the imaginary part Λ that avoids the electromagnetic field to be confined. In addition, we have the relation between ω and the wave number κ, that is,

$$\kappa = \sqrt{\mu\varepsilon\omega} \tag{17}$$

By substituting Eq. (12) into Eq. (10), we have that one of the resonance conditions is that the Fredholm determinant Δ must be zero, that is,

$$\Delta\begin{pmatrix} A & -B \\ B & A \end{pmatrix} = 0 \tag{18}$$

where

$$A = \frac{\sin(\omega-\omega_p)\delta}{(\omega-\omega_p)\delta} - \lambda \tag{19}$$

and

$$B = i\frac{\cos(\omega-\omega_p)\delta}{(\omega-\omega_\mathrm{p})\delta} \tag{20}$$

In Eq. (19), λ^{-1} is the Fredholm eigenvalue.

We can put Eq. (16) into Eqs. (18)–(20) and express the Fredholm determinant as:

$$\Delta = 2\left[\mathrm{K}-\omega_p - i\Lambda\right]^{-2}\delta^{-2}\left[\sin((\mathrm{K}-\omega_p)\delta)\cos(i\Lambda\delta)+\cos((\mathrm{K}-\omega_p)\delta)\sin(i\Lambda\delta)\right]^2$$
$$-2\left[\mathrm{K}-\omega_p - i\Lambda\right]^{-1}\lambda\left[\sin((\mathrm{K}-\omega_p)\delta)\cos(i\Lambda\delta)+\cos((\mathrm{K}-\omega_p)\delta)\sin(i\Lambda\delta)\right] \quad (21)$$
$$+\lambda^2$$

We can explore some of the conditions for the existence of resonances (**Figure 1**); for example, if we take $\mathrm{K}=\omega_p+\frac{n\pi}{2\delta}$, $\lambda=1$, and the condition $\Delta=0$, we obtain the following equation for Λ:

$$\frac{n\pi}{2}\cosh(\Lambda\delta)-((\frac{n\pi}{2})^2+\Lambda^2\delta^2)\lambda=0 \quad (22)$$

or defining

$$x\equiv\Lambda\delta \quad (23)$$

$$2\pi\cosh(x)-4x^2-\pi^2=0 \quad (24)$$

Then, the resonant frequencies will have the following form:

$$\omega_{res}=\omega_p+\frac{n\pi}{2\delta}-i\frac{x}{\delta} \quad (25)$$

Now, we can put realistic values for δ and ω_p taken from reference [1], that is,

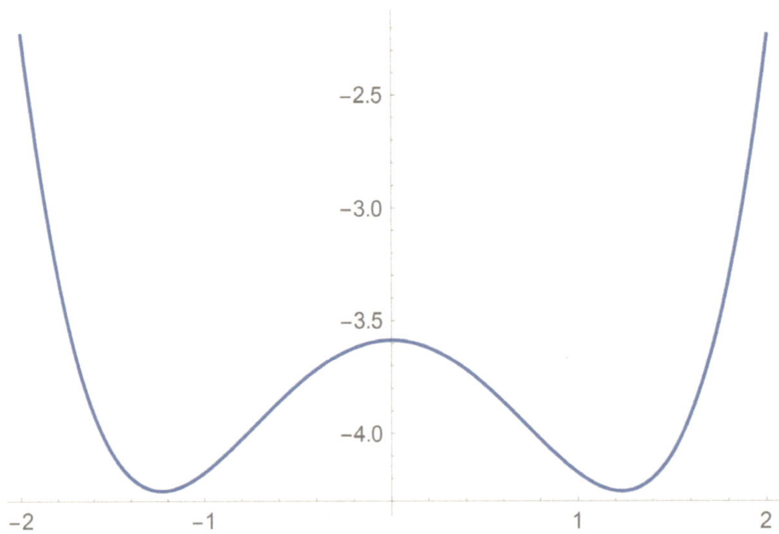

Figure 1. Behavior of Eq. (22) with $n=1$.

$$\delta = 1.68 \times 10^5 Hz \tag{26}$$

and

$$\omega_p = 300 \times 10^6 Hz. \tag{27}$$

So, the first two resonances are

$$\omega_{1,2} \equiv \omega_{\pm} \equiv (3005.1 \pm i(3.778)) \times 10^5 Hz, \tag{28}$$

for $x_{\pm} = \mp 2.2484$.

4. Resonances on a broadcasting process

In the past section, we saw that resonances follow important orthogonal rules. But each resonance has only a unique associated frequency and not a complete band; indeed, the only way for using an individual frequency in a broadcasting process is to emit information in a telegraphic manner; that is, we must have a key and send in the same frequency a succession of intervals of signals with different lengths in time. Fortunately, communication theory (CT) brings us some clues about the problem for sending information [11–16]. First, we recall some statements from this theory, and then we use them. In accordance with these statements, suppose that $f(t)$ is a function that is a member of a set defined in CT as an ensemble and suppose in addition that we are interested on functions that are limited to the band from 0 to W cycles per second, then we have the following theorem [11]:

Theorem II

Let $f(t)$ contain no frequencies over W. Then,

$$f(t) = \sum_{-\infty}^{\infty} X_n \frac{\sin \pi(2Wt - n)}{\pi(2Wt - n)}, \tag{29}$$

where

$$X_n = f\left(\frac{n}{2W}\right). \tag{30}$$

We can see that we have expanded $f(t)$ in terms of orthogonal functions, and the respective coefficients X_n are coordinates in an infinite dimension space.

Theorem (21) can now be taken as a building stone for very special functions with very important properties in the broadcasting processes. We have called these functions as communication packs in previous works. First, we use the cut-off frequency W as a label for distinguishing different packs; second, we use each pack as a new component or coordinate of the signal message $f(t)$ that is, to each W_q frequency corresponds a projection or coordinate

$f_q(t)$, and third, we choose each W_q as a resonant frequency that is $W_q = \omega_R$. Our proposal is that the complete signal $f(t)$ can be recovered by adding its components $f_q(t)$.

5. Why to use communication packs?

We have shown how we can project a signal over different resonant dimensions, but why we must do this. The reason is that theoretically, each resonance is orthogonal to any other resonance, which means that there is no interference between signals traveling over different resonances. Then, we expect that communication packs do not interfere between them because we use different base functions in each pack but also because their defining frequency is a resonant one; that is, we have defined a new space for the broadcasting process and each pack carries a part of the signal over an orthogonal resonant dimension. In addition, we also expect that the infinite sum in Eq. (26) really have a relatively few dominant terms around the resonant frequencies in a manner that we do not need to sum an infinite number of terms for a good approximation. If we want to evaluate the relative broadcasting efficiency between one channel operating with a nonresonant situation and other channel operating with resonant conditions, it is necessary to take into account that a resonant wave cannot live where the precursors lived as we stated above. Therefore, as we have proposed in the abstract, we can provide a specific device, i.e., a pair of circuits, each one with a different response, by selecting the best circuit in any instant for a good reception and avoid the blocking effect in the conventional circuits. In other words, we must remember that resonant solutions vanish at the point sources. Let us take a simple example in which we have only two resonant frequencies and then we can build their respective communication packs with the recipe based on the theorem (29) and explicitly given in another previous work [2–4, 9]:

Suppose that $P(t)$ is the specific signal

$$P(t) = \frac{\sin \pi(2Wt)}{\pi(2Wt)}. \tag{31}$$

Following Theorem II and using the resonances, we get the two communication packs:

$$P_1(t) = \sum_{-\infty}^{\infty} X_{n,1} \frac{\sin \pi(2\omega_1 t - n)}{\pi(2\omega_1 t - n)} \tag{32}$$

$$P_2(t) = \sum_{-\infty}^{\infty} X_{m,2} \frac{\sin \pi(2\omega_2 t - m)}{\pi(2\omega_2 t - m)}, \tag{33}$$

with ω_1 and ω_2 given by Eq. (28):

$$X_{n,1} = P\left(\frac{n}{2\omega_1}\right) \tag{34}$$

$$X_{m,2} = P\left(\frac{m}{2\omega_2}\right). \tag{35}$$

In example of Section 3, we have obtained two resonances so that the two packs are described by Eqs. (32)–(35), but with the numerical values obtained before:

$$X_{n,1} = \frac{\sin \pi\left(2W\left(\frac{n}{2\omega_1}\right)\right)}{\pi\left(2W\left(\frac{n}{2\omega_1}\right)\right)} \tag{36}$$

and

$$X_{m,2} = \frac{\sin \pi\left(2W\left(\frac{m}{2\omega_2}\right)\right)}{\pi\left(2W\left(\frac{m}{2\omega_2}\right)\right)}, \tag{37}$$

That is,

$$X_{n,1} = \frac{\sin \pi \frac{Wn}{\omega_1}}{\pi \frac{Wn}{\omega_1}} \tag{38}$$

and

$$X_{m,2} = \frac{\sin \pi \frac{Wm}{\omega_2}}{\pi \frac{Wm}{\omega_2}} \tag{39}$$

So, the first CP is

$$P_1(t) = \sum_{-\infty}^{\infty} \left(\frac{\sin \pi \frac{Wn}{\omega_1}}{\pi \frac{Wn}{\omega_1}} \right) \frac{\sin \pi(2\omega_1 t - n)}{\pi(2\omega_1 t - n)}, \tag{40}$$

and the second CP is

$$P_2(t) = \sum_{-\infty}^{\infty} \left(\frac{\sin \pi \frac{Wm}{\omega_2}}{\pi \frac{Wm}{\omega_2}} \right) \frac{\sin \pi(2\omega_2 t - m)}{\pi(2\omega_2 t - m)}. \tag{41}$$

Eqs. (40) and (41) can be considered the projections of the real signal (31) over the two dimensions of the resonance space.

6. Concluding remarks

We have shown how we can join several tools that we have developed for the purpose to enhance the broadcasting process; with this aim, we have incorporated the so-called PSM parameters into the algebraic equations (vector-matrix equations) of the VMF searching a way to make communications invulnerable to abrupt changes in the atmospheric conditions. This is very important particularly for high definition channels, which are more sensitive to these abrupt changes, and the PSM (plasma sandwich model) predicts that the mathematical resonances are associated with the delivery of the so-called evanescent waves or to negative values of the refraction index. One of our fundamental proposals is that the atmosphere behaves like a collection of regions with changes from positive to negative (and vice versa) refraction index with unpredictable frequency, and then we can use the PSM to characterize them. On the other hand, we propose the use of the resonant frequencies to overcome the broadcasting barriers by defining a new resonance space created by using the resonances as a new dimension in which the communication packs are the projections of an arbitrary signal. In addition, we suppose that the conventional traveling waves change their regular trajectories when there is a local change in the refraction index sign, so the combined effect of the original paths and the prevalence of the resonant modes make the broadcasting process very difficult without the help of our proposals. By using the results of previous works, we also suggest the use of a device with the possibility for put on and put out of two internal independent circuits each one with a normal (positive refraction index) or resonant (negative refraction index) performance. We underline that communication packs can be constructed even when the current regime is not a resonant.

Author details

Juan Manuel Velázquez-Arcos*, Alejandro Pérez-Ricardez, Ricardo Teodoro Páez-Hernández and Jaime Granados-Samaniego

*Address all correspondence to: jmva@correo.azc.uam.mx

Universidad Autónoma Metropolitana, CDMX, México

References

[1] Kong X-k, Liu S-b, Zhang H-f, Bian B-r, Li H-m, et al. Evanescent wave decomposition in a novel resonator comprising unmagnetized and magnetized plasma layers. Physics of Plasmas. 2013;**20**:043515. DOI: 10.1063/1.4802807

[2] Velázquez-Arcos JM, Granados-Samaniego J. Wave propagation under confinement break. IOSR Journal of Electronics and Communication Engineering, Ver. 1. 2016;**11**(2): 42-48. DOI: 10.9790/2834-1102014248

[3] Velázquez-Arcos JM, Granados-Samaniego J, Navarrete-González TD. Top efficiency on communication theory. IOSR Journal of Electronics and Communication Engineering. 2016;**11**(6):01-11. DOI: 10.9790/2834-1106020111.

[4] Velázquez-Arcos JM. Fredholm's equations for subwavelength focusing. Journal of Mathematical Physics. 2012;**53**(10):103520. DOI: 10.1063/1.4759502

[5] Velázquez-Arcos JM. Fredholm's alternative breaks the confinement of electromagnetic waves. AIP Advances. 2013;**3**:092114. DOI: 10.1063/1.4821336

[6] Velázquez-Arcos JM, Granados-Samaniego J., Vargas CA. The confinement of electromagnetic waves and Fredholm's alternative. In: Electromagnetics in Advanced Applications (ICEAA) International Conference, 9–13 Sept. 2013. doi: 10.1109/ICEAA.2013.6632268; 2013. pp. 411–414

[7] Velázquez-Arcos JM, Pérez-Martínez F, Rivera-Salamanca CA, Granados-Samaniego J. On the application of a recently discovered electromagnetic resonances to communication systems. IJETAE. 2013;**3**(1):466-471 www.ijetae.com, ISSN 2250-2459

[8] Velázquez-Arcos JM, Granados-Samaniego J, Vargas CA. Communication theory and resonances on electromagnetic systems. Electromagnetics in Advanced Applications (ICEAA) 2012 International Conference, 2–7 Sept. 2012, IEEE Cape Town. pp. 392–395. DOI: 10.1109/ICEAA.2012.6328657.

[9] Velázquez-Arcos JM, Granados-Samaniego J. Recent technologies and recording time detection on time reversal signals. International Journal on Recent and Innovation Trends in Computing and Communication. 2014;**2**(11):3447-3450

[10] Hernández-Bautista F, Vargas CA, Velázquez-Arcos JM. Negative refractive index in split ring resonators. Revista Mexicana de Fisica. 2013;**59**(1):139-144 ISSN: 0035-00IX

[11] Shannon CE. A mathematical theory of communication. The Bell System Technical Journal. 1948;**27**:379-423 623-656

[12] Nyquist H. Certain factors affecting telegraph speed. Bell System Technical Journal. 1924; **47**:324

[13] Nyquist H. Certain topics in telegraph transmission theory. Transactions of the A.I.E.E. April 1928;**47**:617

[14] Hartley RVL. Transmission of Information, Bell System Technical Journal, July 1928 pp. 535-563

[15] Wiener N. The Ergodic theorem. Duke Mathematical Journal. 1939;**5**:1-20

[16] Hopf E. On causality statistics and probability. Journal of Mathematical Physics. 1934;**13**(1): 51-102